VORTEX DYNAMICS AND OPTICAL VORTICES

Edited by **Héctor Pérez-de-Tejada**

Vortex Dynamics and Optical Vortices
http://dx.doi.org/10.5772/62608
Edited by Héctor Pérez-de-Tejada

Contributors

Seung-Jae Lee, Carlos Hernandez-Garcia, Laura Rego, Julio San Román, Antonio Picón, Luis Plaja, Fabrizio Bobba, Domenico D'Agostino, Alessandro Scarfato, Maria Iavarone, Goran Karapetrov, Steven Alan Moore, Massimiliano Polichetti, Davide Mancusi, Valentyn Novosad, Volodymyr Yefremenko, Anna Maria Cucolo, Cinzia Di Giorgio, Yangjian Cai, Xianlong Liu, Lin Liu, Yahong Chen, Walter Furlan, İlkay Bakırtaş, Naoto Ohmura, Steven Wang, Jesús Atencia, María Victoria Collados, Iñigo Sola, Julia Marín-Sáez, Warein Holgado, Takashige Omatsu, Katsuhiko Miyamoto, Ryuji Morita, Andrew Lee, Jacques Tempere, Slawomir Mamica, Henk Arnoldus, Xin Li, Zhangjin Xu

Published by InTech
Janeza Trdine 9, 51000 Rijeka, Croatia

Publishing Process Manager Maja Bozicevic
Technical Editor SPi Global
Cover InTech Design team

First published February, 2017
Printed in Croatia
Legal deposit, Croatia: National and University Library in Zagreb

Additional hard copies can be obtained from orders@intechopen.com

Vortex Dynamics and Optical Vortices, Edited by Héctor Pérez-de-Tejada
p. cm.
Print ISBN 978-953-51-2929-5
Online ISBN 978-953-51-2930-1

For EU product safety concerns:
IN TECH d.o.o., Prolaz Marije Krucifikse Kozulić 3, 51000 Rijeka, Croatia,
info@intechopen.com or visit our website at intechopen.com.

Meet the editor

Professor Héctor Pérez-de-Tejada obtained a bachelor's degree in science from the National Autonomous University of Mexico (UNAM) and later received his PhD degree from the University of Colorado (Boulder, Colorado). He has been a full researcher at the Institute of Geophysics of UNAM and participated in various academic functions first at the Ensenada (Baja California) campus and later at the Mexico City campus of that Institution. He was president of the Mexican Geophysical Union and has been a fellow of the National Research System in Mexico (level 3). Currently, he is the head of the Space Science Department of the Institute of Geophysics of UNAM, has participated in numerous research projects including as a guest investigator of the Pioneer Venus Orbiter mission of NASA at Venus, and has been involved in the analysis of plasma data of the Mars and Venus Express spacecraft of ESA. His academic activities have been related to studies of the interaction of the solar wind with planetary ionospheres and he has participated in many scientific publications where he has been the main author in most of them. He has been responsible for activities in the ionospheric sounder station near Mexico City and has been the supervisor of many graduate students at the Ensenada and at the Mexico City campuses of UNAM.

Contents

Preface

The contents of this book are directed to provide updated information on studies of vortex dynamics and optical vortices and on their application to fluid dynamics. The book is intended to serve as a reference guide to scientists, engineers, and students with an interest on the advances in that area of research. The 13 chapters that are presented in the book are related to the analysis of problems conducted in experiments with laser beams that lead to vortices that influence particle motion and also experiments that create vortex structures.

The outline in the contents of the book is to present an overview of experimental results added with computational modeling and theoretical interpretations of the observations. In this manner, the chapters can be summarized as follows: The contributions first refer to a general view of vortices in superfluid atomic gases, followed by a study on the generation of vortex laser beams. An analysis of the dynamic vortex-antivortex in magnetically coupled superconductor magnetic hybrids is then presented with a further study on the contribution of optical vortices illumination to create chiral nanostructures. A discussion of dynamic particle motion in vortex flows is then described together with a simulation of vortex-dominated flows. A study on vortex structures in saturable media is included and also on holographic optical elements in achromatic vortices. A discussion on ultrashort extreme ultraviolet vortices is presented, followed by a study on fractal light vortices, together with that on partially coherent vortex beams. In turn, a study on vortices in electric dipole radiation is included, together with a contribution on spin-wave dynamics under magnetic forces.

The information that is presented in the book provides an updated account on the importance of vortex dynamics and optical vortices in a wide variety of experiments in optics and in computational work, and their description also improves our understanding of important problems in fluid dynamics. Much of what is described is updated information on issues that reveal the value of vortex dynamics and optical vortices to basic matters in science and their application to varied scientific goals. The edition of this book was conducted through the initiative of InTech with the professional work of its staff. The contribution of many scientists and researchers in various areas included in the book provides a substantial amount of innovative information that improves our understanding of vortex dynamics.

Much of the revision of the contents of this book was derived from the participation of the editor on the application of fluid dynamics to various physical problems. The initiative to carry out that effort in response to the invitation by the editorial board of InTech originated from contact with various colleagues who provided useful information in various areas of research in fluid dynamics. Important participation was first contributed by Dr. A. Poveda at UNAM, by Dr. M. Dryer at NOAA, and later by Prof. E. N. Parker at the University of Chicago. Encouragement has been available from Guadalupe Hernandez Ramos and other family members and remains in memory from that of his late parents (Jesus Pérez-de-Tejada and Rogelia Jaime).

Professor Héctor Pérez-de-Tejada
Institute of Geophysics, UNAM
Ciudad Universitaria, Mexico City, Mexico

Chapter 1

Vortex Structures in Ultra-Cold Atomic Gases

Nick Verhelst and Jacques Tempere

Additional information is available at the end of the chapter

http://dx.doi.org/10.5772/67121

Abstract

In this chapter a basic introduction to the theory of vortices in ultra-cold (superfluid) atomic gases is given. The main focus will be on bosonic atomic gases, since these contain the same basic physics, but with simpler formulas. Towards the end of the chapter, the difference between bosonic and fermionic atomic gases is discussed. This discussion will allow the reader to make the conceptual step from bosonic to fermionic gases, while pinpointing the main differences and difficulties when working with fermionic gases rather than bosonic gases. The goal of this chapter is to provide a good and general starting point for researchers, or other interested parties, who wish to start exploring the physics of ultra-cold gases.

Keywords: vortices, Bose-Einstein condensation, superfluidity, ultra-cold atomic gases, quantum turbulence

1. Introduction

Superfluidity was discovered independently by Kapitza [1] and by Allen and Misener [2] in 1938, who found that liquid helium II flows without resistance through capillaries. The theoretical framework to describe superfluids was set up in the next decades, and a central concept in this description is the macroscopic (complex-valued) wave function. A normal fluid, when described at the quantum mechanical level, requires a many-body wave function that depends on the coordinates of all the particles in the fluid. A superfluid, however, can be well described by a macroscopic wave function that depends on a single-position coordinate. This wave function has a straightforward 'hydrodynamic' interpretation: the density of the superfluid is given by the modulus square of the macroscopic wave function, and the velocity field $\mathbf{v}(\mathbf{r},t)$ is proportional to the gradient of the phase $S(\mathbf{r},t)$ of the wave function,

$$\mathbf{v}(\mathbf{r},t) = \frac{\hbar}{m}\nabla S(\mathbf{r},t), \tag{1}$$

with m the mass of the atoms. The single-valuedness of the wave function, in conjunction with Eq. (1) leads to remarkable properties for rotating superfluids. Since the phase can wind only in multiples of 2π, circulation is quantized. Contrary to vortices in classical fluids, in a quantum fluid, vortices carry an integer number of circulation quanta—it is not possible to divide a singly-quantized vortex into smaller units. Upon rotation, a superfluid will not behave as a solid body, but will generate a number of singly-quantized vortices that together carry the total circulation of the rotating fluid. In a (neutral) superfluid, vortices with the same sign of the circulation repel due to kinetic energy of the super-flow. This leads to the formation of a lattice of singly quantized vortices, known as an Abrikosov lattice [3].

In helium-II, the quantization of vorticity and the presence of a lattice of singly-quantized vortices were observed by Vinen [4]. His seminal experiments initiated the study of vortices in liquid helium, which still represents a very active and intense field of research. A thorough review of this field can be found in reference [5]. Liquid helium at non-zero temperature is not a 'pure' superfluid in the sense that thermal and quantum fluctuations lead to a finite viscosity in the bulk liquid, and this complicates the study of superfluid vortices. However, two decades ago, a new superfluid system became experimentally available: ultra-cold atomic quantum gases. These consist of a cloud of typically 10^5 atoms that are optically and/or magnetically trapped, and cooled down to nano-kelvin temperatures. The system is very controllable and versatile: temperature, atom number, trapping geometry and dimensionality, and interaction strength are experimentally tunable. Bosonic atoms were cooled down below the critical temperature for Bose-Einstein condensation (BEC) in 1995 (for a review, see reference [6]), and fermionic atoms were cooled down to a regime where the atoms form superfluid pairs in 2003 (reviewed in reference [7]). The tunable interaction strength allows to bring Fermi gases in the regime where they form Cooper pairs, or in the regime where they form strongly bound molecules that Bose condense, or in the crossover regime between those two limits.

The present chapter aims to familiarize researchers in other fields of vortex physics with the ongoing research on vortices specifically in quantum gases, especially Bose-Einstein condensates. First, we present an (non-exhaustive) overview on the experiments related to these superfluid vortices, and in the second part, we outline the theoretical description of vortices in condensates.

2. Review of experiments on vortices in quantum gases

In the first few years after the initial observation of Bose-Einstein condensation in atomic gases [8–10], it was not clear how to form vortices in these new superfluids. Early experiments with rotating traps failed to produce the vortices (see the contribution of Ketterle and co-workers in reference [6]), and the question was raised on whether the vortex core might be too small to image. The healing length (and core size) is estimated to be of the order of 100 nm. To make observations of quantum gases, the mainstream method [6] is to shine a laser on the cloud, and image the shadow of that cloud. That way, a map of the column density of the cloud can be made, and vortices with their core parallel to the line of sight should appear as holes in the cloud. In this imaging technique, the trap holding the quantum gas is switched off, so the cloud—initially a micron-sized object—can expand to a size large enough to provide a good resolution. The initial difficulties to observe vortices were surprising since theory predicts that when the trap is switched off the vortex core expands faster than the radius of the atomic cloud [11, 12], and this should facilitate the observation, unless the vortex line twists away from the line of sight. An interferometric technique was proposed [13, 14] to overcome possible difficulties with direct imaging. The phase coherence of the superfluid implies matter-wave interference: when two adjacent condensates are allowed to expand, then a series of parallel matter interference fringes appear in the region of overlap of the expanding clouds. The phase pattern of a cloud containing a vortex is different, and this leads to a 'fork dislocation' in the fringe pattern [13]. This idea of interferometry is used in many different branches of physics, e.g. plasma physics to measure the electron density [15, 16] and in the study of optical vortices [17, 18], where the 'fork dislocation' is also found. In both cases, this is done by making an interference pattern of the 'probing beam' (which probes your object) with a 'reference beam'. However, note that the interference pattern of two overlapping condensates is a pure quantum mechanical effect [19], while the interference pattern in conventional interferometry is a pure classical interference effect.

The first successful attempt, in 1999, to create a topological excitation similar to a vortex was realized not by rotating the cloud, but by quantum state manipulation [20]. This requires a multi-component condensate, consisting of atoms in different internal states, typically different hyperfine levels, confined to the same-atom trap. In the experiment given in reference [20], initially the trap contains only a non-rotating one-component BEC. A microwave field is then used to couple two hyperfine states and coherently transform a given number of atoms from the first component into the second, so that a BEC can be grown in the second component. An additional laser beam can be used to imprint a phase profile in the second component in such a way that a single quantum of circulation is present in the second-component, ring-like BEC. The core of the vortex-like structure is filled by the atoms still in the (non-rotating) first component. It can be argued that due to the multi-component nature of the condensate, it should be interpreted as a spinor condensate and the resulting object should consequently be classified rather as a skyrmion than as a 'true' vortex. Indeed, in these cases, the phase can 'untwist' itself, and the excitation is not topologically protected as expected for a quantum vortex [21]. In later experiments using quantum-state manipulation, vortices were created by transferring orbital momentum directly from a laser beam [22]. In other related experiments, the (hyperfine) spin orientation of the atoms in a BEC is manipulated by adiabatically changing an external magnetic field in a position-dependent way to imprint a vortex [23]. In this way, vortices carrying multiple quanta of circulation were also created and their decay into singly quantized vortices was studied [24]. The latest development in vortex physics using internal state manipulation is the possibility to create synthetic magnetic fields [25], i.e. to quantum engineer an effective Lagrangian in which a vector potential term appears, not due to physical rotation or a physical magnetic field, but due to a position-dependent coupling between internal atomic states.

The second class of vortex experiments in quantum gases uses setups in which the condensate is physically rotated. Note that the circulation of a vortex and thus the condensate are quantized. This means that a minimal (critical) frequency is needed in order to have a vortex in a quantum gas. As soon as the critical frequency is reached, the rotational frequency of the condensate snaps to this critical frequency, and starts rotating as a whole. Although initially this method was unsuccessful, it is now the most widely used method to create and study vortices in superfluid quantum gases. The first observation of a 'true' vortex in quantum gases—distinct from the coreless vortex given in reference [20]—can be found in reference [26]. In that experiment a time-dependent optical potential (a 'laser spoon') is used in conjunction with a magnetic trap to stir the condensate, and this proved to be more effective than rotating only the magnetic trap. Below a critical frequency the condensate shakes (surface oscillation modes are present) but it does not rotate (no circulation). Above the critical frequency a singly quantized vortex is nucleated: it arises from a surface mode at the edge of the cloud and moves towards the centre where it remains as long as the cloud is kept in rotation. Using this technique, by rotating faster, an array of vortices was nucleated [27]. Also a 'field-cooled' version of this experiment was performed: rather than rotating a condensate, one cools a rotating normal cloud below the critical temperature [28]. When the temperature drops below the critical temperature, a condensate is nucleated in which the circulation 'snaps' to the nearest quantized value of circulation, with a corresponding number of vortices appearing. A large number of subsequent experiments have further refined the rotation technique for vortex creation—some noteworthy cases are the investigation of vortex nucleation [29], the development of a purely magnetic technique [30] and the use of a rotating optical lattice [31]. These experiments rely largely on the direct imaging technique, in which the vortex core is visible as a hole in the cloud, and the success of these experiments have dispelled the early doubts that the vortex cores might be too small to observe. Nevertheless, the matter-wave interferometric technique was also used to confirm the presence of the phase singularity, and to make sure that there is indeed circulation around the visible hole in the condensate [32, 33].

With techniques to produce vortices in quantum gases well-established, the path was open to study various aspects of vortex physics, such as the properties and dynamics of single vortex lines [34]. In the next three paragraphs we highlight three specific lines of experimental research that have been pursued very intensively: one that focuses on the properties of the vortex lattice, one that looks for spontaneous generation of vortices and anti-vortices, and finally one that focuses on quantum turbulence.

The equilibrium properties of vortex lattices were studied [35], as well as their formation dynamics [36, 37]. The response of the lattice to external perturbations such as shaking and removal of atoms was investigated, and revealed various unusual vortex configurations, such as giant vortices and vortex sheets. This also allowed to excite the 'phonons' or vibrations of the lattice, known as Tkatchenko modes [38]. Increasing the rotation frequency, attempts were made to reach the Landau-level regime, where the number of quantum vortices becomes comparable to the number of particles. Although this regime could not be reached, interesting physics regarding the lattice structure and equilibrium cloud shapes emerged for fast-rotated condensates [39].

By adapting the trapping potential, the atomic cloud can be squeezed into a pancake thin enough to reveal the physics of the two-dimensional (2D) Bose gas—in this case superfluidity is destroyed through the spontaneous appearance of vortices and anti-vortices above a critical temperature, the so-called Kosterlitz-Thouless transition, as given in reference [40]. The creation of vortex-anti-vortex pairs can also be induced in a three-dimensional (3D) cloud by moving an obstacle (such as a blue de-tuned laser beam) through the condensate [41]. The spontaneous generation of vortices and anti-vortices also occurs when a Bose gas is quenched through the phase transition. This quench leads to the nucleation of individual phase-coherent domains with different phases that upon merging lead to phase singularities at the interface of the merged domains. This phenomenon is called the Kibble-Zurek mechanism and with the advent of uniform ('flat box') trapping potentials [42] it has become possible to observe this transition in quantum gases [43].

By tilting the vortex array by a small angle around the axis of symmetry, the precession of the array was observed, and Kelvin modes (transverse oscillations of the vortex line) were seen [44]. The technique of tilting the axes of rotation would later prove useful in the studies on quantum turbulence. This is a regime of turbulent flow characterized by a tangle of vortices, and has been extensively studied in helium [45]. The difficulty in studying vortex tangles in liquid helium is that the core size is very small, so it is hard to directly image the vortices. In quantum gases, vortex cores can be readily imaged, and in addition new *in situ* [46] and tomographic [34] imaging techniques were developed. However, in quantum gases, the ratio between the condensate size and the vortex core is much smaller than in helium (typically it is of the order of 10–50 in BECs), limiting the size of the vortex tangle and inducing edge and finite-size effects. Quantum turbulence was achieved [47] by rotating the condensate around different axes simultaneously, or by rotating in conjunction with shaking. The possibility to flatten the atomic cloud has opened the way to also study quantum turbulence in two dimensions [48]. These techniques offer the potential to resolve many outstanding questions [49] regarding the transition from classical to quantum turbulence, the role of vortex reconnections and the dimensional crossover.

The above examples of recent vortex studies relate to Bose gases. Also in superfluid Fermi gases under rotation, vortex lattices have been observed [50]. The appearance of arrays of singly quantized vortices has been used as a hallmark of superfluidity in these systems. This in turn has been used to investigate how superfluidity breaks down when a population imbalance exists between the two states that combine to form the superfluid pairs [51]. Finally, to conclude this review section, we provide some references to more extensive reviews [52, 53] and a book [54] on vortices in quantum gases.

3. Building a theory of ultra-cold bosonic gases

Before we take a look at vortices, we will briefly discuss and derive the theory that we will use to describe ultra-cold bosonic gases. More elaborated discussions can be found in references [55–57]. The results of this section will be the Gross-Pitaevskii (GP) equation, a hydrodynamic description of the system and the energy of a general ultra-cold, dilute bosonic gas.

3.1. Choosing the right framework: classical or quantum?

In order to give a successful theoretical description of ultra-cold quantum gases, we first need to ask ourselves which framework to use. Is it sufficient to use the classical kinetic theory, or should we resort to quantum theory to get reliable answers? In order to answer this question, we have to look at the de Broglie wavelength[1] (or thermal wavelength):

$$\lambda_{\mathrm{dB}} = \sqrt{\frac{h^2}{2\pi m k_B T}}. \tag{2}$$

As soon as the de Broglie wavelength becomes comparable (or larger) than the typical distance between atoms, we need quantum mechanics. In this case the wavelength associated with a particle will become so large that the quantum mechanical wave functions will overlap.

For typical systems of ultra-cold quantum gases, we deal with clouds of around:

- 10^5 atoms,
- with a scattering length in the order of nm,
- trapped in a magneto-optical trap with a typical length scale of μm,
- at temperatures of the order of nK.

The typical de Broglie wavelength of this system can now be estimated using the above, this results in:

$$\lambda_{\mathrm{dB}} \propto 10^{-5}\mathrm{m} = 10\ \mu\mathrm{m}. \tag{3}$$

The distance between the atoms can be estimated from the inverse particle density (n), which yields:

$$d_{\mathrm{atoms}} \propto \sqrt[3]{n^{-1}} \propto 10^{-8}\mathrm{m} = 10\ \mathrm{nm}. \tag{4}$$

As we see the distance between the atoms is around a factor 1000 smaller than the de Broglie wavelength, which means that we should use quantum mechanics.

Since we are working with a gas of N atoms (many body system), we will start in second quantization [59, 60] since it is then easier to write down (and simplify) our Hamiltonian. The most general Hamiltonian for a system with two-body interactions is [61]:

$$\hat{H} = \int d\mathbf{r} \hat{\psi}^{\dagger}(\mathbf{r}) \left[-\frac{\hbar^2 \nabla^2}{2m} + V_1(\mathbf{r}) \right] \hat{\psi}(\mathbf{r}) + \frac{1}{2} \int d\mathbf{r} \int d\mathbf{r}' \hat{\psi}^{\dagger}(\mathbf{r}) \hat{\psi}^{\dagger}(\mathbf{r}') V_2(\mathbf{r}-\mathbf{r}') \hat{\psi}(\mathbf{r}') \hat{\psi}(\mathbf{r}), \tag{5}$$

[1]This wavelength is derived by comparing the kinetic energy of a free particle with its thermal energy, using the de Broglie hypothesis $p = \hbar k$. This wavelength also emerges naturally when looking at ideal gases, see reference [58].

http://dx.doi.org/10.5772/67121

where $V_1(\mathbf{r})$ is a global potential (e.g. a trapping potential) and $V_2(\mathbf{r}-\mathbf{r}')$ describes the interactions.[2] The operators $\hat{\psi}(\mathbf{r})$ are bosonic operators, and the extension to fermionic atomic gases is discussed towards the end of the chapter. Determining the two potentials will fix the system, allowing the calculation of the dynamics of the system.

3.2. Simplifying by assumptions: ultra-cold dilute bosonic gases

For ultra-cold gases the main contribution to the partial wave expansion will come from the s-wave scattering. Since the range of the inter-atomic potential is much smaller than the inter-atomic distance (dilute gases), the potential can be approximated as a contact potential. Using partial wave analysis [62], the two-body potential can now be simply written as [63, 64]:

$$V_2(\mathbf{r}-\mathbf{r}') = g\delta(\mathbf{r}-\mathbf{r}') \qquad \text{with} \qquad g = \frac{4\pi\hbar^2 a_s}{m}, \tag{6}$$

with a_s the (boson-boson[3]) s-wave scattering length. Physically this s-wave scattering length can be interpreted as the effective radius a_s of our interacting particles if they would be approximated as hard spheres.[4]

A second approximation can be made if we assume that we have a Bose-Einstein condensate (BEC). The mechanism behind Bose-Einstein condensation can be easily seen when we look at the Bose-Einstein distribution:

$$n(\varepsilon;\mu,T) = \left[\exp\left(\frac{\varepsilon-\mu}{k_B T}\right)-1\right]^{-1} \qquad \text{with} \qquad \varepsilon > 0 \qquad \text{and} \qquad \mu \leq 0, \tag{7}$$

where the energy was chosen in such a way that the lowest energy level is situated in $\varepsilon = 0$. This distribution has a divergence in $\varepsilon = \mu$ and drops exponentially for large values of ε. This means that we can store an infinite amount of particles in the lowest energy level, while we can only have a finite amount in the higher energy levels. The total number of particles in our system of bosons is given by:

$$N_{\text{tot}} = N_0 + \underbrace{\int_0^\infty d\varepsilon\, g(\varepsilon) n(\varepsilon;\mu,T)}_{N_{\text{exc}}}, \tag{8}$$

where N_0 is the number of particles in the ground state, N_{exc} the number of particles in the excited state and $g(\varepsilon)$ the density of states. When we start putting more and more particles in our system, we will see at a certain point that the number of particles in excited states will start saturating (the chemical potential becomes maximal). At the point where the

[2] Of course in reality three (or more) body interactions can be a possibility. Since ultra-cold atomic gases are usually dilute gases, interactions between more than two particles are negligible.

[3] Also sometimes noted as a_{BB}. For now this does not matter, when you study fermions this will become more important.

[4] The larger the spheres (value of a_s), the more interactions, and the larger the value of g.

number of excited states is saturated, every new particle will be placed in the ground state. If we keep increasing the number of particles, then at a certain point the number of particles in the ground state will become macroscopically large, dominating the many-body wave function. This means that for a BEC all particles are in the same quantum state and hence the many-particle wave function $\psi(\mathbf{r})$ can be approximated by a macroscopic wave function $\Psi(\mathbf{r})$.

Notice that in order to achieve a BEC we want as many particles as possible, while for our approximation of the interaction potential we want a dilute gas. How can we combine this? The answer is to make the gas ultra-cold, this way the number of particles required to get a BEC becomes small enough to make the gas dilute. If our atomic gas is not diluted, higher order interactions will play a more important role. These higher order interactions will lead to particle being expelled from the atomic gas.

Combining both assumptions, it is now possible to simplify our Hamiltonian to make it equal to:

$$\hat{H} = -\frac{\hbar^2\nabla^2}{2m} + V_1(\mathbf{r}) + \frac{1}{2}g\Psi^*(\mathbf{r})\Psi(\mathbf{r}). \tag{9}$$

In order to get our equation of motion for our system (9), we use the Heisenberg equation of motion, resulting in:

$$i\hbar\frac{\partial\Psi(\mathbf{r},t)}{\partial t} = \left(-\frac{\hbar^2\nabla^2}{2m} + V_1(\mathbf{r}) + g|\Psi(\mathbf{r},t)|^2\right)\Psi(\mathbf{r},t). \tag{10}$$

The above equation (10) is known as the time dependent Gross-Pitaevskii equation (or the nonlinear Schrödinger equation). Note that we substituted the operator $\hat{\psi}(\mathbf{r},t)$ by the macroscopic field value $\Psi(\mathbf{r},t)$. Mathematically this substitution is not done since we go from an operator on Hilbert space to a function in space and time. However from a physical point of view this substitution makes sense. In order to see why the substitution of the operator is sensible, we look at the Penrose-Onsager criterion for BEC. The Penrose-Onsager criterion for BEC states that when the ground state becomes macroscopically occupied (which is the case), our one-particle density matrix should have an eigenvalue that becomes macroscopically large. This allows us to approximate the density matrix by its biggest contribution (the macroscopic wave function):

$$\rho_1(\mathbf{r};\mathbf{r}') = \langle\hat{\psi}^\dagger(\mathbf{r})\hat{\psi}(\mathbf{r}')\rangle \approx \Psi^*(\mathbf{r})\Psi(\mathbf{r}'). \tag{11}$$

Since all expectation values for our system can be rewritten in terms of the density matrix ρ_1, we can replace all operators by the macroscopic field. With the two above Eqs. (9) and (10), we are ready to describe our condensate of bosons.

Note that if you want to minimize your energy (9) to find the ground state of your system with a given normalization, you explicitly have to use Lagrange multipliers to fix your normalization.

This is because of the fact that our Hamiltonian and thus equations of motion are non-linear. For the ground state this means that you define the Hamiltonian:

$$\hat{H}_\mu = -\frac{\hbar^2\nabla^2}{2m} - \mu + V_1(\mathbf{r}) + \frac{1}{2} g\Psi^*(\mathbf{r})\Psi(\mathbf{r}), \tag{12}$$

which now contains the chemical potential as a Lagrange multiplier for the number of particles. Doing a functional minimization then yields the differential equation:

$$\mu\Psi_\mu(\mathbf{r},t) = \left(-\frac{\hbar^2\nabla^2}{2m} + V_1(\mathbf{r}) + g|\Psi_\mu(\mathbf{r},t)|^2\right)\Psi_\mu(\mathbf{r},t). \tag{13}$$

So depending on whether you work with a fixed number of particles or a fixed chemical potential you use respectively, Eqs.(12) and (13) or Eqs.(9) and (10). The relation between both views can be easily established by comparison. For the energy we get the known Legendre transform from statistical mechanics:

$$\hat{H} = \hat{H}_\mu + \mu\hat{N}. \tag{14}$$

For the macroscopic wave function, we have an additional phase rotation:

$$\Psi(\mathbf{r},t) = \exp\left(\frac{i\mu t}{\hbar}\right)\Psi_\mu(\mathbf{r},t). \tag{15}$$

With the above two relations it is possible to switch between fixed number of particles or fixed chemical potential if needed.

3.3. The hydrodynamic description

In order to simplify the calculations one usually works in the hydrodynamic description. This means that our complex field $\Psi(\mathbf{r},t)$ will be written as the product of two real fields $n(\mathbf{r},t)$ and $S(\mathbf{r},t)$ as:

$$\Psi(\mathbf{r},t) = \sqrt{n(\mathbf{r},t)}e^{iS(\mathbf{r},t)}. \tag{16}$$

The field $n(\mathbf{r},t)$ is known as the superfluid density (field), whereas $S(\mathbf{r},t)$ is known as the phase field.

If we now take the Gross-Pitaevskii equation, Eq. (10) and treat it canonically[5] we get, after simplification:

---

[5]Multiply (10) by the complex conjugate field $\Psi^*(\mathbf{r},t)$, and subtract the product $\Psi^*\partial_t\Psi$ by the complex conjugate of the whole $\Psi\partial_t\Psi^*$. In other words, calculate $\Psi^*\partial_t\Psi - \Psi\partial_t\Psi^*$.

$$\frac{\partial|\Psi(\mathbf{r},t)|^2}{\partial t} + \nabla\cdot\left(\frac{\hbar}{2mi}[\Psi^*(\mathbf{r},t)\nabla\Psi(\mathbf{r},t) - \Psi(\mathbf{r},t)\nabla\Psi^*(\mathbf{r},t)]\right) = 0. \tag{17}$$

This is a result known from standard quantum mechanics, where it is seen as a continuity equation since $|\Psi(\mathbf{r},t)|^2 = n(\mathbf{r},t)$ is interpreted as a (probability) density. If (17) is a continuity equation, then the part with a divergence is given by a current $\mathbf{j}(\mathbf{r},t) = n(\mathbf{r},t)\mathbf{v}(\mathbf{r},t)$. Using the fact that the argument of the divergence is a current, we can define our velocity field as:

$$\mathbf{v}(\mathbf{r},t) = \frac{\hbar}{2mi}\frac{\Psi^*(\mathbf{r},t)\nabla\Psi(\mathbf{r},t) - \Psi(\mathbf{r},t)\nabla\Psi^*(\mathbf{r},t)}{|\Psi(\mathbf{r},t)|^2}. \tag{18}$$

Of course we want to work in the hydrodynamic equations so we will substitute the hydrodynamic description (16) in the continuity equation, Eq. (17) and the definition for the velocity field, Eq. (18). For the continuity equation, Eq. (17) this yields (as expected) that:

$$\frac{\partial n(\mathbf{r},t)}{\partial t} + \nabla\cdot[n(\mathbf{r},t)\mathbf{v}(\mathbf{r},t)] = 0, \tag{19}$$

which is known from classical hydrodynamics. For the velocity field (18), we find that

$$\mathbf{v}(\mathbf{r},t) = \frac{\hbar}{m}\nabla S(\mathbf{r},t), \tag{20}$$

so the velocity field is determined entirely by the phase field. Note that our velocity field is also irrotational:

$$\nabla\times\mathbf{v}(\mathbf{r},t) = \nabla\times\left(\frac{\hbar}{m}\nabla S(\mathbf{r},t)\right) = 0, \tag{21}$$

meaning that we will only have irrotational flow patterns.

Of course since we have two real fields $n(\mathbf{r},t)$ and $S(\mathbf{r},t)$, we also need two equations of motion. To find these equations of motion, we need to substitute the hydrodynamic description (16) into the Gross-Pitaevskii equation, Eq. (10). After the substitution, the Gross-Pitaevskii equation, Eq. (10) splits up into real and imaginary parts which can be considered as two independent equations since both fields are real-valued. The real and imaginary parts after minor simplifications are given by:

$$\hbar\frac{\partial\sqrt{n(\mathbf{r},t)}}{\partial t} = -\frac{\hbar^2}{2m}[2\left(\nabla\sqrt{n(\mathbf{r},t)}\right)\cdot\left(\nabla S(\mathbf{r},t)\right) + \sqrt{n(\mathbf{r},t)}\left(\nabla^2 S(\mathbf{r},t)\right)], \tag{22}$$

$$-m\frac{\partial}{\partial t}\left(\frac{\hbar}{m}S(\mathbf{r},t)\right) = -\frac{\hbar^2}{2m}\frac{\nabla^2\left(\sqrt{n(\mathbf{r},t)}\right)}{\sqrt{n(\mathbf{r},t)}} + \frac{m}{2}\left(\frac{\hbar}{m}\nabla S(\mathbf{r},t)\right)^2 + V_1(\mathbf{r}) + gn(\mathbf{r},t), \tag{23}$$

http://dx.doi.org/10.5772/67121

where the first line (22) is the imaginary part and the second line (23) is the real part of the Gross-Pitaevskii equation, Eq. (10). The imaginary part (22) can be simplified and rewritten in terms of the velocity field and density field, this will yield the continuity equation, Eq. (19). The real part (23) can also be rewritten in terms of the velocity and density fields, this is most easily done by taking the gradient of the equation (23). After simplifying the gradient of the real part (23) we get:

$$m\frac{\partial \mathbf{v}(\mathbf{r},t)}{\partial t} + m\left(\mathbf{v}(\mathbf{r},t)\cdot\nabla\right)\mathbf{v}(\mathbf{r},t) + \nabla\cdot\left(-\frac{\hbar^2}{2m}\frac{\nabla^2\left(\sqrt{n(\mathbf{r},t)}\right)}{\sqrt{n(\mathbf{r},t)}} + V_1(\mathbf{r}) + gn(\mathbf{r},t)\right) = 0, \tag{24}$$

which is the same as the Euler equation (or frictionless Navier-Stokes equation) from hydrodynamics. In other words, in a BEC we have an irrotational, frictionless flow. By looking at the Euler equation (24), we can identify three kinds of pressure that will occur in our system:

Mechanical pressure: The first pressure we can recognize is the one particle potential $V_1(\mathbf{r})$. This pressure is called the mechanical pressure, and this is the pressure used to keep the condensate together.

Interaction pressure: The second pressure is given by the interaction term $gn(\mathbf{r},t)$ and is known as the interaction pressure. This pressure is a consequence of the collisions between the atoms, leading to a pressure that is proportional to the density.

Quantum pressure: The remaining term

$$-\frac{\hbar^2}{2m}\frac{\nabla^2\left(\sqrt{n(\mathbf{r},t)}\right)}{\sqrt{n(\mathbf{r},t)}} \tag{25}$$

is known as the quantum pressure. This pressure is a consequence of quantum mechanics where your kinetic energy is proportional to the second derivative of the wave function (the curvature). Note that this pressure is a pure quantum mechanical effect, indeed if we take the limit $h \to 0$ this term becomes equal to zero.

3.4. Summary

So for our ultra-cold bosonic gases forming a BEC, we get one macroscopic wave function $\Psi(\mathbf{r},t)$ which is subject to the Gross-Pitaevskii equation, Eq. (10). This complex wave function $\Psi(\mathbf{r},t)$ can be rewritten in terms of two real valued fields (16): a density field $n(\mathbf{r},t)$ and a phase field $S(\mathbf{r},t)$. The description in terms of the real fields is called the hydrodynamic description, where one usually uses the velocity field (20) instead of the phase field. The relations for the density field and velocity field are given by the continuity equation, Eq. (19) and the Euler equation, Eq. (24), where the quantum mechanics enters through the quantum pressure. From the hydrodynamic description we concluded that a BEC typically describes a frictionless, irrotational flow.

4. Basic properties of a vortex in a superfluid

Now that we know the basic physics for a BEC, we can look at vortices in ultra-cold gases. Note that the results discussed here will be general for both vortices in superfluid Bose gases as well as in superfluid Fermi gases.[6] In this section we will not look at the structure of the vortex (this is done in the next section), but we will rather define what we mean with a vortex and what this implies for our vortex flow. By thoroughly studying the definition of a vortex we can already derive a few basic properties of a vortex. The rest of the section is devoted to two analytically solvable systems of a BEC and on how vortices are detected in experiments.

4.1. Definition of a (quantum) vortex

In order to study a vortex, we first have to explain what we mean when we talk about a vortex. Just as in classical fluid dynamics, we start from the definition of the circulation[7] κ, defined as:

$$\kappa = \oint_C d\mathbf{s} \cdot \mathbf{v}(\mathbf{r},t), \tag{26}$$

where C is a closed contour containing the vortex core. For the circulation we then have three possibilities: counter-clockwise rotation ($\kappa > 0$), clockwise rotation ($\kappa < 0$) and no rotation $\kappa = 0$. The value of the circulation κ should be the same for *any contour* C that encloses the vortex core.[8] Given the velocity field $\mathbf{v}(\mathbf{r},t)$, it is always possible to define a vorticity field as:

$$\boldsymbol{\omega}(\mathbf{r},t) = \nabla \times \mathbf{v}(\mathbf{r},t), \tag{27}$$

which is a quantity telling us how the fluid rotates locally. The vorticity follows the vorticity equation which can be derived by taking the curl of the Euler equation, Eq. (24). Applying Stokes' theorem in Eq. (26), we can relate the circulation κ in terms of the vorticity (24), this results in:

$$\kappa = \int_S d\mathbf{S} \cdot \omega(\mathbf{r},t), \tag{28}$$

making the circulation equal to the flux of vorticity. Note that we can use Stokes' theorem as long as S is a surface without any holes.

Using the above two formulas (26) and (28) we can calculate the circulation of a BEC. To calculate the circulation using the velocity field, we will use the definition of the velocity field in terms of the phase field (20). The circulation then becomes:

---

[6]This is because of the fact that fermions have to pair up and become bosons in order to become superfluid. So the effective superfluid behaves like a BEC.

[7]In fluid dynamics the circulation is usually denoted by Γ.

[8]This is because of the fact that we assume that the different fields (density field, velocity field,...) are meromorphic functions and hence the singularities are localized. According to Morera's theorem we can freely deform the contour (as long as we do not cross a singularity) and still have the same contour integral.

$$\kappa = \frac{\hbar}{m}\oint_C \nabla S(\mathbf{r},t) = \frac{\hbar}{m}\left(S(\mathbf{r}_e,t) - S(\mathbf{r}_b,t)\right), \tag{29}$$

where the gradient theorem was used. In (29) $\mathbf{r}_b$ is the point where the closed loop begins and $\mathbf{r}_e$ is the point where the loop ends, although the physical positions are the same the points $\mathbf{r}_b$ and $\mathbf{r}_e$ can be different. An example of two points with a different coordinate but the same physical location is given for example when we describe the phase in polar coordinates $S(\rho,\phi,z,t)$. Making one complete circle in the angular coordinate then yields for the start and end points that:

$$S(\mathbf{r}_b,t) = S(\rho,\phi,z,t) \quad \text{and} \quad S(\mathbf{r}_e,t) = S(\rho,\phi + 2\pi,z,t). \tag{30}$$

Since we expect to find the same physics again when we make one round, we also expect our quantum mechanical wave function to be the same up to a global phase that is a multiple of 2π. The fact that we expect to find the same physics again tells us that the phase difference in (29) has to be given by:

$$S(\mathbf{r}_e,t) - S(\mathbf{r}_b,t) = n2\pi \quad \text{with} \quad n \in \mathbb{Z}. \tag{31}$$

Using the above restriction (31) on the phase difference the equation for the circulation (29) then yields:

$$\kappa = n\frac{h}{m} \quad \text{with} \quad n \in \mathbb{Z}, \tag{32}$$

in other words the circulation is quantized in multiples of h / m.

The circulation can now also be calculated by using the formula containing the vorticity field (28), yielding

$$\kappa = \int_S d\mathbf{S} \cdot \omega(\mathbf{r},t) = \int_S d\mathbf{S} \cdot \left[\nabla \times \left(\frac{\hbar}{m}\nabla S(\mathbf{r},t)\right)\right] = 0, \tag{33}$$

where we got zero since the curl of a gradient field is always zero. The fact that our two results for the circulation Eqs. (32) and (33) are in contradiction means we did something (mathematically) wrong. Since Eq. (26) is the definition and the restriction Eq. (31) is exact our result, Eq. (32) should be the right one. The only assumption that we made to go from the definition of the circulation Eq. (26) to the formula with the vorticity field, Eq. (28) is our condensate (and thus the surface we integrate over) has no holes. The fact that the results are different means that it was wrong to assume that the condensate has no holes. Or in other words, having a vortex (so a non-vanishing circulation: $n \neq 0$ in Eq. (32)) will create a hole in the condensate. We can even go further and say that it is impossible for a vortex to begin and end in two different points *inside* of the BEC, and this can be illustrated by contradiction. Suppose we have a vortex that ends (or begins) inside of the BEC, then we can choose a surface S belonging to the contour C in such a way that the surface does not intersect with the vortex, an example is given in **Figure 1**. If we can choose our surface as illustrated in **Figure 1**, then Stokes' theorem should hold, leading to a contradiction when calculating the circulation using the two formulas (26) and (28). What we can conclude from this is that vortices extend through the entire BEC,[9] with end and begin points at the end of the condensate.

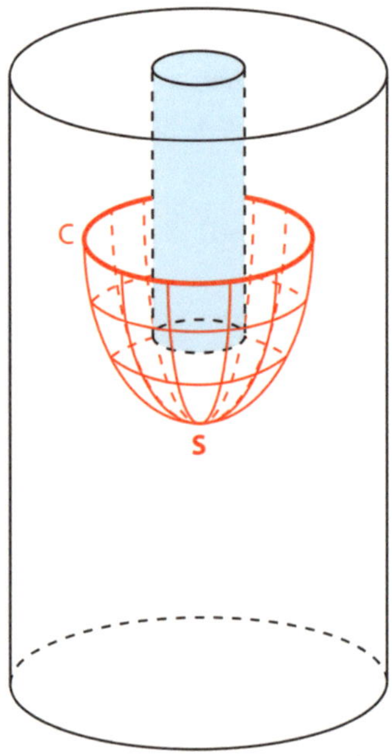

Figure 1. In blue a cylindrical vortex that ends/begins inside the BEC. In red the contour C and its surface S is chosen in such a way that the surface does not intersect with the vortex core.

From the definition of the circulation we can now make three general conclusions about vortices in ultra-cold gases:

- Vortices will make *a hole* in the BEC.
- The *circulation is quantized* with quanta h / m.
- Vortices *cannot have a beginning or ending within* the BEC.

The second property is also the reason why vortices in ultra-cold gases are called quantum vortices, and throughout this chapter, we will however drop the prefix quantum. The last property also tells us that vortices have to be formed at the edges of the BEC, since they cannot end the start inside the BEC. This can be seen in experiments where a condensate that is removing its vortices just pushes them to the outside of the condensate, where they leave.

[9]The exception to this is vortex rings which are closed rings with a non-vanishing circulation. But since a ring has no beginning or ending this is actually no exception to the rule.

http://dx.doi.org/10.5772/67121

In this chapter we will look at the simplest kind of vortices: vortices which have a cylindrically symmetric. This means that we will be looking at straight vortices with through the centre the (rotational) symmetry axis, for which we will choose the z axis. If our macroscopic wave function $\Psi(\mathbf{r},t)$ has a cylindrical symmetry, it is possible to rewrite the wave function as:

$$\Psi(\mathbf{r},t) = f(\rho,z,t)e^{in\phi}, \tag{34}$$

where $f(\rho,z,t)$ is a real-valued structure function. Using the symmetric form of the wave function, Eq. (34), we can calculate the velocity field using Eq. (20), this yields:

$$\mathbf{v}(\mathbf{r},t) = \frac{\hbar}{m}\left(\frac{1}{\rho}\frac{\partial(n\phi)}{\partial\phi}\mathbf{e}_\phi\right) = n\frac{\hbar}{m\rho}\mathbf{e}_\phi. \tag{35}$$

What we see is that the velocity field lies along the ϕ direction, which is indeed a spinning motion in the counter-clockwise direction[10] for positive values of n (vortices). If n is a negative number, the singularity is called an anti-vortex, and we have a clockwise rotation of the condensate. We also see that the velocity field drops as $\propto 1/\rho$ as we move away from the vortex core. Using the velocity field Eq. (35) we can calculate the angular momentum, this becomes equal to:

$$\mathbf{L}_z(\mathbf{r},t) = \mathbf{r}\times m\mathbf{v}(\mathbf{r},t) = m\rho v(\mathbf{r},t) = n\hbar. \tag{36}$$

This tells us that the angular momentum is quantized, every time that we make a vortex the condensate gets an angular momentum kick of the size $n\hbar$. Finally we can also look at the circulation (26) of the wave function $\Psi(\mathbf{r},t)$, this yields:

$$\kappa = \oint_C d\mathbf{s}\cdot\mathbf{v}(\mathbf{r},t) = \int_0^{2\pi}\rho d\phi v(\mathbf{r},t) = \int_0^{2\pi}\rho d\phi n\frac{\hbar}{m\rho}\mathbf{e}_\phi = n\frac{h}{m}. \tag{37}$$

Comparing the above result (37) with the circulation of a BEC (32), we can indeed see that the wave function, Eq. (34) can describe a cylindrically symmetric vortex. The number n in the phase factor of the wave function, Eq. (34) is the number of circulation quanta $h\,/\,m$ the vortex will have.

With now the macroscopic wave function (34) as a starting point, we can now start our study of vortices in a BEC. In order to determine the vortex structure, we should determine the functional dependence of $f(\rho,z,t)$ either analytically (by using a variational model) or numerically, this is done in the next section. To conclude this section we will look at two analytic cases: the homogeneous condensate, and the condensate near a hard wall. The first analytic case will tell us how the condensate behaves far away from the vortex, and this asymptotic behaviour will come in handy

[10] Since we are working in a right-handed coordinate system (ρ,ϕ,z), the direction of rotation along ϕ is given by the right-hand rule. This results in a counter-clockwise direction for the velocity field of the condensate when viewed from above (along the $-\mathbf{e}_z$ direction).

both analytically and numerically. The second analytic case tells us how the condensate heals if you make a hole in it, which is the basis of a commonly used variational model.

4.2. A stationary homogeneous BEC

The first system that can be solved analytically is the homogeneous BEC. For this system we expect that the density is given by:

$$|\Psi(\mathbf{r})|^2 = n_\infty, \tag{38}$$

with n_∞ the particle density of the BEC. If we want to work with a fixed number of N particles, we need to determine the chemical potential. Substituting the wave function, Eq. (38) into the Gross-Pitaevskii equation for a fixed number of particles, Eq. (13) then yields:

$$gn_\infty \Psi(\mathbf{r}) = \mu \Psi(\mathbf{r}). \tag{39}$$

Solving the above equation then yields for the chemical potential:

$$\mu = gn_\infty. \tag{40}$$

The energy of the condensate can be found by taking the expectation value of the Hamiltonian (9), resulting in:

$$E_N[\Psi] = \frac{1}{2} g \int_V d^3\mathbf{r} |\Psi(\mathbf{r})|^4 = \frac{1}{2} g n_\infty^2 V = \frac{1}{2} g \frac{N^2}{V} = \frac{1}{2} \mu N, \tag{41}$$

with V the volume of the BEC and N the total number of particles in the BEC. The above formula for the energy (41) tells us that a BEC likes to spread out and maximize its volume (this decreases the energy). This means that in experiments we will need an external potential $V_1(\mathbf{r})$ to keep the BEC together. Usually this is a harmonic trapping potential of the form

$$V_1(\mathbf{r}) = \frac{1}{2} m \omega_\perp^2 \rho^2 + \frac{1}{2} m \omega_\parallel^2 z^2, \tag{42}$$

which follows from the magneto-optical trap. This harmonic trapping potential leads to cigar-shaped condensates. Another type of confinement potential is the hard wall potential, this trapping potential is realized by using a hollow laser [65], this leads to a potential of the form:

$$V_1(\mathbf{r}) = \Theta(R - \rho). \tag{43}$$

In this chapter, we will mainly focus on the latter hard-wall potential (43) because this yields the simplest calculations and still contains all of the physics. For the harmonic-trapping potential, it is also possible to make simple approximations in order to get easier calculations. The most common approximation made when working with a harmonic trapping potential is the Thomas-Fermi approximation. In the Thomas-Fermi approximation, it is assumed that the kinetic energy is negligible, and this is the case if the number of particles is large enough. In typical experiments this is usually the case.

http://dx.doi.org/10.5772/67121

4.3. A stationary BEC near a hard wall and the healing length

The next analytic case that will be considered is if we put a hard wall edge in the BEC. In other words we will punch a hole in the BEC and see how it heals. This analytic case is important because it will give us the length scale over which the condensate heals, also called the healing length. This healing length will be useful to use as a typical length scale of the system in order to make numerical equations dimensionless. To introduce the healing length, one usually looks at the Gross-Pitaevskii equation, Eq. (10) and fills in the length scale for which the kinetic energy term becomes equal to the interaction term. When substituting the second derivative in Eq. (10) by $1/\xi^2$ this then yields the equation:

$$\frac{\hbar^2}{2m\xi^2} = gn_\infty, \tag{44}$$

for the healing length ξ this solves to:

$$\xi = \frac{\hbar}{\sqrt{2mn_\infty g}} = \frac{1}{\sqrt{8\pi n_\infty a_s}}. \tag{45}$$

Given the healing length, we can now again look at a free BEC (in $x > 0$) in which we place a hard wall potential at $x = 0$. Filling in the Gross-Pitaevskii equation, Eq. (10) for a 1D BEC now yields:

$$-\frac{\hbar^2}{2m}\frac{d^2\Psi(x)}{dx^2} + g|\Psi(x)|^2\Psi(x) = \mu\Psi(x). \tag{46}$$

In the limit $x \to \infty$ we get our homogeneous BEC back, meaning that the chemical potential is given by Eq. (40). If we now define the structure function $f(x)$ as

$$\Psi(x) = \sqrt{n_\infty} f(x) \tag{47}$$

and we fill in the chemical potential, then the Gross-Pitaevskii equation can be written as:

$$\xi^2 \frac{d^2 f(x)}{dx^2} - [f(x)^2 - 1] f(x) = 0, \tag{48}$$

where we divided the Gross-Pitaevskii equation by gn_∞ in order to get the healing length. The above non-linear differential equation is solved by substituting a hyperbolic tangent. The solution for the condensate wave function with a hard wall potential is then given by:

$$\Psi(x) = \sqrt{n_\infty}\tanh\left(\frac{x}{\sqrt{2}\xi}\right). \tag{49}$$

What we see is that the condensate 'heals' from a hole over a typical length given by the healing length ξ, and this is also illustrated in **Figure 2**.

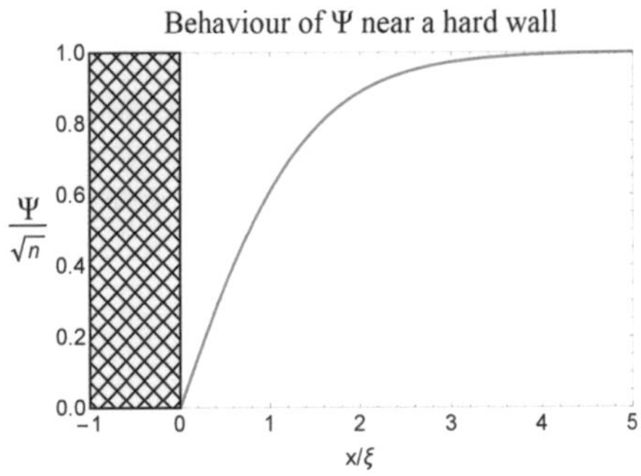

Figure 2. The healing of a BEC near a hard wall over a typical length of ξ.

This solution for a hard wall edge of the BEC will play an important role when we will be studying vortices. As said at the beginning of this section, a vortex will make a hole in the condensate from which the condensate also heals. It is to be expected that when the condensate heals from a hole of a vortex, it happens in an analogous way as when it heals from a hole due to a hard wall. This is because of the fact that both holes follow the same condensate physics, with the same boundary condition: $\Psi(x = 0) = 0$, where $x = 0$ is either the centre of the vortex or the location of the hard wall. The difference for the vortex is that there is an extra rotation of the condensate, this rotation tends to 'smear out' the hole of the vortex, resulting in a larger length over which the condensate heals from a vortex: $\alpha\xi$ (with $\alpha \geq 1$). Thanks to the analogous behaviour as a vortex, the hard wall solution will seem to be a very good variational model to study vortices in ultra-cold gases, even for fermionic gases [66].

4.4. Detection of vortices in superfluids

In essence, every experiment on atomic clouds is done by scattering probe laser light off it. So essentially experimenters will take a 'picture' of the condensate by using a laser in either a destructive or non-destructive manner. This 'picture' that is obtained is in essence the plot of the superfluid density field $n(\mathbf{r},t)$ of the hydrodynamic description given in Eq. (16) at a given time. A phase can also be measured by the way the condensate interacts with either the laser or another condensate.

This means that the main source of information is given by visual information. For the detection of vortices this is problematic since the healing length is of the order of 10 nm (if we fill in the quantities of the third section). To get an idea of how the healing lengths compare to the scattering lengths for several elements that are used to make a BEC, we have given a few values in **Table 1** for typical condensate atoms.

From **Table 1**, we see that the healing length is indeed a very small value, which makes vortices hard to detect (and even measure their structure). The common method to detect vortices in a superfluid is to turn off the confining potential. As we know a BEC likes to maximize its volume, so turning off the confinement will lead to an expanding cloud of atoms

Condensate atom	Scattering length a_s	Healing length ξ	a_s / ξ
Cesium (^{133}Cs)	300	3.646×10^2	1.214
Rubidium (^{87}Rb)	100	6.31×10^2	6.31
Sodium (^{23}Na)	50	8.92×10^2	17.84
Lithium (^{7}Li)	30	1.15×10^3	38.3
Hydrogen (^{1}H)	1	6.31×10^3	6310

Both lengths are given in units of $a_0 = 10^{-8}$cm (or 0.1 nm), for a typical particle density equal to $n_\infty = 10^{15}$cc^{-1} (or 10^{21}m^{-3}). Note that although the healing length is very small, it is always considerably larger than the s-wave scattering length.

Table 1. The approximate scattering length a_s and approximate healing length ξ (44) for a few typical atoms that undergo BEC [55].

(also called 'Ballistic expansion' by experimenters). Since the BEC expands, the cores of the vortices will start expanding as well, making the vortex cores easier to detect. Typical images that we get from this kind of experiments are given in **Figure 3**. As we see the vortices become visible after expansion. The two things that we can see in an experiment are how many vortices we have and how they are ordered.

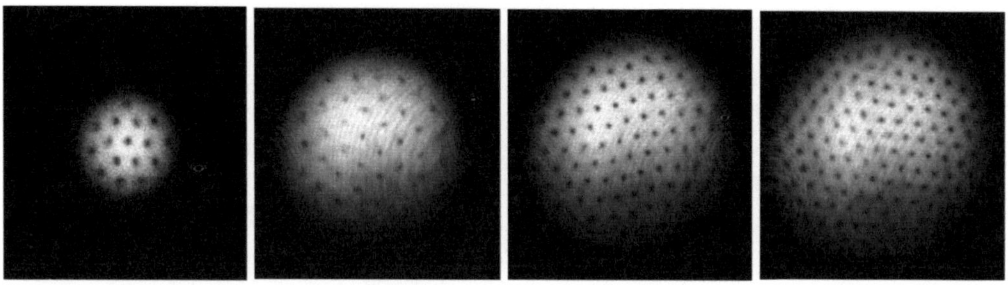

(a) $N = 16$ (b) $N = 32$ (c) $N = 80$ (d) $N = 130$

Figure 3. Four experimental outcomes for vortices, depending on how fast the condensate is stirred. From left to right (a–d), the stirring frequency increases, leading to the formation of more vortices in the BEC. The diameter of the cloud in figure (d) is 1 mm after the ballistic expansion, which is a magnification of 20 times. Figures taken from Ref. [67].

During ballistic expansion, we can also extract extra information by using interference of matter waves (condensates). When we make two BECs overlap they will give rise to a matter interference pattern. This means that if we take a reference BEC (without any vortices) and make this interfere with a BEC that contains a vortex, the effect of the vortex can be seen in the interference pattern. The vortex shows up as a fork dislocation, an example of a BEC with one vortex overlapping with a BEC with no vortices is given in **Figure 4**. As seen in the figure, the presence of the vortex leads to an irregularity in the interference pattern that is known as a fork dislocation. The fork dislocation was also already known in optical systems where optical vortices are also a field of study [69].

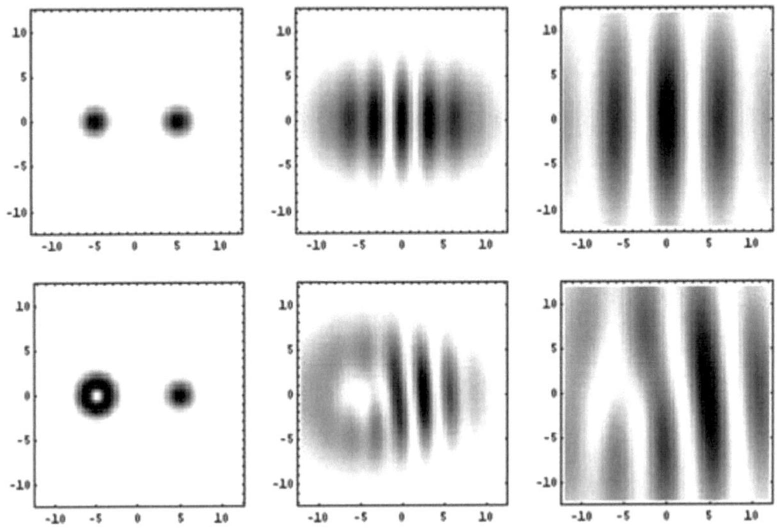

Figure 4. Two BECs in a trap, when the trap is turned off the condensates will expand and eventually overlap leading to an interference pattern. The top figures show two condensates without a vortex, in the bottom figures the left condensate has a vortex. In the interference pattern with the vortex, the fork dislocation becomes apparent. Figure taken from Ref. [68].

The additional information that we get with respect to the simple ballistic expansion is the fact that we can now also see the number of vortex quanta (or circulation quanta) that are present in the vortex. The more circulation quanta a vortex has, the more fringes will be involved in the fork dislocation. The interference pattern of two condensates can also tell us something about the sign of the circulation. A vortex (velocity field gives a counter-clockwise rotation) will have a fork dislocation which is upside down compared to the fork dislocation of an anti-vortex (spinning clockwise).

5. Vortex formation in Bose-Einstein condensates

With two basic solutions for ultra-cold gases and the general properties of a vortex from the previous section we can now continue. In this section we will first look at how vortices appear in a condensate. Afterwards, we will look at the structure of a single vortex and two simple variational models.

5.1. Energy of a rotating BEC

Since vortices classically appear when we start stirring our system, a rotating condensate might seem a good setting for a vortex system. In atomic gases this rotating system is achieved by stirring the condensate using a (blue de-tuned) laser which acts like a spoon. We will look at a system that rotates with a rotational velocity equal to Ω. Looking back at the previous section it only makes sense that we should be looking for vortices in a rotating condensate. The only

http://dx.doi.org/10.5772/67121

way to get a rotating BEC (and hence a non-zero circulation in Eq. (26)) is by having a vortex in the BEC.[11]

From the third section the energy of a BEC is already known and given by the expectation value of Eq. (9). This leads to three different energy contributions given by:

$$E_{\text{kin}}[\Psi(\mathbf{r})] = \frac{-\hbar^2}{2m}\int_V d^3\mathbf{r}\Psi^*(\mathbf{r})\nabla^2\Psi(\mathbf{r}) = \frac{\hbar^2}{2m}\int_V d^3\mathbf{r}|\nabla\Psi(\mathbf{r})|^2, \tag{50}$$

$$E_{\text{trap}}[\Psi(\mathbf{r})] = \int_V d^3\mathbf{r}\Psi^*(\mathbf{r})V_1(\mathbf{r})\Psi(\mathbf{r}), \tag{51}$$

$$E_{\text{int}}[\Psi(\mathbf{r})] = \frac{1}{2}g\int d^3\mathbf{r}|\Psi(\mathbf{r})|^4, \tag{52}$$

where the second equality in Eq. (50) follows from partial integration and the fact that the boundary term drops out.[12] The time dependency was explicitly left out since we will be looking at stationary solutions for the vortex in the remainder of the chapter.

Since for vortices we will be working in the hydrodynamic description, we will substitute the hydrodynamic form of the wave function in Eq. (16) in the different energy contributions in Eqs. (50)–(52), this yields:

$$E_{\text{kin}}[\Psi(\mathbf{r})] = \frac{\hbar^2}{2m}\int_V d^3\mathbf{r}\left(\nabla\sqrt{n(\mathbf{r})}\right)^2 + \frac{m}{2}\int_V d^3\mathbf{r}[n(\mathbf{r})v^2(\mathbf{r})], \tag{53}$$

$$E_{\text{trap}}[\Psi(\mathbf{r})] = \int_V d^3\mathbf{r}\, n(\mathbf{r})V_1(\mathbf{r}), \tag{54}$$

$$E_{\text{int}}[\Psi(\mathbf{r})] = \frac{1}{2}g\int d^3\mathbf{r}\left(n(\mathbf{r})\right)^2, \tag{55}$$

The only missing ingredient is the Hamiltonian in a rotating frame of reference, this Hamiltonian is given by:

$$\hat{H}' = \hat{H} - \mathbf{\Omega}\cdot\hat{\mathbf{L}}. \tag{56}$$

This additional term $-\mathbf{\Omega}\cdot\hat{\mathbf{L}}$ is nothing else but the rotational energy (also known from classical mechanics). What we see is that in order to form a vortex the profit in energy when rotating $\mathbf{\Omega}\cdot\hat{\mathbf{L}}$ should be larger than the energy cost of making the vortex. In other words, a vortex will be formed if in Eq. (56) we have that $\langle\hat{H}'\rangle$ with a vortex is smaller than $\langle\hat{H}\rangle$ without a vortex. In order to give an estimate of the minimal required rotational velocity Ω needed to form a vortex

[11] A condensate with the same amount of vortices and anti-vortices does not need to be rotating since these circulations cancel each other out.

[12] We either will look at an infinite condensate or a condensate with a hard-wall edge. Both cases yield a zero valued boundary term

we will need a model for the vortex core. To get a fairly rough, but close estimate we introduce the simplest possible model for a vortex: the cylindrical vortex hole. This model will allow us to easily make a few statements about vortex stability and interactions.

5.2. Simplest model for a vortex

The simplest model for a vortex that we can imagine is to represent the vortex by a hollow cylindrical hole. The radius of the vortex hole is chosen to be equal to the healing length[13] and the height of the vortex given by H. As we know the velocity field of the vortex should be given by Eq. (35), only inside the vortex core there is no condensate and we are free to choose our velocity field. In order to have a correct circulation (even within the vortex core) we assume the velocity field to take a constant value. The simple vortex model is depicted in **Figure 5**.

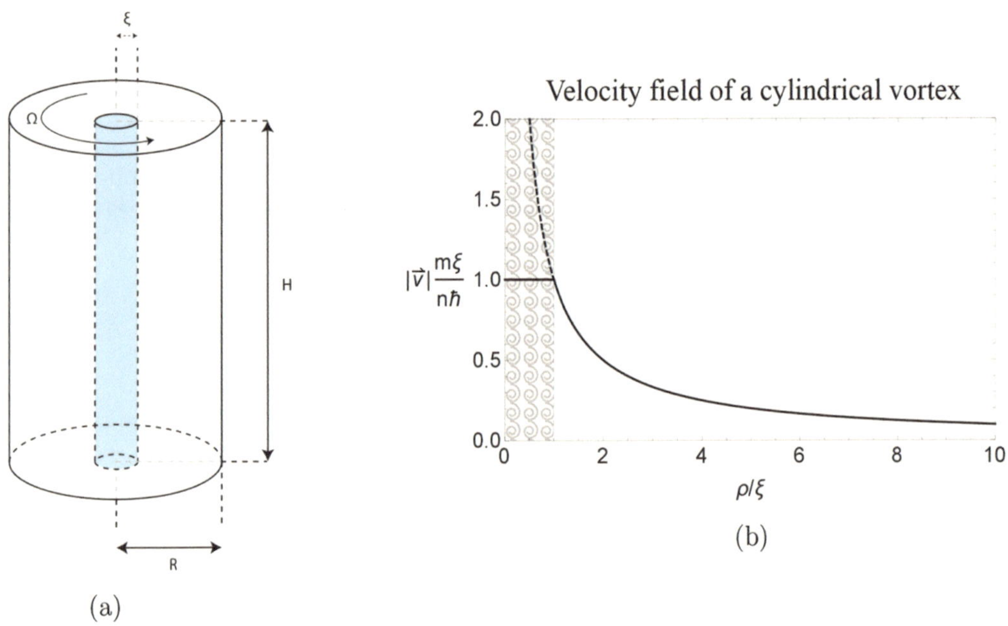

Figure 5. The simplest possible vortex model: a hollow cylindrical vortex core with radius ξ in a cylindrical condensate with height H and radius R, shown in figure (a). Figure (b) shows the corresponding vortex velocity field.

In formulas this means that the simple vortex model is given by:

$$n(\mathbf{r}) = n_{\infty}\Theta(\rho-\xi)\Theta(R-\rho), \tag{57}$$

$$\mathbf{v}(\mathbf{r}) = n\frac{\hbar}{m\rho}\mathbf{e}_{\phi}\Theta(\rho-\xi)\Theta(R-\rho) + n\frac{\hbar}{m\xi}\mathbf{e}_{\phi}\Theta(\xi-\rho). \tag{58}$$

With this model (and the above formulas) we can now calculate the different energies as given in Eqs. (41–44) for the rotating condensate.

---

[13]You can change this by some effective length R_{cylinder}, this will not change the calculations or results

http://dx.doi.org/10.5772/67121

5.3. Formation of one and more vortices

Since only the energy difference compared to the homogeneous (vortex free) state is relevant, we will only calculate the energy differences between the homogeneous and vortex state. We will look at a condensate confined by an infinite potential well, so $V_1(\mathbf{r}) = 0$ within the condensate. This means that we have two possible energy contributions, one for the velocity field and one for the interactions.

If we have a vortex in our condensate, then we will get a velocity field of the form (58), which will give an extra kinetic energy. The difference kinetic energy (41) then becomes[14]:

$$\Delta E_{\text{kin}} = \int_V d^3\mathbf{r}\left(\frac{m}{2}n(\mathbf{r})v^2(\mathbf{r})\right) = 2\pi\frac{1}{2}mn_\infty H\int_\xi^R \rho d\rho\left(n\frac{\hbar}{m\rho}\right)^2 = n^2\pi n_\infty\frac{\hbar^2}{m}H\ln\left(\frac{R}{\xi}\right). \tag{59}$$

We also have a contribution in energy of the interaction energy, namely the energy that is needed to punch a hole in the BEC. This needed energy is given by

$$\Delta E_{\text{int}} = \frac{1}{2}g\int_V d^3\mathbf{r}\Big(n(\mathbf{r})\Big)^2 = \frac{1}{2}gV_{\text{hole}}, \tag{60}$$

with

$$V_{\text{hole}} = \pi\xi^2 H, \tag{61}$$

the volume of the hole that has been made in the condensate. If we now assume (for simplicity) that[15] $R >> \xi$ (the condensate is much larger than the volume of the hole), then the energy contribution of Eq. (60) will be negligible. The total cost of energy for making a vortex is thus given by:

$$\Delta E = \Delta E_{\text{kin}} = n^2\pi n_\infty\frac{\hbar^2}{m}H\ln\left(\frac{R}{\xi}\right). \tag{62}$$

In order to be advantageous, the energy cost of making a vortex from Eq. (62) should be smaller as the energy reduction of making a vortex rotate. In order to calculate the energy reduction of rotating the condensate, we need to calculate the second term of Eq. (56), for this we need the expected angular momentum $\langle\hat{L}\rangle$. Within our simple model the expected angular momentum $\langle\hat{L}\rangle$ can be easily calculated:

$$\langle\hat{L}\rangle = \langle\hat{r}\times m\mathbf{v}(\mathbf{r})\rangle = \int_V d^3\mathbf{r}\left[(\rho\mathbf{e}_\rho)\times\left(mn(\mathbf{r})\frac{n\hbar}{m\rho}\mathbf{e}_\phi\right)\right] = n\pi\hbar n_\infty H(R^2 - \xi^2)\mathbf{e}_z. \tag{63}$$

With Eq. (63) the energy reduction of rotating the condensate is given by:

---

[14]The first term drops out since we have a homogeneous condensate and thus $\nabla\left(\sqrt{n(\mathbf{r})}\right) = 0$.

[15]This is not a bad assumption since in most condensates we have that $R / \xi \approx 30$.

$$\Delta E_{\mathrm{rot}} = \langle \boldsymbol{\Omega} \cdot \hat{\boldsymbol{L}} \rangle = n\pi\hbar n_{\infty} H(R^2 - \xi^2)\mathbf{e}_z, \tag{64}$$

where we assumed that $\boldsymbol{\Omega}$ is parallel to the vortex core. The fact that $\boldsymbol{\Omega}$ is parallel to the vortex core is also needed because otherwise we can have circulation in the BEC without having a vortex, in the previous section we showed this was not possible.

A vortex will be formed in the BEC if it is energetically favourable; this means that the cost of making a vortex as in Eq. (62) is smaller than (or equal to) the energy profit or rotating the condensate as in Eq. (64). Setting the energy cost (62) equal to the energy profit (64) we can now calculate the critical frequency:

$$\Omega_{\mathrm{cr}} = n \frac{\hbar}{m} \frac{\ln(R/\xi)}{R^2 - \xi^2}. \tag{65}$$

The critical frequency yields the minimal rotational frequency that is required in order to get a vortex. The critical frequency is plotted as a function of the condensate size in **Figure 6**.

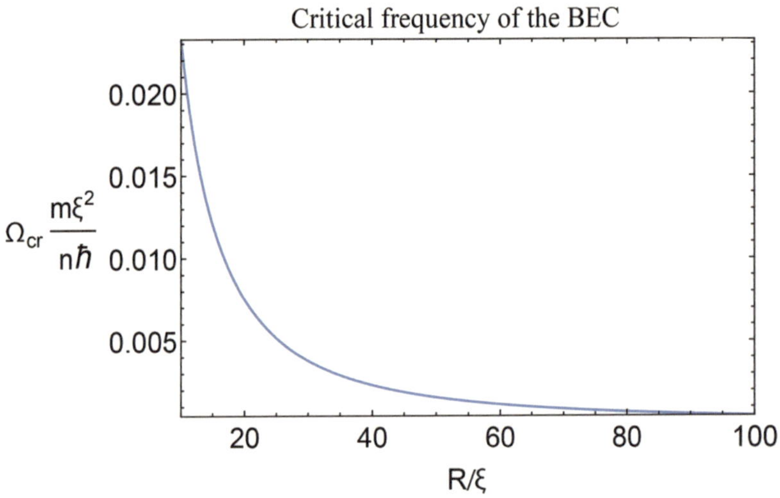

Figure 6. The critical frequency in Eq. (65) as a function of R / ξ.

A first thing that can be noted is that the critical frequency given in Eq. (65) is directly proportional to the number of circulation quanta present in the vortex. This means that a $n = 1$ vortex (one circulation quantum) is the first possible vortex that can be formed. If we now rotate twice as fast we can in principle, a vortex can be formed with $n = 2$ (two circulation quanta) *or* we can make two $n = 1$ vortices. Which one of the two that will be formed depends on the relative energy of both vortex configurations. In order to see which configuration will be beneficial, it is worthwhile to look at how the energy cost (62) and energy reduction (64) depend on the number of circulation quanta n. For the energy cost (62) we see that $\Delta E_{\mathrm{kin}} \propto n^2$, while the energy reduction (64) goes as $\Delta E_{\mathrm{rot}} \propto n$. This means that as your number of circulation quanta rises, the energy cost will rise faster than the energy reduction. Indeed if we have two $n = 1$ vortices, we see that the energy cost will rise with a factor two, and the energy reduction will also rise with a factor two. While if we have a $n = 2$ vortex the energy cost will rise with a factor four, while the energy reduction will still remain a factor two. This means that for the condensate it is more advantageous to make more vortices as you rotate faster, rather than making one big vortex. This is also seen in experiments (for example **Figure 3**), due to interactions the vortices will form lattices (Abrikosov lattices) in the BEC. Note that for both cases (many vortices of one big vortex) the critical frequency (65) will be the same.

A second aspect that can be noticed is that from our simple derivation it follows that two vortices will repel each other, while a vortex and anti-vortex will attract each other. The fact that two vortices repel each other can be seen from the fact that if we try to push them in the same place, they will behave like one vortex with the double amount of circulation quanta. As we have seen one vortex with the double amount of circulation quanta is energetically less advantageous than two separate vortices. This means that two vortices will repel each other in order to avoid an energetically less advantageous situation. For a vortex-anti-vortex pair this is different, these will attract in order to reduce the energy. If we have a vortex and an anti-vortex the total circulation equals zero, meaning that the condensate as a whole is not rotating. The fact that the condensate is not rotating means that there is no energy reduction (64), but only an energy cost (62) for making the holes. By pushing the vortex and anti-vortex together, their velocity fields will cancel each other out, leading to an annihilation of both vortices. Interactions between vortex pairs and vortex-anti-vortex pairs will be studied in more detail in the next section.

The results about vortex formation in a BEC can be summarized as:

- Energetically is more favourable to *only* make vortices with *one* circulation quantum ($n = 1$).
- Two vortices will repel each other.
- A vortex and anti-vortex will attract each other.

Since for a condensate it is advantageous to only make $n = 1$ vortices, we will assume for the remainder of the chapter that $n = 1$. So we will only look at vortices with one circulation quantum.

5.4. The exact structure of a single vortex

For our derivation of the vortex formation we assumed that our vortex core was a cylindrical hole. As it will turn out this is not a bad approximation as long as $R >> \xi$, which was exactly what we assumed in our previous derivation for the critical frequency. Of course we should always check whether our assumptions are good, in order to do this we will determine the *exact* structure of a $n = 1$ vortex.

To determine the vortex structure we start from the Gross-Pitaevskii equation, Eq. (13), where we substitute the Laplacian in cylindrical coordinates (ρ,ϕ,z):

$$-\frac{\hbar^2}{2m}\left[\frac{1}{\rho}\frac{\partial}{\partial\rho}\left(\frac{\partial\Psi(\mathbf{r})}{\partial\rho}\right)+\frac{1}{\rho^2}\frac{\partial^2\Psi(\mathbf{r})}{\partial\phi^2}+\frac{\partial^2\Psi(\mathbf{r})}{\partial z^2}\right]+V_1(\mathbf{r})\Psi(\mathbf{r})+g|\Psi(\mathbf{r})|^2\Psi(\mathbf{r})=\mu\Psi(\mathbf{r}). \quad (66)$$

In the Gross-Pitaevskii equation, Eq. (66) we can now substitute the general cylindrical symmetric wave function (34). After eliminating the phase factor $e^{i\phi}$ we find that our equation of motion (66) becomes equal to:

$$-\frac{\hbar^2}{2m}\left[\frac{1}{\rho}\frac{\partial}{\partial\rho}\left(\rho\frac{\partial f(\rho,z)}{\partial\rho}\right)+\frac{\partial^2 f(\rho,z)}{\partial z^2}\right]+\left[\frac{\hbar^2}{2m\rho^2}+V_1(\rho,z)\right]f(\rho,z)+gf^3(\rho,z)=\mu f(\rho,z). \quad (67)$$

The extra term next to the potential is the rotational energy due to the vortex flow, note that this term fill force $f(\rho,z)$ to zero as we approach the vortex core in $\rho = 0$.

The Gross-Piteavskii equation, Eq. (67) will now be solved for the uniform BEC, so $V_1(\rho,z) = 0$. Since there is no more z dependency in the equation of motion (67) for the chosen potential, the structure of the vortex will also no longer depend on z. For a uniform BEC the Gross-Pitaevskii equation then becomes:

$$-\frac{\hbar^2}{2m}\frac{1}{\rho}\frac{\partial}{\partial\rho}\left(\rho\frac{df(\rho)}{d\rho}\right)+\frac{\hbar^2}{2m\rho^2}f(\rho)+gf^3(\rho)=\mu f(\rho). \quad (68)$$

We are now left with an ordinary non-linear differential equation (68) that needs to be solved. This means that the solution will be most likely given by a numerical solution. In order to easily solve the equation of motion (68) numerically, it should be made dimensionless. If we look at the limit $\rho \to \infty$ (far away from the vortex core) we find that the solution to the differential equation is given by:

$$f(\rho\to\infty)=\sqrt{\frac{\mu}{g}}=\sqrt{n_\infty}, \quad (69)$$

where the last equality follows from the solution of a homogeneous BEC in Section 4. We now have two relevant quantities to make the differential equation dimensionless, namely the value at infinity f_0 and the healing length ξ:

$$f_0=\sqrt{\frac{\mu}{g}}=\sqrt{n_\infty} \quad \text{and} \quad \xi=\frac{1}{\sqrt{8\pi n_\infty a_s}}=\frac{1}{f_0}\frac{\hbar}{\sqrt{2gm}}. \quad (70)$$

Using the asymptotic value and healing length (70), we can define a dimensionless radial coordinate x and a wave function $\chi(x)$ as:

$$x=\frac{\rho}{\xi} \quad \text{and} \quad \chi(x)=\frac{f(x\xi)}{f_0}. \quad (71)$$

Substituting our dimensionless quantities (70) and variables (71) into the differential equation (68) yields:

$$-\frac{1}{x}\frac{d}{dx}\left(x\frac{d\chi(x)}{dx}\right)+\frac{\chi(x)}{x^2}+\chi^3(x)-\chi(x)=0. \quad (72)$$

This remaining differential equation, Eq. (72) can now be solved. In order to solve the differential equation, Eq. (72) for the vortex structure, we impose the boundary conditions:

$$\chi(x=0)=0 \quad \text{and} \quad \chi(x\to\infty)=1. \quad (73)$$

The last boundary condition can also be replaced by

$$\frac{d\chi(x)}{dx}|_{x\to\infty} = 0, \tag{74}$$

depending on which numerical algorithm is used. In the latter case one should check whether the stable value equals one. In order to solve the equation of motion, Eq. (72) together with the boundary conditions (73) the shooting method can be used. The shooting method is a fast and popular way to solve non-linear ODEs with boundary conditions. Using the shooting method one starts in one of the boundary conditions and then 'shoots' out paths at different slopes. In order for a solution to become accepted, it should end within a given tolerance near the other boundary condition. At first the possible paths for $\chi(x)$ are shot at various large slopes, the range of slopes under which the paths are shot out reduces in each step. The reduction of the range of slopes is done by choosing the slopes the two closest paths near the other boundary condition. The great advantage of the shooting method is that it reduces a problem with boundary conditions to a problem with initial conditions, which is more easily solved. The shooting method is demonstrated in **Figure 7** for our equation of motion, Eq. (58). The big problem with the shooting method is that for unstable (non-linear) problems (as our vortex structure for example) the algorithm is also fairly unstable. A trick to overcome this instability is to start at a low x value (e.g. at $x = 3$ in **Figure 7**) for comparing with the boundary condition. At lower values of x, the value of $\chi(x)$ is smaller and hence also the value of $\chi^3(x)$, leading to a more stable program. Each time the interval of slopes becomes narrower, the x-value at which the solutions are compared to the boundary value increases since the solution then becomes more stable.

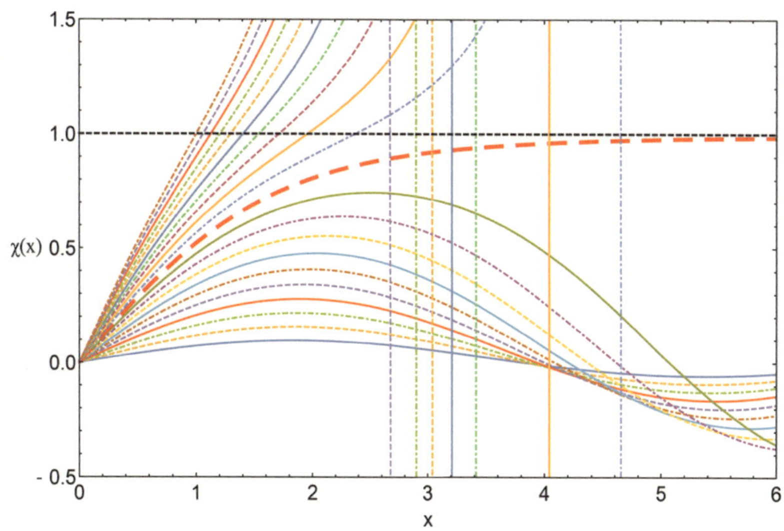

Figure 7. The shooting method solution of the ODE (72) with boundary conditions (73). The eventual vortex structure is plotted as a thick red dashed line.

As we see from the above discussion solving a non-linear ODE is quite involved and unstable. A more stable method to find the vortex structure is to numerically minimize the energy functional. In polar coordinates the total energy of the BEC (expectation value of Eq. (9)) is given by:

$$E[f(\rho)] = \int_V d^3\mathbf{r}\left[\frac{\hbar^2}{2m}\left(\frac{df(\rho)}{d\rho}\right)^2 + \frac{\hbar^2}{2m}\frac{f^2(\rho)}{\rho^2} + \frac{g}{2}f^4(\rho)\right]. \tag{75}$$

Since we have no ϕ of z dependency in the energy (75), these can be integrated out yielding an energy per height H of the condensate:

$$\frac{E[f(\rho)]}{H} = \int_0^R 2\pi\rho d\rho\left[\frac{\hbar^2}{2m}\left(\frac{df(\rho)}{d\rho}\right)^2 + \frac{\hbar^2}{2m}\frac{f^2(\rho)}{\rho^2} + \frac{g}{2}f^4(\rho)\right]. \tag{76}$$

One thing that can already be noted is the fact that for large values of ρ we have that $f(\rho)$ becomes constant, this will yield a logarithmic divergence of the integral. In order to get sensible results we can either renormalize the integral by subtracting the $\rho \to \infty$ behaviour, or (as is the case in Eq. (76)) keeping the integration interval finite. In order to simplify the numerics, the energy of the uniform BEC is subtracted from the energy of the cylindrical vortex (76). For a homogeneous BEC we know from Section 4 the energy of a homogeneous BEC (41), per unit length this becomes:

$$\frac{E_0[\Psi]}{H} = \frac{1}{2}gn_\infty^2\frac{V}{H} = \frac{1}{2}gn_\infty^2\pi R^2 = \frac{1}{2}gf_0^4\pi R^2, \tag{77}$$

with V the volume of the vortex core. Since the number of particles should be preserved when we go from a uniform BEC to a BEC containing a vortex, the number of particles per unit length N / H should also be preserved. The number of particles per unit length (per cross section of the cylindrical condensate) can be written as ν, related to n_∞ as:

$$n_\infty = \frac{\nu}{\pi R^2}. \tag{78}$$

Since in the system with the vortex the number of particles per unit length ν should be the same as in the uniform case, we can now calculate ν using the vortex solution:

$$\nu = \frac{1}{H}\int_V d^3\mathbf{r} f^2(\rho) = \int_0^R 2\pi\rho d\rho f^2(\rho) = \pi R^2 f_0^2 - \int_0^R 2\pi\rho d\rho\left(f_0^2 - f^2(\rho)\right), \tag{79}$$

where we added and subtracted the uniform solution in the last step. Using both Eqs. (78) and (79), the energy of a homogeneous BEC per unit length (77) can be rewritten as:

$$\frac{E_0[\Psi]}{H} = \frac{1}{2}gf_0^4\pi R^2 - gf_0^2\int_0^R 2\pi\rho d\rho\left(f_0^2 - f^2(\rho)\right), \tag{80}$$

where the term

$$\frac{g}{\pi R^2}\left(\int_0^R 2\pi\rho d\rho\left(f_0^2 - f^2(\rho)\right)\right)^2 \tag{81}$$

was dropped. The term that was dropped is of the order

$$\frac{f_0^4 g \xi^4}{R^2}, \tag{82}$$

Since we assume that $R >> \xi$, this term is negligible. If we now subtract the energy of the homogeneous BEC (80) from our total energy (76) we get:

$$\frac{E[f(\rho)]}{H} = \int_0^R 2\pi\rho d\rho \left[\frac{\hbar^2}{2m}\left(\frac{df(\rho)}{d\rho}\right)^2 + \frac{\hbar^2}{2m}\frac{f^2(\rho)}{\rho^2} + \frac{g}{2}\left(f_0^2 - f^2(\rho)\right)^2\right]. \tag{83}$$

The advantage of writing the energy as Eq. (83) is that the last term now vanishes as $\rho \to \infty$ together with the first term. The form (83) is a convenient form for numerical calculations of the vortex energy. By using our dimensionless quantities (70) and variables (71), the energy (83) can be made dimensionless yielding:

$$\frac{E[\chi(x)]}{H} = \frac{\pi\hbar^2}{m} n_\infty \int_0^{R/\xi} x dx \left[\left(\frac{d\chi(x)}{dx}\right)^2 + \frac{\chi^2(x)}{x^2} + \frac{1}{2}\left(1 - \chi^2(x)\right)^2\right]. \tag{84}$$

Note that the upper boundary of the energy integral equals R/ξ, since $R >> \xi$ this upper boundary is usually (with variational models) approximated by ∞. If we now integrate to a very large value of R / ξ then the second term in (84) will have a dominant contribution. As a first approximation for the energy of the vortex we get:

$$\frac{E[\chi(x)]}{H} \approx \frac{\pi\hbar^2}{m} n_\infty \ln\left(\frac{R}{\xi}\right). \tag{85}$$

Note that this is the energy of the cylindrical vortex (62). To get a numerically even nicer form for the energy (83) the first (asymptotic) approximation (85) can be pulled out of the integral. After rewriting the numerically most optimal form is given by:

$$\begin{aligned}\frac{E[\chi(x)]}{H} &= \frac{\pi\hbar^2}{m} n_\infty \ln\left(\frac{R}{\xi}\right) \\ &+ \frac{\pi\hbar^2}{m} n_\infty \int_0^{R/\xi} x dx \left[\left(\frac{d(1-\chi(x))}{dx}\right)^2 - \frac{1-\chi^2(x)}{x^2} + \frac{1}{2}\left(1-\chi^2(x)\right)^2\right].\end{aligned} \tag{86}$$

The main advantage of Eq. (86) is that the integral integrates over $(1-\chi(x))$ which drops to zero as $x \to \infty$, this makes Eq. (86) numerically more stable than Eq. (84). This numerical stability is because of the fact that the logarithmic divergence was analytically removed in Eq. (86), allowing for an integration up to ∞. We can substitute the vortex exact vortex solution (shown in **Figure 7**) in the energy, the dependency as a function of R / ξ is shown in **Figure 8**. By making an asymptotic fit of the form $\ln(a \times x)$ the asymptotic expansion for the exact solution was found:

$$\frac{E[\chi_{\text{exact}}]}{H} = \frac{\pi\hbar}{m} n_\infty \ln\left(1.466\frac{R}{\xi}\right). \tag{87}$$

As we see in **Figure 8**, the asymptotic solution of Eq. (87) becomes quite accurate quite fast (already around $R/\xi = 2$). The fact that the asymptotic solution is accurate so fast supports the claim at the beginning of the section that the cylindrical vortex already gives quite a good estimate. In **Figure 8**, we in fact see that for condensates that are large enough and do not contain too many vortices, the cylindrical solution will give good results.

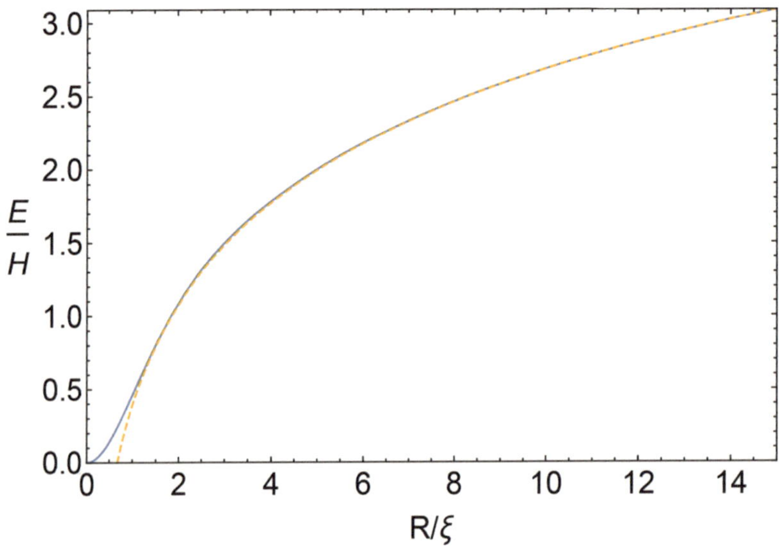

Figure 8. The energy in Eq. (86) as a function of R / ξ for the exact solution. The dashed line represents the asymptotic expansion (87).

5.5. Variational model for the vortex core profile

In the previous we solved for the vortex structure exactly. Once we start looking at more complicated situations this will become quite a time-consuming activity. In order to reduce the numerical work and allow for more analytical work, it is possible to use a variational model. A good variational model should be subjected to the boundary conditions (73). Three possible candidates are given by:

$$\chi_1(x) = \frac{x}{\sqrt{\alpha_1 + x^2}}, \tag{88}$$

$$\chi_2(x) = \tanh\left(\frac{x}{\sqrt{2}\alpha_2}\right), \tag{89}$$

$$\chi_3(x) = 1\text{-}\exp\left(-\frac{x}{\alpha_3}\right). \tag{90}$$

In order to make the calculations as easy as possible, it is wise to choose a variational function consisting of analytical functions, so either Eq. (89) or Eq. (90). Remember from Section 4 that a condensate heals from a hole like a hyperbolic tangent (49), this might be a possible argument to choose (89) as the variational vortex function. In order to determine the values of $\{\alpha_1,\alpha_2,\alpha_3\}$ the variational models are substituted into the energy functional (86) and this energy is minimized with respect to $\{\alpha_1,\alpha_2,\alpha_3\}$. The values for that are found for the variational parameters are given by:

$$\alpha_1 = 2, \tag{91}$$

$$\alpha_2 = \sqrt{\frac{3}{4\ln(2)-1}}\approx 1.3, \tag{92}$$

$$\alpha_3 = \frac{12}{\sqrt{89}}\approx 1.27. \tag{93}$$

Given these variational parameters we can now see how they compare to the exact solution. The three variational solutions for Eqs.(88)–(90) are plotted with the exact solution in **Figure 9**.

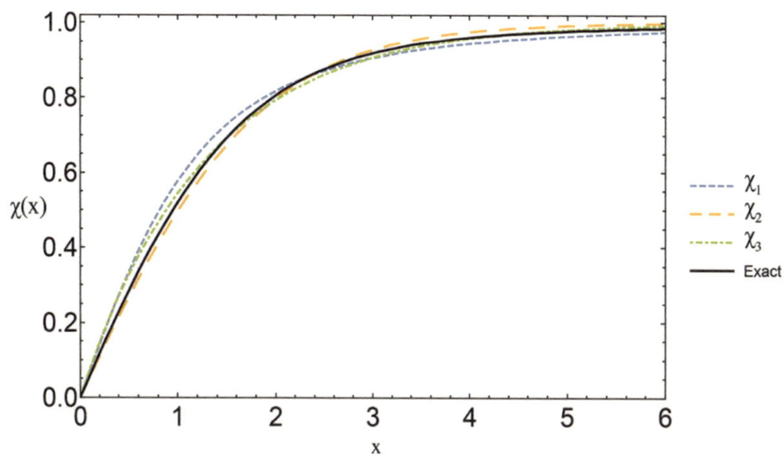

Figure 9. The exact solution for the vortex structure together with the three variational solutions (88)–(90).

For the remainder of this chapter we will work with either the cylindrical vortex core (57) or the hyperbolic tangent (89). This choice of variational model is made because for the hyperbolic tangent it is already proven that it gives good results for the vortex core structure [88]. For larger values of R / ξ, the hyperbolic tangent reduces to the cylindrical vortex, as it should since the exact solution does the same, in order to show that the hyperbolic tangent indeed reduces to the cylindrical case.

6. Several basic vortex configurations

Now that we know how reliable our variational models are and their range of applicability, we can look at three different vortex configurations:

- A vortex-antivortex system
- An off-centre vortex

- A vortex-vortex system

This will give us an idea of how stable vortices are in a BEC and how they interact. The interactions between vortices happen because the vortices feel each other's velocity field, which leads to a Magnus force. For the first part, the easiest variational model (cylindrical hole) is used. Once the forces are known, we can also look at the energy (and thus potential) of the system in a variational manner. For this system, the variational solutions using the hyperbolic tangent are presented.

Note that although these methods might seem specific to vortices and hydrodynamics, there are a lot of other applications. An example is given by the application of the Magnus force to model the interaction of the solar wind plasma with the upper layers of planetary atmospheres. In the absence of particle-particle interactions between both plasma media (solar wind and atmosphere), the kinetic momentum of the solar wind will be transferred to the planetary particles through wave-particle interactions. These wave-particle interactions will produce vortex motion, which will cause these Magnus force interactions [70, 71].

6.1. A vortex in a velocity field: Magnus force

The interaction of vortices is because they feel each other's velocity field. If we have a rotating system (vortex) that moves with a relative velocity with respect to a fluid, it will feel a Magnus force. A nice qualitative picture of the Magnus force is given in **Figure 10**. From the image on the right side, it is intuitively clear how the Magnus force will emerge. The rotating cylinder will bend the indecent velocity field. In order to bend the velocity field, an acceleration is needed, which results in a force (second law of Newton). This force will both act on the fluid as well as on the cylinder.

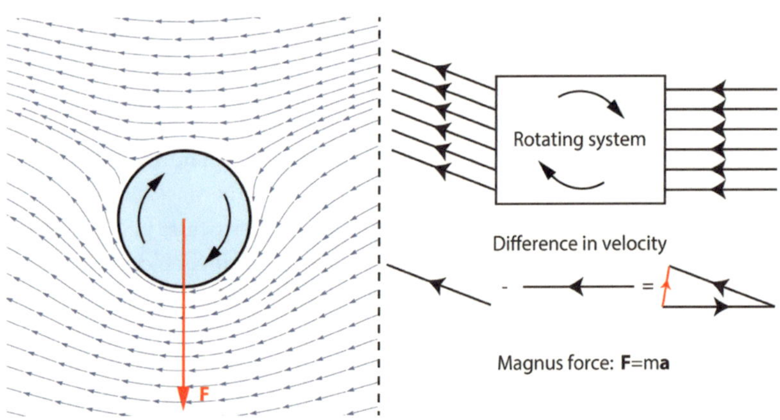

Figure 10. Left: a flow (velocity field) around a rotating (circular) body together with the resulting Magnus force (red), right: the simple qualitative picture.

In order to calculate the force on a vortex, we will use the variational model of the cylindrical vortex core. In this case the force can even be analytically calculated. The starting point for the Magnus force is the Bernoulli equation

$$p(\mathbf{r}) + \frac{1}{2}n(\mathbf{r})v^2(\mathbf{r}) + \phi(\mathbf{r}) = \text{const}, \tag{94}$$

where $\phi(\mathbf{r})$ is a potential term. The Bernoulli equation (94) can be derived from the Euler equation ((24)) by integration. Note that since our vortex flow is irrotational the Bernouilli equation, Eq. (94) is valid everywhere in the BEC and not only along the same stream line. For simplicity we also look in a homogeneous BEC (no potentials) so that our potential term $\phi(\mathbf{r})$ in the Bernoulli equation drops out. Taking our reference values far away from the vortex core (at infinity), the Bernoulli equation becomes:

$$p(\mathbf{r}) + \frac{1}{2}n(\mathbf{r})v^2(\mathbf{r}) = p_\infty + \frac{1}{2}n_\infty v_\infty^2. \tag{95}$$

Note that $\mathbf{v}_i nfty$ is equal to the translational velocity v_{trans} of the vortex. Since the velocity field $\mathbf{v}(\mathbf{r})$ and density field $n(\mathbf{r})$ of the vortex are known the Bernoulli equation, Eq. (95) gives us a prescription for the pressure. Given the pressure we can calculate the resulting force as:

$$\mathbf{F} = -\oint_S d\mathbf{S}p(\mathbf{r}) = -\int_V d^3\mathbf{r}\left(\nabla p(\mathbf{r})\right), \tag{96}$$

where the Gauss integration theorem was used. To calculate the Magnus force we now fill in the pressure using the Bernoulli equation, Eq. (95). The density field of our cylindrical vortex is known (57), the velocity field is given by [72]:

$$\mathbf{v}(\mathbf{r}) = \left(2v_{\text{trans}}\sin(\phi) + \frac{\kappa}{2\pi\xi}\right)\mathbf{e}_\phi, \tag{97}$$

which is a combination of a doublet flow with a vortex flow in a uniform velocity field. For the Bernoulli equation, Eq. (95) we need the square of the velocity, leading to

$$v^2(\mathbf{r}) = 4v_{\text{trans}}^2\sin^2(\phi) + \left(\frac{\kappa}{2\pi\xi}\right)^2 + 2\frac{\kappa}{\pi\xi}v_{\text{trans}}\sin(\phi). \tag{98}$$

In order to calculate the Magnus force (96) we will use the first formula, expanding this in the three parts of the cylinder this gives:

$$\mathbf{F} = -\oint_S d\mathbf{S}p(\mathbf{r}) = -(\int_{\text{Upper plane}} + \int_{\text{Lower plane}} + \int_{\text{Mantle}})d\mathbf{S}p(\mathbf{r}). \tag{99}$$

The first two integrals in Eq. (99) will be equal to zero since the upper and lower parts of the cylindrical vortex are not inside of the BEC, yielding:

$$\mathbf{F} = -\int_{\text{Mantle}} d\mathbf{S}p(\mathbf{r}) = H\int_0^{2\pi} R\cos(\phi)d\phi\mathbf{e}_x + H\int_0^{2\pi} R\sin(\phi)d\phi\mathbf{e}_y, \tag{100}$$

the $\mathbf{e}_z$ component is zero since this comes from the lower and upper planes which are zero. Note that constant terms (p_∞,...) do not contribute since they gets averaged out in the integration. Substituting the pressure $p(\mathbf{r})$ using the Bernoulli equation (95) the Magnus force becomes (100):

$$\mathbf{F} = -\oint_S d\mathbf{S}\left[-\frac{1}{2}n_\infty\left(4v_{\text{trans}}^2 \sin^2(\phi) + \frac{2\kappa}{\pi\xi}v_{\text{trans}}\sin(\phi)\right)\right] = H\xi n_\infty \frac{\kappa}{\xi}v_{\text{trans}}\mathbf{e}_y. \tag{101}$$

The Magnus force (per unit of length) can now be rewritten as[16]

$$\frac{\mathbf{F}_{\text{Magnus}}}{H} = n_\infty \kappa v_{\text{trans}}\mathbf{e}_y = n_\infty \kappa \times \mathbf{v}_{\text{trans}}, \tag{102}$$

where $\kappa = \kappa\mathbf{e}_z$ and $\mathbf{v}_{\text{trans}} = v_{\text{trans}}\mathbf{e}_x$. Note that this means that the z-axis should point along the vortex core. Given the above formula for the Magnus force, we can now look at our three different vortex configurations.

6.2. Force between a vortex-anti-vortex

The first structure that we will study is a vortex and anti-vortex that can feel each other's velocity field. This is the simplest system that we can have since for this system the condensate does not have to be rotating (total circulation is zero). We look at a vortex-anti-vortex pair where the two vortices are separated by a distance d, a top-view sketch of this system is given in **Figure 11**.

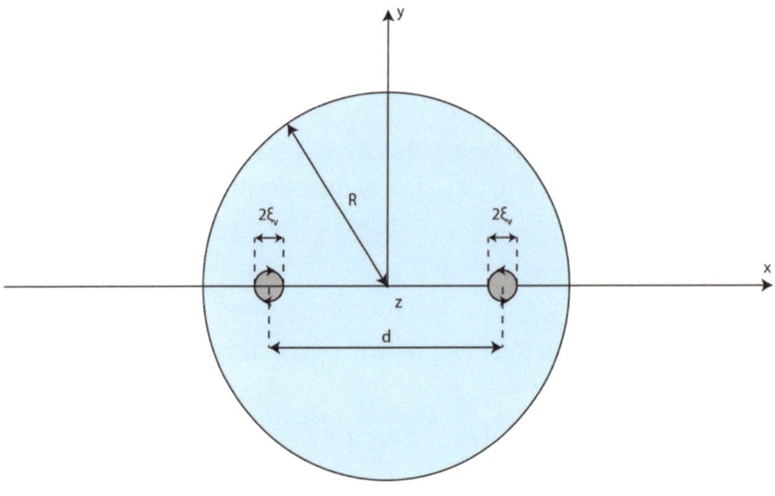

Figure 11. A top-view sketch of the vortex-anti-vortex system.

In order to determine the force, we need to fill in the formula of the Magnus force (102). What we need to know is the circulation and the translational velocity at which the vortex is moving. The circulation is known since we only look at vortices with one vortex quantum. For the

[16]Note that in aerodynamics this equation is also called the Kutta-Joukowski theorem, which gives a good estimate for the lift of the vortex flow.

http://dx.doi.org/10.5772/67121

translational velocity we say that the vortices are moving in each other's velocity field (35), this allows us to estimate $\mathbf{v}_{\text{trans}}$ by the velocity of the other vortex. Using the above the Magnus force (102) on a vortex can be calculated as:

$$\frac{\mathbf{F}_{\text{vortex}}}{H} = n_\infty(\kappa \mathbf{e}_z)\times\left(\frac{\kappa}{2\pi d}\right)e_\phi = -\frac{n_\infty \kappa^2}{2\pi d}\mathbf{e}_\rho. \tag{103}$$

As we see from the force on both vortices in the vortex-anti-vortex pair (103), the force points along $-\mathbf{e}_\rho$. The fact that the force points along $-\mathbf{e}_\rho$ means that both the vortex and anti-vortex are being pulled towards the centre of the condensate. Once both vortices meet in the middle, they will annihilate each other. This again verifies the fact that vortices and anti-vortices will attract each other, this result was derived in the previous section based on energy statements. With Eq. (103) we now have an idea on the factors influencing the force between the vortices.

6.3. Force on an off-centre vortex

Given the previous vortex case, we can now look at a vortex that is displaced from the origin with a distance d. A top-view sketch of the system is given in **Figure 12**.

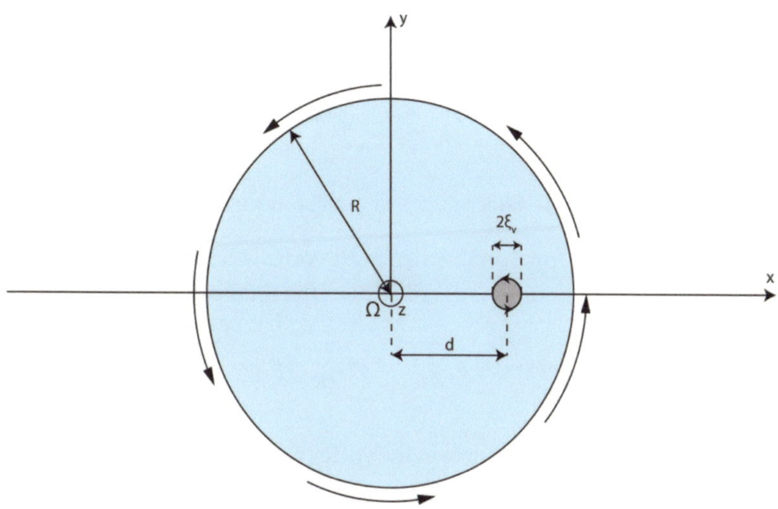

Figure 12. A top-view sketch of our single vortex system.

Now the translational velocity of the off-centre vortex is given by the rotational velocity of the trapping potential $\mathbf{v} = \mathbf{\Omega}\times\mathbf{r}$. Substituting the translational velocity into the formula for the Magnus force then yields that:

$$\frac{\mathbf{F}_{\text{vortex}}}{H} = n_\infty \boldsymbol{\kappa}\times(\mathbf{\Omega}\times\mathbf{r}) = n_\infty\kappa\Omega\rho[\mathbf{e}_z\times(\mathbf{e}_z\times\mathbf{e}_\rho)] = -n_\infty\kappa\Omega\rho\mathbf{e}_\rho. \tag{104}$$

As we see from Eq. (104) the off-centre vortex is being pushed towards the centre of the vortex due to the rotational velocity.

Now there is still one problem with our setup (**Figure 12**) that comes from the velocity field (35). Since we are working in a cylindrical trap with a hard wall potential, our velocity at the edge of the condensate cannot have a radial component (flow in or out the condensate). In order to get the correct velocity field, we should in principle solve the equations of motion again, given the new boundary conditions. However, since the velocity equation is a linear equation in this case we can apply the method of images. In order to get the right velocity field, that follows the boundary conditions, an anti-vortex is added at a distance $b > R$ from the centre of the trap. The full situation with the image vortex is shown in **Figure 13**.

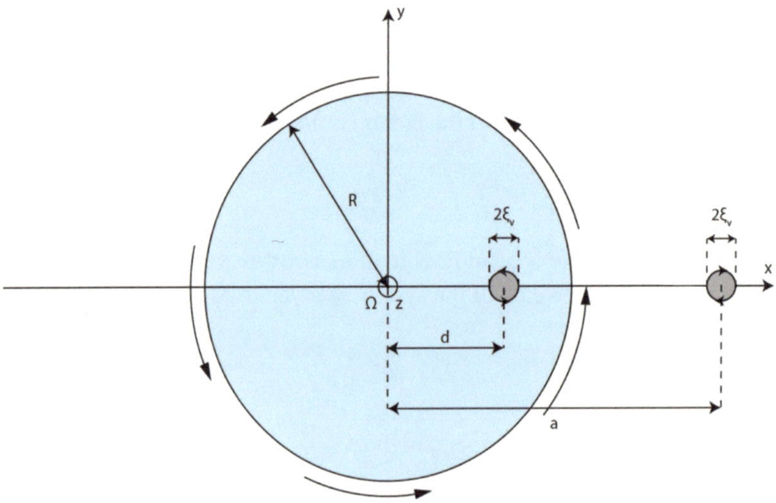

Figure 13. A top-view sketch of our single vortex system with an anti-vortex image.

To get a correct estimate of the force on the displaced vortex, we need to determine the position b of the image anti-vortex. To do this, we write down the total velocity field given by:

$$\mathbf{v}_{\text{tot}}(\mathbf{r}) = \frac{\kappa}{2\pi}\frac{1}{\rho'}\boldsymbol{e}'_{\phi} - \frac{\kappa}{2\pi}\frac{1}{\rho''}\boldsymbol{e}''_{\phi}. \tag{105}$$

Here, the notation ρ' and ρ'' are the distances to the vortex and anti-vortex, respectively. The vectors $\boldsymbol{e}'_{\phi}$ and $\boldsymbol{e}''_{\phi}$ are the basis vectors $\mathbf{e}_{\phi}$ with the vortex and anti-vortex, respectively, in the origin of the polar coordinate system. Rewriting the above in the chosen system of coordinates (x,y,z) in **Figure 13** yields:

$$\begin{aligned} \mathbf{v}_{\text{tot}}(x,y) = {} & \frac{\kappa}{2\pi}\frac{1}{\sqrt{(x-d)^2+y^2}}\left[\frac{-y}{\sqrt{(x-d)^2+y^2}}\mathbf{e}_x + \frac{x-d}{\sqrt{(x-d)^2+y^2}}\mathbf{e}_y\right] \\ & -\frac{\kappa}{2\pi}\frac{1}{\sqrt{(x-b)^2+y^2}}\left[\frac{-y}{\sqrt{(x-b)^2+y^2}}\mathbf{e}_x + \frac{x-b}{\sqrt{(x-b)^2+y^2}}\mathbf{e}_y.\right] \end{aligned} \tag{106}$$

Our Boundary condition (no radial flow) is given by

http://dx.doi.org/10.5772/67121

$$\mathbf{v}_{\text{tot}}(x,y)\cdot\mathbf{e}_\rho = 0. \tag{107}$$

Writing out the left-hand side of the boundary condition Eq. (73) using Eq. (106) then yields:

$$\begin{aligned}\mathbf{v}_{\text{tot}}(x,y)\cdot\mathbf{e}_r &= \frac{\kappa}{2\pi}\frac{1}{(x-d)^2+y^2}\left[\frac{-xy}{\sqrt{x^2+y^2}}+\frac{(x-d)y}{\sqrt{x^2+y^2}}\right]\\ &-\frac{\kappa}{2\pi}\frac{1}{(x-b)^2+y^2}\left[\frac{-xy}{\sqrt{x^2+y^2}}+\frac{(x-b)y}{\sqrt{x^2+y^2}}\right].\end{aligned} \tag{108}$$

For the boundary condition Eq. (108) should equal zero, using the fact that on the boundary $x^2+y^2=R^2$ yields that:

$$0 = db^2-[R^2+d^2]b+dR^2. \tag{109}$$

Solving the second order equation in b then yields two possible solutions:

$$b_+ = d, \tag{110}$$

$$b_- = \frac{R^2}{d}. \tag{111}$$

The first solution is an anti-vortex in the same place as the vortex, hence an annihilation of the vortex. The second solution is a vortex outside of the condensate (since $d\leq R$). The resulting velocity field then becomes:

$$\begin{aligned}\mathbf{v}_{\text{tot}}(x,y;d,R) &= \frac{\kappa}{2\pi R}\frac{1}{(x/R-d/R)^2+(y/R)^2}\left[-\frac{y}{R}\mathbf{e}_x+\left(\frac{x}{R}-\frac{d}{R}\right)\mathbf{e}_y\right]\\ &-\frac{\kappa}{2\pi R}\frac{1}{(x/R-R/d)^2+(y/R)^2}\left[-\frac{y}{R}\mathbf{e}_x+\left(\frac{x}{R}-\frac{R}{d}\right)\mathbf{e}_y\right].\end{aligned} \tag{112}$$

The size of the velocity field $v_{\text{tot}} = \sqrt{\mathbf{v}_{\text{tot}}\cdot\mathbf{v}_{\text{tot}}}$ is given by:

$$v_{\text{tot}}(x,y;d,R) = \frac{\kappa}{2\pi R}\sqrt{\frac{(d/R-R/d)^2}{[(x/R-d/R)^2+(y/R)^2][(x/R-R/d)^2+(y/R)^2]}}. \tag{113}$$

Note that in the limit $d\to 0$ we get our single vortex case back, and in the case $d\to R$ the vortex and anti-vortex annihilate yielding a zero-velocity field.

Using the above result, we see that in order to get a correct velocity field, we should add an anti-vortex. The addition of the anti-vortex means that our Magnus force gets an extra contribution, pulling the vortex out of the condensate. Since the distance between the vortex and anti-vortex is given by:

$$a = \frac{R^2}{d}-d = \frac{R^2-d^2}{d}, \tag{114}$$

the extra term in the Magnus force becomes equal to:

$$\frac{\mathbf{F}_{\text{Vortex}}}{L} = \frac{n_\infty \kappa^2}{2\pi a} \mathbf{e}_x = \frac{n_\infty \kappa^2}{2\pi} \frac{d}{R^2 - d^2} \mathbf{e}_x. \tag{115}$$

Combining both forces on the vortex (rotating condensate and anti-vortex), the total force on the vortex becomes equal to:

$$\frac{\mathbf{F}_{\text{total}}}{L} = \left(\frac{n_\infty \kappa^2}{2\pi} \frac{d}{R^2 - d^2} - n_\infty \kappa \Omega d \right) \mathbf{e}_x = n_\infty \kappa d \left(\frac{\kappa}{2\pi} \frac{1}{R^2 - d^2} - \Omega \right) \mathbf{e}_x. \tag{116}$$

In the total force we see a competition between the two forces, by looking for the positions of the vortex d where the force equals zero, we can find the equilibrium positions. Setting Eq. (116) to zero then yields two solutions:

$$d_1 = 0, \tag{117}$$

$$d_2 = R\sqrt{1 - \frac{\kappa}{2\pi R^2 \Omega}}. \tag{118}$$

The first solution yields a stable position at the centre of the vortex, while the second solution gives an equilibrium outside of the centre. Note that the second solution only exists if the condensate is rotating *fast* enough. The second solution is thus some kind of maximum (barrier) needed to keep the vortex inside of the condensate when it is rotating. The minimal velocity needed to keep the vortex inside is given by:

$$\Omega_{\min} = \frac{\hbar}{mR^2}. \tag{119}$$

Note that this frequency (119) is almost the same as the critical frequency (65). The big difference with the critical frequency is that there we have a length scale (the healing length) that also plays a role. The reason for this difference in frequency is because of the fact that when the Magnus force (102) is calculated, the typical length scale of the vortex (healing length) gets eliminated. This means that the obtained results are excellent as long as the vortices are far enough away from each other. When the vortices do get close to each other, for example if the off centre vortex reaches the edge, we need a more precise (variational or exact) calculation that also takes the vortex structure into account.

Using the total force on the vortex (116) a potential can also be defined as:

$$\frac{U}{L} = -\int_0^d \left(\frac{\mathbf{F}_{\text{total}}}{L} \right) d(d) \mathbf{e}_\rho = n_\infty \kappa \left(\frac{\kappa}{4\pi} \ln\left(\frac{R^2 - d^2}{R^2} \right) + \Omega \frac{d^2}{2} \right). \tag{120}$$

This potential (120) is also plotted in **Figure 14**. As we see in **Figure 14**, as long as the condensate spins fast enough, a vortex will always be pushed towards the centre.

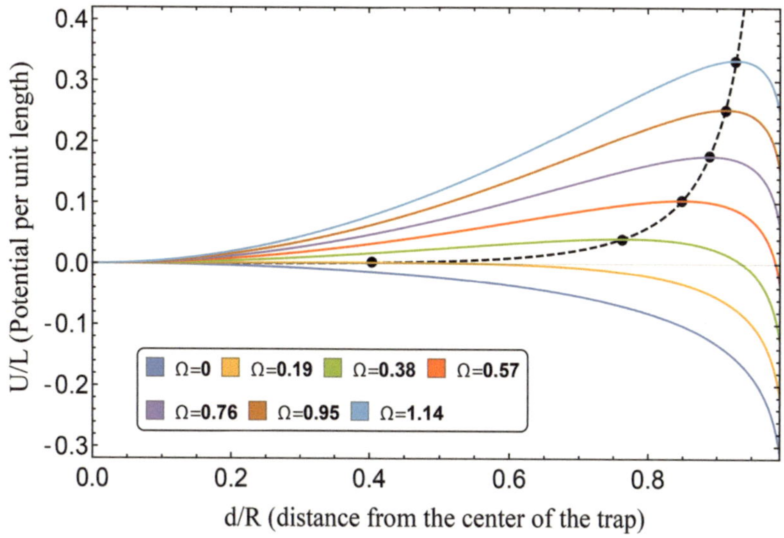

Figure 14. The resulting potential for the off-centre vortex. The black dashed line and dots yield the second stationary position of the off-centre vortex (118).

6.4. Vortex-vortex

The last situation we can look at is the vortex-vortex interaction. As we know from the previous section two vortices will repel each other, meaning that now the centre of the trap will not yield a stable solution. As we know from the off-centre vortex, we want our boundary condition to hold Eq. (107), which means that each vortex should have an anti-vortex. A sketch of the two-vortex system is given in **Figure 15**.

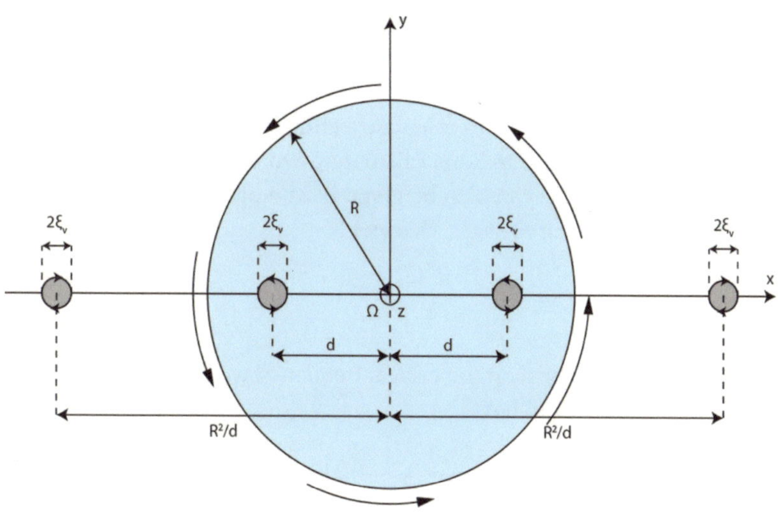

Figure 15. A top-view sketch of the two vortex system with the corresponding anti-vortices.

As earlier, we can now write down the Magnus force of the three different vortices on one of the vortices in the condensate. Since the system is symmetric, it does not matter which vortex is chosen for the calculation; we chose the right one, yielding:

$$\frac{\mathbf{F}_{\text{vortex}}}{L} = \frac{n_\infty \kappa^2}{2\pi}\left(\frac{d}{R^2-d^2} + \frac{1}{2d} - \frac{d}{R^2+d^2}\right)\mathbf{e}_x = \frac{n_\infty \kappa^2}{2\pi}\frac{R^4+3d^4}{2d(R^4-d^4)}\mathbf{e}_x. \tag{121}$$

As we see from Eq. (121) the vortices push each other out of the condensate. Since the vortices give a total circulation to the condensate, we should also have a rotation of the superfluid.[17] Adding the attractive rotational force then yields for the total force on the (right) vortex that:

$$\frac{\mathbf{F}_{\text{total}}}{L} = n_\infty \kappa \left(\frac{\kappa}{2\pi}\frac{R^4+3d^4}{2d(R^4-d^4)} - \Omega d\right)\mathbf{e}_x. \tag{122}$$

Again we can look for stable positions of the vortex pair by solving the equation

$$0 = \frac{\kappa}{2\pi}\frac{R^4+3d^4}{2d(R^4-d^4)} - \Omega. \tag{123}$$

This equation (123) is analytically solvable by substituting the variable $y = d^2$ and solving the remaining equation of third degree by Cardano's method. This will however lead to a large (non-insightful) equation. For simplicity we assume that $d << R$ (vortices close to the centre), this then yields for the equilibrium positions that:

$$\frac{\kappa}{4\pi} = \Omega d^2 \Rightarrow d = \sqrt{\frac{\kappa}{4\pi\Omega}} = \sqrt{\frac{\hbar}{2m\Omega}}. \tag{124}$$

The equilibrium position, both exact and the above approximation (124), is plotted in **Figure 16**. As we see, the faster the condensate rotates, the more the vortices are pushed together. In the limit $d << R$ a very high rotational frequency is needed in order to get accurate results. The limit $d << R$ corresponds to ignoring the two image vortices, and these get pushed to infinity as d decreases. In principle we can again calculate a critical frequency as before, but this would require the full formula and some tedious calculating work that is left for the reader. A nice estimate of the critical frequency can also be given by the plot in **Figure 16** by looking at the frequency for which the solid curve stops, this yields:

$$\frac{R^2\Omega}{\kappa} \approx 0.351. \tag{125}$$

Notice that this is slightly higher than the critical frequency we would expect from the discussion in Section 5, where we would have:

$$\frac{R^2\Omega}{\kappa} = \frac{1}{\pi} \approx 0.318. \tag{126}$$

This overestimation is again an artefact of the fact that the vortex core structure is ignored.

---

[17]Note that as before this rotation is needed to keep the vortices inside of the condensate.

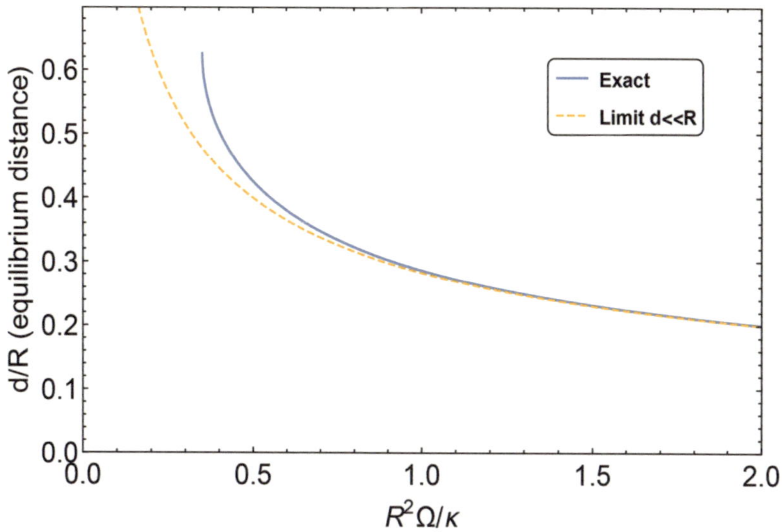

Figure 16. The equilibrium position of a vortex-vortex pair.

Again it is possible to define a potential which yields:

$$\frac{U}{L} = -\int_{R/2}^{d} \left(\frac{\mathbf{F}_{\text{total}}}{L}\right) d(d)\mathbf{e}_x = n_\infty \kappa \left[\frac{\kappa}{4\pi} \ln\left(\frac{R/2}{d} \frac{R^4 - d^4}{R^4 - (R/2)^4}\right) + \frac{\Omega}{2}\left(d^2 - \left(\frac{R}{2}\right)^2\right)\right]. \tag{127}$$

This potential is plotted for two different rotational frequencies in **Figure 17**. In **Figure 17**, we see that the vortices indeed repel each other but stay in the trap if the trapping frequency is high enough.

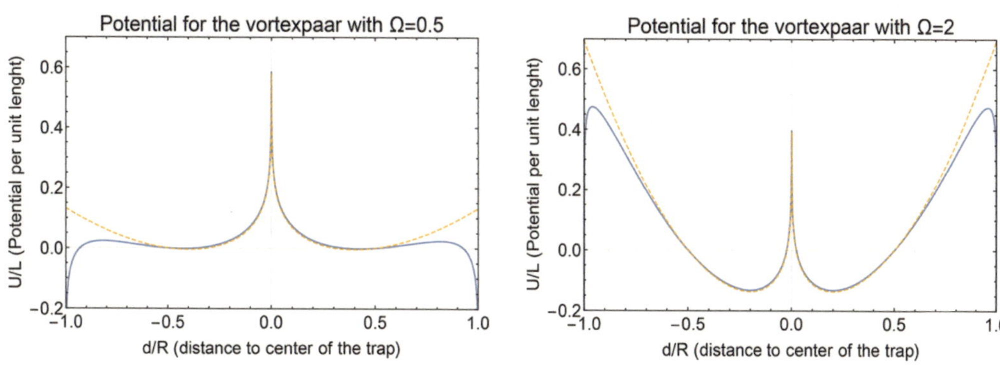

Figure 17. The potential for a vortex-vortex pair for two different frequencies. The solid curves yield the full solution, the dashed curves are the limit $d \ll R$.

6.5. Variational calculations for the energy functional

From the previous sections we now know four variational models for the vortex structure $\chi(\mathbf{r})$ given by Eqs. (57), (88)–(90). Given one of these variational, we can write the variational macroscopic wave function for a vortex or anti-vortex as:

$$\Psi(\mathbf{r}) = \sqrt{n_\infty}\chi(\mathbf{r})e^{iS(\mathbf{r})}. \tag{128}$$

The above macroscopic wave function (128) is not limited to one-vortex states. We can also construct a structure function $\chi(\mathbf{r})$ for a multi-vortex state. Assume we have a total of N_v vortices, then the variational structure function is written as[18]

$$\chi(\mathbf{r}) = \prod_{j=1}^{N_v}\chi_j(\mathbf{r}), \tag{129}$$

where $\chi_j(\mathbf{r})$ are the one-vortex structure functions. Usually one takes the hyperbolic tangent (89) for the one-vortex structure, resulting for Eq. (129) in:

$$\chi(\mathbf{r}) = \prod_{j=1}^{N_v}\tanh\left(\frac{|\mathbf{r}-\mathbf{r}_j|}{\alpha\sqrt{2}\xi}\right) \qquad \text{with} \qquad \alpha = \sqrt{\frac{3}{4\ln(2)-1}}\approx 1.3, \tag{130}$$

with $\mathbf{r}_j$ the positions for the vortices. The velocity field for one vortex in Eq. (35) can be rewritten in terms of a circulation vector $\boldsymbol{\kappa} = \kappa\mathbf{e}_z$ and a position vector $\mathbf{r} = \rho\mathbf{e}_\rho$. Note that the z-axis is oriented parallel to the rotational velocity $\boldsymbol{\Omega}$. Using the circulation vector κ and position vector $\mathbf{r}$ the velocity field (27) for a vortex or anti-vortex becomes:

$$\mathbf{v}(\mathbf{r}) = \frac{\pm\hbar}{m\rho}\mathbf{e}_\phi = \frac{\kappa}{2\pi\rho}\mathbf{e}_\phi = \frac{\kappa}{2\pi\rho}(\mathbf{e}_z\times\mathbf{e}_\rho) = \frac{\boldsymbol{\kappa}\times\mathbf{r}}{2\pi r^2}. \tag{131}$$

where the sign of κ depends on whether you have a vortex $\kappa > 0$ or anti-vortex $\kappa < 0$. If we look at a multi-vortex structure with N_v vortices, then the velocity field is given by[19]:

$$\mathbf{v}(\mathbf{r}) = \sum_{j=1}^{N_v}\frac{\boldsymbol{\kappa}\times(\mathbf{r}-\mathbf{r}_j)}{2\pi|\mathbf{r}-\mathbf{r}_j|^2}. \tag{132}$$

Since we are going to look at stability of the system, we will again add the rotational term according to Eq. (56). For the off-centre vortex and vortex-vortex structures, this will be needed. For the vortex-anti-vortex this will not be needed since the condensate will not be rotating then.

Finally since we will be working in polar coordinates, our structure functions $\chi(\mathbf{r})$ and velocity field should be converted to polar coordinates. Since we will be looking at one- and two-vortex systems we can make the calculations simpler by choosing the x-axis of the Cartesian

[18]The reason that we have to multiply the vortex solutions for the structure rather than add them has several reasons. First of all the particle density is given by the macroscopic wave function (and thus of the structure function), if we have a sum here this will yield extra terms. Second the Gross-Pitaevskii equation, Eq. (10) is a non-linear equation, which means that a sum of two solutions is *not* a solution of the differential equation. By multiplying the structure functions we create *one* state (solution) with multiple vortices.

[19]In the case of velocity fields we take the sum. This also has two reasons. First of all we know that each vortex hole gives rise to such a velocity field (35). Second in the case of the cylindrical hole we have an irrotational and incompressible flow for which the velocity follows a linear differential equation. So for the cylindrical vortex hole we can add velocity solutions thanks to linearity. For other vortex structures this variational model is again approximate, but appropriate for the chosen variational model for the density (129).

http://dx.doi.org/10.5772/67121

coordinate system such that all of the vortices lie on the x-axis. For the structure function with the hyperbolic tangent given in Eq. (89) we get in polar coordinates for a vortex in $(x,y) = (d,0)$:

$$\begin{aligned} \tanh\left(\frac{|\mathbf{r}-d\mathbf{e}_x|}{\alpha\sqrt{2}\xi}\right) &= \tanh\left(\frac{\sqrt{(\rho\cos(\phi)-d)^2 + (\rho\sin(\phi))^2}}{\alpha\sqrt{2}\xi}\right) \\ &= \tanh\left(\frac{\sqrt{\rho^2-2\rho d\cos(\phi) + d^2}}{\alpha\sqrt{2}\xi}\right) \end{aligned} \tag{133}$$

Of course in order to do a variational calculation we need an energy functional, this is given by:

$$E[n(\mathbf{r}),\mathbf{v}(\mathbf{r})] = \int_V d^3\mathbf{r}\left(\frac{1}{2}mn(\mathbf{r})v^2(\mathbf{r}) + n(\mathbf{r})V_{\text{trap}}(\mathbf{r}) + gn^2(\mathbf{r})\right). \tag{134}$$

In this chapter we look at a homogeneous condensate ($V_{\text{trap}}(\mathbf{r})$). For the variational calculations, we also assumed that the vortices are small compared to the size of the condensate ($R >> \xi$), if that is the case then our interaction term is constant and can be neglected. Finally we need to add the rotational energy, doing this results in the following form for the rotational energy:

$$E[n(\mathbf{r}),\mathbf{v}(\mathbf{r})] = \int_V d^3\mathbf{r}\left(\frac{1}{2}mn(\mathbf{r})v^2(\mathbf{r})\right)-\mathbf{\Omega}\cdot\int_V d\mathbf{r}\left(\mathbf{r}\times(mn(\mathbf{r})\mathbf{v}(\mathbf{r}))\right). \tag{135}$$

Choosing the rotational velocity along the z-axis is possible to do the z integration resulting in:

$$\frac{E[n(\mathbf{r}),\mathbf{v}(\mathbf{r})]}{mL} = \frac{1}{2}\int_S d^2\mathbf{r}\left(n(\mathbf{r})v^2(\mathbf{r})\right)-\Omega\int_S d^2\mathbf{r}|\mathbf{r}|n(\mathbf{r})|\mathbf{v}(\mathbf{r})|. \tag{136}$$

The above energy can be used for further variational calculations; the calculations themselves are left to the reader.

6.6. Variational results for the energy of the systems

Given the variational energy (136) it is possible to look at our three different vortex cases again. To conclude the section the different energies for the three-vortex systems above will be calculated variationally. For the different calculations a trap of the size $R = 30$ was chosen with a healing length of $\xi = 1$. As a variational model, the hyperbolic tangent was chosen.

For the vortex pair the energy as a function of the distance between the vortices is given in **Figure 18**. As we see the vortex-anti-vortex pair attracts as well in the variational model so our results from the simple calculations on the Magnus force still hold. For the off-centre vortex and vortex pair, a rotating condensate is needed. This means that the two integrals of Eq. (136) should be integrated. The first integral yields the energy of the stationary condensate, the second integral yields the energy reduction of rotating the condensate. The total energy is then given by a competition between both energies.

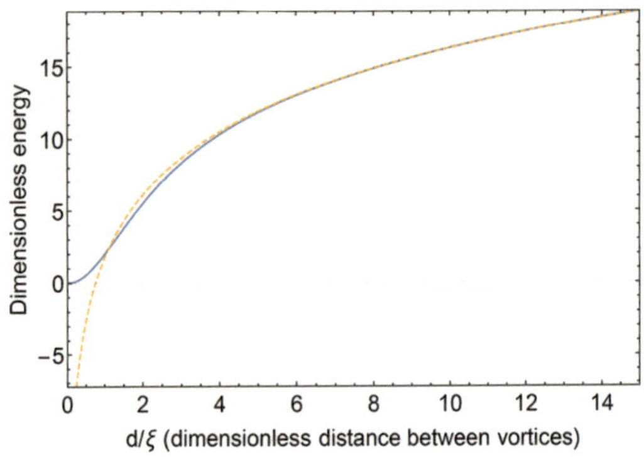

Figure 18. The variational energy for a vortex-anti-vortex pair the solid line gives the full solution, whereas the dashed line yields the asymptotic solution.

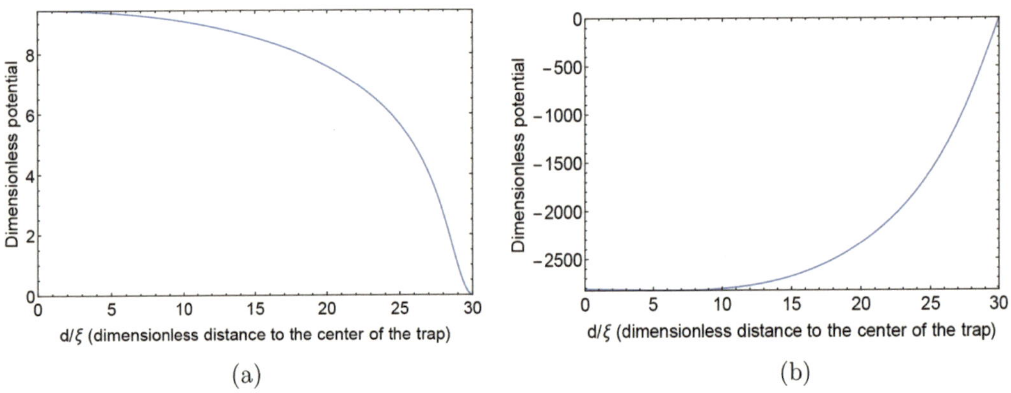

Figure 19. The kinetic (a) and rotational (b) energy for the off-center vortex.

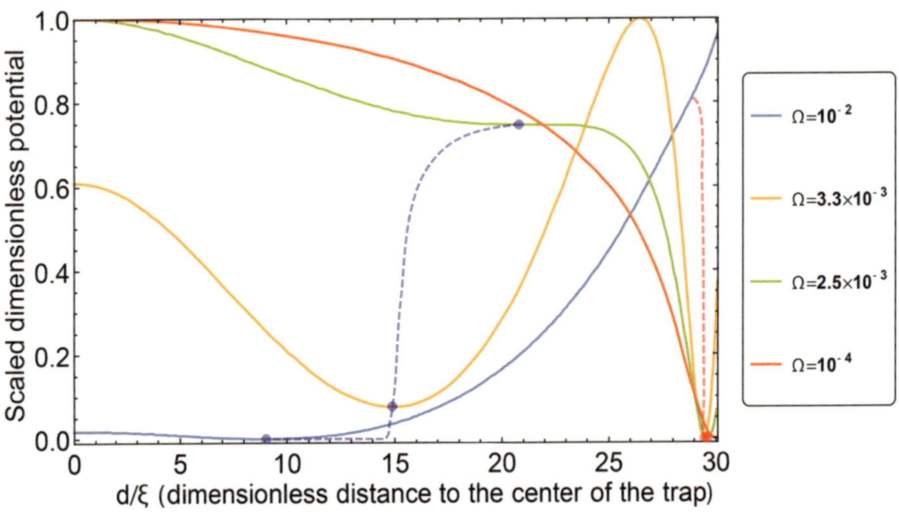

Figure 20. The variational energy for the off-centre vortex for several values of Ω. The dashed lines trace the minima and maxima of the potential.

The potentials for the off-centre vortex and vortex-vortex systems are given in **Figure 19**. As seen on **Figure 19**, a repulsive and attractive energy are in competition. Depending on the values of the rotational frequency we will have different situations. The situation for the off-centre vortex is given for a few values of the rotational frequency Ω in **Figure 20**. What we can notice in **Figure 20** is that for a single vortex we find a non-centred vortex in the stable situation. This is because of the fact that the variational velocity field of the cylindrical model was used for the hyperbolic tangent, yielding small deviations. For the vortex-vortex case the same can be done, resulting in the variational energies as given in **Figure 21**. Here we again see that as the rotational velocity increases, the vortices get pushed more towards the centre of the trap, while if the rotational velocity decreases below a certain value the vortices get pushed out of the trap.

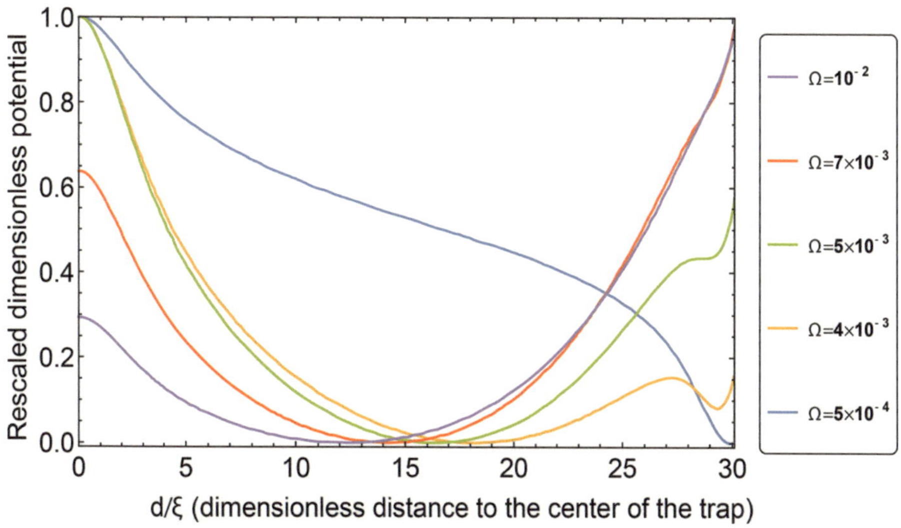

Figure 21. The variational energy for the vortex-vortex for several values of Ω. The dashed lines trace the minima and maxima of the potential.

7. The Berezinskii-Kosterlitz-Thouless phase transition

To conclude the discussion about vortices in BECs we will look at a vortex induced phase transition, the Berezinskii-Kosterlitz-Thouless (BKT) phase transition (Nobel Prize in Physics 2016).

7.1. Vortex entropy

Vortices carry entropy. The entropy of a given state can be computed from the free energy,

$$S = \left.\frac{\partial F}{\partial T}\right|_{V,N}, \tag{137}$$

but in this simplified presentation we will use some qualitative arguments to estimate the entropy. The ' size' of a vortex is given by its healing length ξ, and the number of ways that an object of this size can be placed in a condensate of radius R is estimated by ξ^2/R^2, where we consider either a two-dimensional condensate, or we enforce the vortex lines to be parallel. An illustration of a possible microstate—a way of placing vortices—is shown in **Figure 22**. With this simple argument we can estimate the entropy of one or more vortices and anti-vortices, again by counting the number Ω of microstates, and using Boltzmann's formula $S = k_B \log\Omega$. For example, the entropy of a single vortex becomes:

$$S_{1\text{vortex}} = k_B \ln(\Omega_{1\text{vortex}}) \approx k_B \ln\left[\left(\frac{R}{\xi}\right)^2\right] = 2k_B \ln\left(\frac{R}{\xi}\right). \tag{138}$$

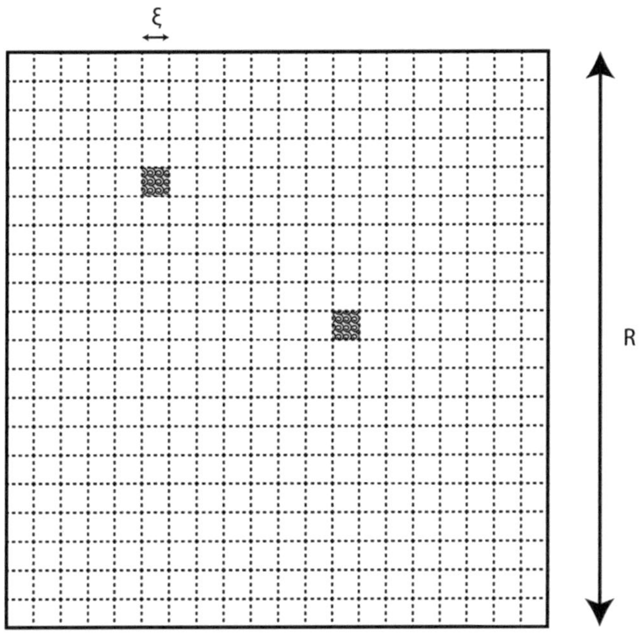

Figure 22. A schematic view of a possible microstate contributing to the vortex positional entropy.

When two vortices are present, or a vortex and an anti-vortex are present, we consider them to be independent. From the previous section, it is clear that this is an approximation that fails when they are close together. However, the number of microstates where this happens is small, and we can still assume the entropy is extensive in the number of vortices:

$$S_{2\text{vortices}} = 2S_{1\text{vortex}} = 4k_B \ln\left(\frac{R}{\xi}\right). \tag{139}$$

When information (or order) is added to the system, the entropy must decrease. For example, we might know that the vortex and anti-vortex are bound, and must be in adjacent patches of **Figure 22**. Then the number of microstates that are compatible with this constraint is again proportional to ξ^2/R^2, so that up to a constant, negligible for $R \gg \xi$,

http://dx.doi.org/10.5772/67121

$$S_{\text{vort exp air}} \approx 2k_B \ln\left(\frac{R}{\xi}\right) = \frac{1}{2} S_{\text{2vortices}}. \tag{140}$$

7.2. BKT temperature

In a three-dimensional Bose gas, superfluidity vanishes above the critical temperature for Bose-Einstein condensation. In the normal state, phase correlations vanish exponentially fast as a function of distance, and phase coherence is absent. In the BEC regime, the phase is locked by long-range off-diagonal order [73], and the phase correlation function is a constant as a function of distance. However, this is not the case for a two-dimensional Bose gas. Due to quantum fluctuations, there is no phase transition to the Bose-Einstein condensation (more accurately: the critical temperature is zero). Nevertheless, this does not mean that phase coherence is lost altogether: the phase correlation function gets a power-law decay rather than the exponentially fast decay in the normal state [74]. This regime is called 'quasi-condensation' and allows for superfluidity.

Obviously, the mechanism whereby superfluidity is destroyed must be different. Berezinskii [75] and independently Kosterlitz and Thouless [76, 77] proposed that phase coherence gets destroyed by the unbinding of vortex-anti-vortex pairs, leading to a sudden proliferation of vortices and anti-vortices that scramble the phase coherence. Here, the results for the interaction energy between vortices and anti-vortices calculated above will be of use. The binding energy of a vortex-anti-vortex pair can be defined as the energy it takes to move a vortex and an anti-vortex that form a pair, away from each other to infinity.

Heuristically, the Berezinskii-Kosterlitz-Thouless (BKT) transition can be thought of as a competition between the internal energy and the entropy for vortices. The realized state is that which minimizes the free energy

$$F = U - TS. \tag{141}$$

When the temperature is low $T \approx 0$, the internal energy (here, the binding energy of the vortex-anti-vortex pair) dominates. In this case, the pairs remain bound and no uncompensated phase singularities perturb the coherence. Indeed, as can be seen from **Figure 18**, the internal energy is minimal when the vortex and anti-vortex are co-located. The phase windings then cancel out, and $U = 0$, $F = 0$. When the temperature is increased, the entropy term TS can become large enough to dominate the internal energy, leading to a transition to the more disordered state. Focusing on the vortex-anti-vortex pair, the disordered state (the state with highest entropy) is that in which the vortex and the anti-vortex are unbound and can be placed freely anywhere in the available volume. In that disordered state, we will also take U to be equal to twice the kinetic energy as in Eq. (62)—the number of microstates where the vortex is close to the anti-vortex is negligible in comparison to the number of microstates when they are well separated, so we can neglect the interaction. This allows to approximate the internal energy of the disordered state by

$$U = 2\pi n_\infty \frac{\hbar^2}{m} H \ln\left(\frac{R}{\xi}\right). \tag{142}$$

Comparing the free energy of the state with bound pairs ($F = 0$) with the free energy of the disordered state yields the critical temperature where superfluidity vanishes:

$$0 = U - T_C S. \tag{143}$$

Using our earlier results (142) for U and (139) for S, the critical temperature is found to be

$$T_C = \frac{2\pi n_\infty \frac{\hbar^2}{m} H \ln(R/\xi)}{4k_B \ln(R/\xi)} = \frac{\pi n_\infty \hbar^2 H}{2mk_B}. \tag{144}$$

In the units used for (143), we can see that the right-hand side indeed describes a temperature, and that it is determined by the superfluid density n_∞. Above this temperature, it is advantageous for the free energy to spontaneously create and unbind vortex-anti-vortex pairs that scramble the phase and turn the gas into the normal state. As mentioned in the introduction, the BKT transition was observed experimentally [40]. The observed critical phase space density differs from the back of the envelope estimate given here by a factor of about four, mostly due to the fact that the experimental condensate does not have a uniform density, and also due to finite size effects.

8. Concluding section

With this chapter an introduction to vortices in ultra-cold gases was given. The main focus of the discussion was aimed towards atomic gases of bosons. However the derived (qualitative) results are also applicable to fermionic quantum gases. In this concluding section the main results of the chapter are summarized together with a first look at a condensate of fermionic atoms.

8.1. Conclusions

Concerning vortices in superfluid atomic gases, the following conclusions can be made:

- A vortex will always make a hole that extends throughout the entire condensate.
- The circulation of a vortex is quantized in multiples of h / m see Eq. (32).
- Vortices in a BEC will always carry one quantum of circulation, this is energetically more favourable.
- For vortices a minimal rotational frequency (critical frequency) is required as given in Eq. (65).
- Two vortices will repel each other.
- A vortex and anti-vortex will attract each other.
- Rotating a BEC faster will push vortices to the centre of the trap more, for a single vortex.
- A single vortex in a trap will be most stable in the centre of the trap, for a single vortex this means that it is stable in the centre of the trap.
- For a 2D quasi-condensate superfluidity is destroyed by the unbinding of vortex-anti-vortex pairs that destroy the overall phase coherence. This happens at a certain critical temperature T_C.

http://dx.doi.org/10.5772/67121

8.2. One step further: fermionic gases

This chapter was restricted to superfluidity in Bose gases since these already (qualitatively) describe the behaviour of vortices in general superfluids very well. If we are working with fermionic gases a few things change, these are briefly discussed in this section. More elaborate discussions about fermionic quantum gases can be found in references [78–81].

First of all when we have fermions, we have to take the 'spin'. With spin we label the different quantum states of our atom, in fermionic atomic gases these will be two different hyperfine states of the considered atom. These spins will be called 'up' and 'down' as an analogy with the spin known from standard quantum mechanics. In order to fix the number of particles in both spin states, we need two chemical potentials labelled $\mu_\uparrow$ and $\mu_\downarrow$. If we now say that we have an ultra-cold gas and thus have *s*-wave scattering, the Pauli exclusion principle tells us that only opposite spins can interact. Since the *s*-scattering wave function is symmetric, the spin wave function should be the anti-symmetric singlet state. Applying this to our Hamiltonian we get:

$$\hat{H} = \int d^3\mathbf{r}\left(\sum_{\sigma=\uparrow,\downarrow}\hat{\psi}_\sigma^\dagger(\mathbf{r},t)\left[-\frac{\hbar^2\nabla^2}{2m}-\mu_\sigma\right]\hat{\psi}_\sigma(\mathbf{r},t)+\frac{1}{2}\sum_{\sigma=\uparrow,\downarrow}g\hat{\psi}_\sigma^\dagger(\mathbf{r},t)\hat{\psi}_{\neg\sigma}^\dagger(r',t)\hat{\psi}_{\neg\sigma}(r',t)\hat{\psi}_\sigma(\mathbf{r},t)\right). \quad (145)$$

Given the above Hamiltonian (144) atomic Fermi gases can be studied. In order to simplify calculations, one usually defines the average chemical potential μ and polarization ζ as:

$$\mu = \frac{\mu_\uparrow + \mu_\downarrow}{2}, \quad (146)$$

$$\zeta = \frac{\mu_\uparrow - \mu_\downarrow}{2}. \quad (147)$$

Of course the question we can ask now is 'what is the criterion for superfluidity now'? In Section 3, this criterion was given by the Penrose-Onsager criterion which states that the one-body density matrix should have a macroscopic eigenvalue, indicating that there was one state being occupied macroscopically. For fermions, the criterion for bosons was expanded by C.N. Yang, hence named the Penrose-Onsager-Yang criterion. Yang stated that for fermions the second order density matrix should have a macroscopically large eigenvalue, indicating a condensation of fermion pairs. These fermion pairs are also referred to as Cooper pairs, named after L.N. Cooper who first proposed this for electrons [82]. Once this is assumed it is possible to rewrite the entire system in terms of a macroscopic two-body wave function $\Psi(\mathbf{y},\mathbf{x})$. This macroscopic wave function is usually rewritten by assuming the following product form:

$$\Psi(\mathbf{y},\mathbf{x}) = \psi(\mathbf{R})\chi(\boldsymbol{\rho}), \quad (148)$$

where $\mathbf{R}$ is the centre of mass coordinate and $\boldsymbol{\rho}$ is the relative position. So in the end $\psi(\mathbf{R})$ will describe superfluidity of the Cooper pairs and $\chi(\boldsymbol{\rho})$ will describe the interactions between the atoms in a different spin state.

The study of superfluid fermionic atoms opens up a few extra parameters to play with in experiments, which can help explore the physics of ultra-cold gases. The two most prominent are:

Spin-imbalance: A first parameter to play with is the number of 'spin-up' and 'spin-down' atoms. If the numbers are unequal (or if the polarisation ζ in Eq. (146) is non-zero), then you will have excess atoms when the condensation occurs. These excess atoms will need to go somewhere. Usually excess atoms are pushed towards the edge of the condensate, or placed inside holes in the condensate, for example vortices. When the number of 'spin-up' and 'spin-down' atoms is unequal one also talks about pair frustration since the unbound (frustrated atoms) will still try to pair up with other paired atoms.

BEC-BCS crossover: A second parameter that can be played with is the interaction strength a_s, this can also be done with bosons but is more interesting to do with fermions. With fermions it is possible to change the sign of the interaction from negative (repulsive) to positive (attractive). By doing this it is possible to go from a gas of interacting fermions (Cooper pairs or BCS), to a gas of interacting bosons (BEC) by making the interactions so strong that the particles are quasi bosons. The point at which the interaction changes in sign is also referred to as unitatiry. A qualitative picture of the transition form a repulsive interaction (BCS) to an attractive interaction (BEC) is given in **Figure 23**.

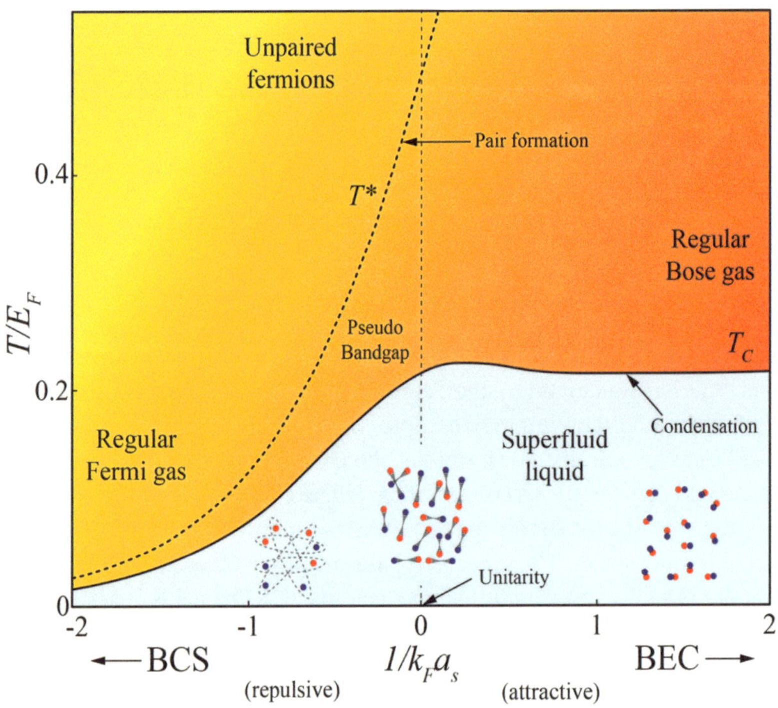

Figure 23. A qualitative picture of the BEC-BCS crossover, figure adapted from Ref. [83].

Acknowledgments

The authors gratefully acknowledge the extensive, weekly discussions with S.N. Klimin, G. Lombardi, W. Van Alphen and J.P.A. Devreese. Financial support was provided by the Fund for Scientific Research Flanders, through FWO projects: G042915N (Superfluidity and superconductivity in multi-component quantum condensates) and G011512N (Quantum turbulence in atomic and solid state Bose-Einstein condensates). One of us (N.V.) acknowledges a Ph.D. fellowship of the University of Antwerp (UAntwerp).

Author details

Nick Verhelst[1] and Jacques Tempere[1,2]*

*Address all correspondence to: Jacques.Tempere@uantwerpen.be

1 Theory of Quantum systems & Complex Systems, University of Antwerp, Antwerp, Belgium

2 Lyman Laboratory of Physics, Harvard University, Cambridge, USA

References

[1] P. Kapitza, Nature **141**, 74 (1938).

[2] J.F. Allen, A.D. Misener, Nature **141**, 75 (1938).

[3] A.A. Abrikosov, J. Phys. Chem. Solids **2**, 199 (1957).

[4] W.F. Vinen, Proc. R. Soc. A **260**, 218 (1961).

[5] R.J. Donnelly, *Quantized Vortices in Helium II* (Cambridge University Press, Cambridge,1991).

[6] M. Inguscio, S. Stringari, C. Wieman (eds.), *Bose-Einstein Condensation in Atomic Gases*, Proceedings of the International School of Physics, " Enrico Fermi", vol. CXL (IOS Press, Amsterdam, 1999).

[7] M. Inguscio, W. Ketterle, C. Salomon (eds.), *Ultra-Cold Fermi Gases*, Proceedings of the International School of Physics, " Enrico Fermi," vol. CLXIV (IOS Press, Amsterdam, 2008)

[8] M.H. Anderson, J.R. Ensher, M.R. Matthews, C.E. Wieman, E.A. Cornell, Science **269**, 198 (1995).

[9] K.B. Davis, M.-O. Mewes, M.R. Andrews, N.J. van Druten, D.S. Durfee, D.M. Kurn, W. Ketterle, Phys. Rev. Lett. **75**, 3969 (1995).

[10] C.C. Bradley, C. A. Sackett, J.J. Tollett, R.G. Hulet, Phys. Rev. Lett. **75**, 1687 (1995).

[11] E. Lundh, C.J. Pethick and H. Smith, Phys. Rev. A **58**, 4816 (1998).

[12] F. Dalfovo and M. Modugno, Phys. Rev. A **61**, 023605 (2000).

[13] J. Tempere, J.T. Devreese, Solid State Comm. **108**, 993 (1998).

[14] E.L. Bolda, D.F. Walls, Phys. Rev. Lett. **81**, 5477 (1998).

[15] C.O. Weiss and K. Teuber, Optics Comm. **16**, 399 (1976).

[16] P.M. Bellan, *Fundamentals of Plasma Physics* (Cambridge University Press, Cambridge, 2008).

[17] J.F. Nye and M.V. Berry, Proc. R. Soc. Lond. **336**, 165 (1974).

[18] L. Marrucci, C. Manzo and D. Paparo, Phys. Rev. Lett. **96**, 163905 (2006).

[19] M.R. Andrews, C.G. Townsend, H.-J. Miesner, D.S. Durfee, D.M. Kum, W. Ketterle, Science **275**, 637 (1997).

[20] M. R. Matthews, B. P. Anderson, P. C. Haljan, D. S. Hall, C. E. Wieman, E. A. Cornell, Phys. Rev. Lett. **83**, 2498 (1999).

[21] M. R. Matthews, B. P. Anderson, P. C. Haljan, D. S. Hall, M. J. Holland, J. E. Williams, C. E. Wieman, E. A. Cornell, Phys. Rev. Lett. **83**, 3358 (1999).

[22] M. F. Andersen, C. Ryu, Pierre Cladé, Vasant Natarajan, A. Vaziri, K. Helmerson, W. D. Phillips, Phys. Rev. Lett. **97**, 170406 (2006).

[23] A. E. Leanhardt, A. Görlitz, A. P. Chikkatur, D. Kielpinski, Y. Shin, D. E. Pritchard, W. Ketterle, Phys. Rev. Lett. **89**, 190403 (2002).

[24] T. Isoshima, M. Okano, H. Yasuda, K. Kasa, J. A. M. Huhtamäki, M. Kumakura, Y. Takahashi. Phys. Rev. Lett. **99**, 200403 (2007).

[25] Y.-J. Lin, R. L. Compton, K. Jimenez-Garcia, J. V. Porto, I. B. Spielman, Nature **462**, 628 (2009).

[26] K. W. Madison, F. Chevy, W. Wohlleben, J. Dalibard, Phys. Rev. Lett. **84**, 806 (2000).

[27] K.W. Madison, F. Chevy, W. Wohlleben, J. Dalibard, J. Mod. Opt. **47**, 2715 (2000).

[28] P. C. Haljan, I. Coddington, P. Engels, E. A. Cornell, Phys. Rev. Lett. **87**, 210403 (2001).

[29] C. Raman, J. R. Abo-Shaeer, J. M. Vogels, K. Xu, W. Ketterle, Phys. Rev. Lett. **87**, 210402 (2001).

[30] E. Hodby, G. Hechenblaikner, S. A. Hopkins, O. M. Maragò, C. J. Foot, Phys. Rev. Lett. **88**, 010405 (2001).

[31] R. A.Williams, S. Al-Assam, C. J. Foot, Phys. Rev. Lett. **104**, 050404 (2010).

[32] S. Inouye, S. Gupta, T. Rosenband, A.P. Chikkatur, A. Görlitz, T.L. Gustavson, A.E. Leanhardt, D.E. Pritchard, W. Ketterle, Phys. Rev. Lett. **87**, 080402 (2001).

[33] F. Chevy, K. W. Madison, V. Bretin, J. Dalibard, Phys. Rev. A **64**, 031601 (2001).

[34] P. Rosenbusch, V. Bretin, J. Dalibard, Phys. Rev. Lett. **89**, 200403 (2002).

[35] I. Coddington, P.C. Haljan, P. Engels, V. Schweikhard, S. Tung, E.A. Cornell, Phys. Rev. A **70**, 063607 (2004).

[36] J. R. Abo-Shaeer, C. Raman, W. Ketterle, Phys. Rev. Lett. **88**, 070409 (2002).

[37] P. Engels, I. Coddington, P. C. Haljan, E.A. Cornell, Phys. Rev. Lett. **89**, 100403 (2002).

[38] I. Coddington, P. Engels, V. Schweikhard, E.A. Cornell, Phys. Rev. Lett. **91**, 100402 (2003).

[39] S. Stock, B. Battelier, V. Bretin, Z. Hadzibabic, J. Dalibard, Laser Phys. Lett. **2**, 275 (2005).

[40] Z. Hadzibabic, P. Krüger, M. Cheneau, B. Battelier, J. Dalibard, Nature **441**, 1118 (2006).

[41] T.W. Neely, E.C. Samson, A.S. Bradley, M.J. Davis, B.P. Anderson, Phys. Rev. Lett. **104**, (2010).

[42] A.L. Gaunt, T.F. Schmidutz, I. Gotlibovych, R.P. Smith, Z. Hadzibabic, Phys. Rev. Lett. **110**, 200406 (2013).

[43] L. Chomaz, L. Corman, T. Bienaimé, R. Desbuquois, C. Weitenberg, S. Nascimbène, J. Beugnon, J. Dalibard, Nat. Commun. **6**, 6172 (2015).

[44] N. L. Smith,W. H. Heathcote, J. M. Krueger, and C.J. Foot, Phys. Rev. Lett. **93**, 080406 (2004).

[45] C. F. Barenghi, R. J. Donnelly, W. F. Vinen (Eds.), *Quantized Vortex Dynamics and Superfluid Turbulence* (Springer Verlag, Berlin, 2001).

[46] N. Gemelke, X. Zhang, C.-L. Hung, and C. Chin, Nature **460**, 995 (2009).

[47] E.A.L. Henn, J.A.Seman, G.Roati, K.M.F. Magalhaes, V.S. Bagnato, Phys. Rev. Lett. **103**, 045301 (2009).

[48] K. E. Wilson, E. C. Samson, Z. L. Newman, T. W. Neely, B. P. Anderson, Ann. Rev. Cold At. Mol. **1**, 261 (2013).

[49] M.C. Tsatsos, P.E.S. Tavares, A. Cidrim, A.R. Fritsch, M.A. Caracanhas, F.E.A. dos Santos, C.F. Barenghi, V.S. Bagnato, Phys. Rep. **622**, 1 (2016).

[50] M.W. Zwierlein, J.R. Abo-Shaeer, A. Schirotzek, C.H. Schunk, W. Ketterle, Nature **435**, 1047 (2005).

[51] M.W. Zwierlein, A. Schirotzek, C.H. Schunck, W. Ketterle, Science **311**, 492 (2006).

[52] B.P. Anderson, J. Low. Temp. Phys. **161**, 574 (2010).

[53] A.L. Fetter, Rev. Mod. Phys. **81**, 647 (2009).

[54] A. Aftalion, *Vortices in Bose-Einstein Condensates* (Birkhäuser, Boston, 2006).

[55] C.J. Pethick, H. Smith, *Bose-Einstein Condensation in Dilute Gases* (Cambridge University Press, Cambridge UK, 2008).

[56] S. Stringari, L.P. Pitaevskii, *Bose-Einstein condensation* (Clarendon Press, Oxford, 2003).

[57] J.F. Annett, *Superconductivity, superfluids and condensates* (Oxford Master Series in Condensed Matter Physics, Oxford University Press, New York, 2004).

[58] C. Kittel and H. Kroemer, *Thermal physics (2 ed)* (W.H. Freeman, New York, 1980).

[59] W. Greiner, J. Reinhardt, *Field Quantization* (Springer-Verlag, Berlin, 1996).

[60] F. Mandel, G. Shaw, *Quantum Field Theory* (John Wiley & Sons, West-Sussex UK, 2010).

[61] W.H. Dickhoff, D. Van Neck *Many-Body Theory Exposed!: Propagator description of quantum mechanics in many-body systems* (World Scientific, Singapore, 2005).

[62] J.D. Griffiths, *Introduction to Quantum Mechanics* (Pearson Prentice Hall, New Jersey, 2013).

[63] L.D. Landau, E.M. Lifshitz, *Quantum Mechanics* (Pergamon, New York, 1987).

[64] I. Bloch, J. Dalibard, W. Zwerger, *Many-body physics with ultracold gases*, Rev. Mod. Phys. **80**, 885 (2008).

[65] I. Manek, Yu.B. Ovchinnikov, R. Grimm, *Generation of a Hollow Laser Beam for Atom Trapping Using an Axicon*, Opt. Commun. 147, 67 (1998).

[66] N. Verhelst, S.N. Klimin, J. Tempere, *Verification of an analytic fit for the vortex core profile in superfluid Fermi gases*, Physica C (to be published).

[67] W. Ketterle, *The Magic of Matter Waves*, MIT Physics annual (2001).

[68] J. Tempere, J.T. Devreese, *Vorticity in Bose-Einstein condensed atomic gases*, Physica C **369**, 28 (2002).

[69] N.R. Heckenberg, R. McDuff, C.P. Smith, A.G. White, *Generation of optical phase singularities by computer-generated holograms*, Opt. Lett. **17**, 3, 221 (1992).

[70] H. Pérez-de-Tejada, Magnus force in the Venus ionosphere, J. Geophys. Res. **111**, A11105 (2006).

[71] R. Lundin, S. Barabash, Y. Futaana, M. Holmström, H. Perez-de-Tejada J.-A. Sauvaud, *A large-scale flow vortex in the Venus plasma tail and its fluid dynamic interpretation*, Geophys. Res. Lett. **40**, 1273 (2013).

[72] J.D. Anderson, *Fundamentals of Aerodynamics* (McGraw Hill, New York, 2007).

[73] C.N. Yang, Rev. Mod. Phys. **34**, 694 (1962).

[74] Yu. Kagan, B.V. Svistunov, G.V. Shlyapnikov, Sov. Phys. JETP **66**, 314 (1987).

[75] V. L. Berezinskii, Sov. Phys. JETP **32**, 493 (1971).

[76] J. M. Kosterlitz, D.J. Thouless, J. Phys. C **6**, 1181 (1973).

[77] J. M. Kosterlitz, J. Phys. C **7**, 1046 (1974).

[78] S. Giorgini, L.P. Pitaevskii, S. Stringari, *Theory of ultracold atomic Fermi gases*, Rev. Mod. Phys. **80**, 1216–1268 (2008).

[79] A.J. Leggett, *Quantum Liquids: Bose-Einstein condensation and Cooper pairing in condensed matter systems* (Oxford Graduate Texts, Oxford University Press, New York, 2006).

[80] M. Inguscio, W. Ketterle, C. Salomon, *Ultra-cold Fermi gases* (IOS Press, Amsterdam, The Netherlands, 2007).

[81] H.T.C. Stoof, K.B. Gubbels, D.B.M. Dickerscheid, *Ultracold Quantum Fields* (Springer, The Netherlands, 2009).

[82] L.N. Cooper, *Bound electron pairs in a degenerate Fermi gas*, Phys. Rev. **104**, 1189 (1956).

[83] C.A.R. Sá de Melo, *When fermions become bosons: Pairing in ultracold gases*, Phys. Today **61**, No. 10, 45 (2008).

Chapter 2

Direct Generation of Vortex Laser Beams and Their Non-Linear Wavelength Conversion

Andrew James Lee and Takashige Omatsu

Additional information is available at the end of the chapter

http://dx.doi.org/10.5772/66425

Abstract

Vortex laser beams are a technology that has revolutionised applications in micro- and nano-manipulation, micro-fabrication and super-resolution microscopy, and is now heralding advances in quantum communication. In order to service these, and emergent applications, the ability to generate powerful vortex laser beams with user-controlled spatial and wavefront properties, and importantly wavelength, is required. In this chapter, we discuss methods of generating vortex laser beams using both external beam conversion methods, and directly from a laser resonator. We then examine the wavelength conversion of vortex laser beams through non-linear processes of stimulated Raman scattering (SRS), sum-frequency generation (SFG), second harmonic generation (SHG) and optical parametric oscillation. We reveal that under different types of non-linear wavelength conversion, the spatial and wavefront properties of the vortex modes change, and in some cases, the spatial profile also evolve under propagation. We present a theoretical model which explains these dynamics, through decomposition of the vortex mode into constituent Hermite-Gaussian modes of the laser resonator.

Keywords: vortex laser, Lagurre-Gaussian modes, optical non-linear conversion, stimulated Raman scattering, sum-frequency generation, second harmonic generation, optical parametric oscillation, topological charge

1. Introduction

Vortex laser beams are characterised as having an annular spatial profile (with a central dark spot), and a wavefront which spirals along the direction of propagation of the beam (like a corkscrew). This is in contrast to a conventional Gaussian beam that has a plane wavefront [1–4]. The spatial, phase and wavefront properties of a Gaussian beam and vortex laser beams are shown in **Figure 1**.

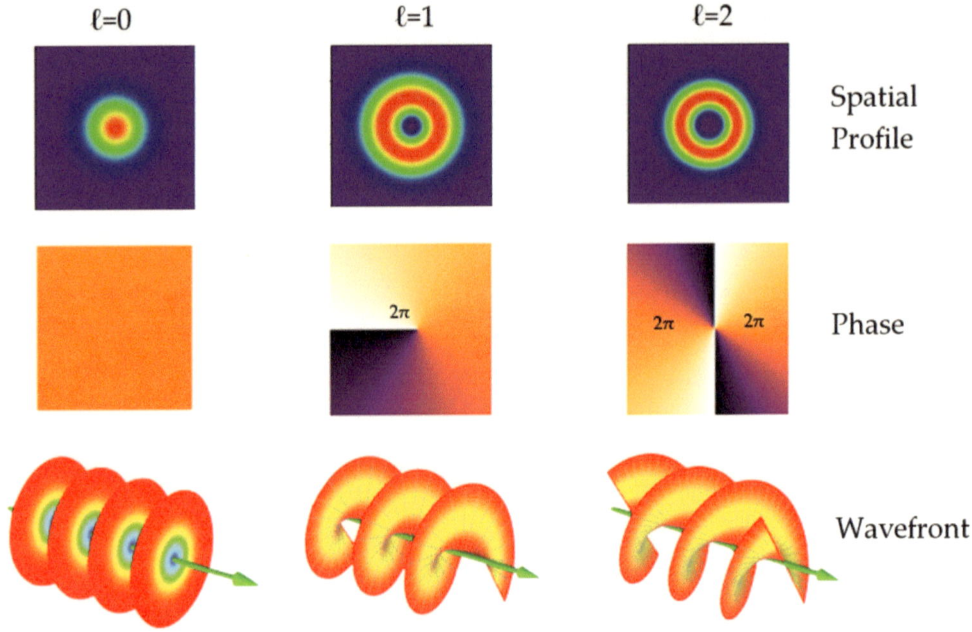

Figure 1. Comparison of Gaussian and vortex laser beams showing spatial and wavefront profiles (Image modified, credit: E-karimi, Creative Commons Attribution-Share Alike 3.0 unported).

The topological charge (denoted l) of a vortex beam is equal to the integer number of a 2π phase change of the wavefront on any closed path around the propagation axis. Equivalently, one round trip around the phase surface leads to the next (or prior) coil with a pitch of $l\lambda$, where λ is the wavelength of the vortex beam [4]. Due to this spiral wavefront, vortex laser beams inherently carry orbital angular momentum (OAM), with each photon carrying orbital angular momentum of $\ell\hbar$ (where $\hbar$ is the reduced Planck constant). The magnitude of a vortex beam topological charge also impacts the size of the central dark spot relative to the overall size of the annular beam, with higher topological charge resulting in a larger spot. The direction of the wavefront, spiralling clockwise or anti-clockwise along the axis of propagation also indicates the sign of the topological charge, with positive (+) topological charge having a wavefront which spirals clockwise, and a negative (−) topological charge having a wavefront spiralling anti-clockwise [4].

It is not surprising that these unique characteristics of vortex laser beams have resulted in their application in a diverse range of applications. Indeed, vortex beams have revolutionised applications ranging from optical tweezing, in which micro- to nano-scale objects can be trapped and manipulated [3, 5], to fabrication of chiral nano-structures [6, 7], through to quantum communication [8]. Perhaps one of the most significant application of optical vortex beams has been in super resolution microscopy based on stimulated emission depletion (STED), a microscopy technique which has gained recognition with the Nobel Prize in Chemistry (2014). In STED microscopy, the annular profile of a vortex laser beam is used to selectively deplete the outer region of a fluorescing particle being imaged. This effectively

http://dx.doi.org/10.5772/66425

allows the resolution of fluorescence microscopy to exceed the diffraction limit [9, 10]. This host of applications therefore not only exploit the orbital angular momentum properties of the vortex beam, but also its annular spatial profile.

With the diversity of applications for vortex laser beams, comes the requirement for beams of the appropriate power and beam quality, topological charge, and importantly wavelength. As with 'standard' laser beams, there is continual drive to develop sources which can offer a diverse range of laser powers, beam quality and wavelengths. In this chapter, we detail methods of generating vortex laser beams, with a focus on methods of directly generating vortex beams from solid-state lasers (some of the most robust laser architectures currently available); and methods of wavelength-tuning vortex laser beams using non-linear optical processes. Here, we outline the dynamics of the processes by which these vortex beams can be manipulated, and the diversity of outputs that can be accessed.

2. Generation of vortex laser beams

There are a host of methods of generating vortex laser beams, however they can be broadly classified into two groups, those acting as convertors— taking a Gaussian beam laser beam and converting it into a vortex beam; and the direct generation methods whereby a vortex beam is produced directly from a laser resonator. The most commonly applied methods are those of the former group involving conversion of a Gaussian laser beam into a vortex beam.

2.1. Mode convertors

Conversion of a Gaussian laser beam into a vortex laser beam involves transformation of a plane wavefront into a spiral wavefront. This can be achieved using a number of methods including:

Phase plates: These are specially fabricated plates where the region through which the Gaussian beam propagates gradually increases in thickness (by stepping) in a spiral fashion. This transforms the plane wavefront of the Gaussian beam into a spiral wavefront by introducing a phase shift [11]. Through appropriate design of the phase plate, it is possible to use it to generate vortex laser beams with a desired topological charge.

Holograms/spatial light modulators (SLM): Much like phase plates, these elements act on the phase of the incident light to generate a spiral wavefront. Spatial light modulators also enable control over the order and sign of topological charge of the beam as they can be readily programmed [12, 13].

Cylindrical lens pairs: These act on higher-order Gaussian beams, typically Hermite-Gaussian beams. The cylindrical lens pair converts Hermite-Gaussian beams of a particular order, to their Laguerre-Gaussian counterparts [14]. As will be examined further in this chapter, Laguerre-Gaussian beams of the form LG_{0l} are vortex beams with topological charge l.

These methods are effective in converting Gaussian laser beams into vortex laser beams, however they are additional components which must be aligned precisely in addition to the actual laser source itself. Phase plates and SLMs, in particular, suffer from relatively low conversion efficiency, and can be easily damaged at high laser powers. This limits their capacity to generate vortex laser beams with high power and high quality.

The ability to also generate vortex laser beams of a desired wavelength using these conversion methods is predicated on the initial Gaussian laser beam undergoing conversion, already being of the desired wavelength. In the following section, we detail methods of directly generating vortex laser beams from all solid-state laser systems, and demonstrate that these systems are capable of producing very high-power beams with excellent beam quality, without the need for additional, bulky and easily damaged beam converting components.

2.2. Direct generation methods

Methods of generating vortex laser beam output directly from a laser cavity rely on suppression of the lowest order Gaussian mode (TEM_{00}) and preferential oscillation of Lagurre-Gaussian modes of the form $LG_{0\ell}$. These Lagurre-Gaussian modes are vortex modes with topological charge (ℓ) [15, 16]. Suppression of the lowest order Gaussian mode can been achieved in a number of ways, including pumping the laser gain medium with an annular pump spot [17, 18], using intra-cavity defects [19–23] and thermal lensing [24, 25].

In this chapter, we focus on the generation of vortex laser beams using two of these methods, using an intra-cavity defect spot and thermal lensing. Intra-cavity defect spots are an effective means of very simply suppressing oscillation of the lowest-order Gaussian mode, to promote oscillation of an LG_{01} mode, and generate vortex laser output from an end-pumped solid-state laser. In these systems, the intra-cavity field is very intense, and by making use of this, we can readily investigate non-linear conversion of vortex beams. The use of thermal lensing to produce a vortex laser beam is also investigated in the context of a side-pumped slab laser. As will be detailed, this configuration is a very effective way of directly generating very high-power vortex laser beams, something which cannot be readily achieved using extra-cavity conversion components owing to component damage.

2.2.1. Intra-cavity defect spot

The use of an intra-cavity defect to suppress the lowest order oscillating mode (generally TEM_{00}) within a laser cavity has been achieved using a number of different laser systems, including He-Ne lasers [20] and solid-state lasers [19, 20–23]. The common feature of each of these systems is that an intra-cavity defect is used; such as engineered damage spots on resonator mirrors. Key to the effectiveness of this method is the ability for the defect to force oscillation of LG modes of the form $LG_{0\ell}$. The defect spot in effect breaks the symmetry of the laser resonator to enable oscillation of these LG modes. This requires accurate mode matching between the lowest order Gaussian mode (TEM_{00}) and the defect spot. As reported in the literature, the defect spot to cavity mode ratio is generally within the range ~0.15–0.2 for solid-state lasers [19–23]. The operating principle of this technique is shown schematically in **Figure 2**.

Here, we investigate generation of vortex laser emission from a continuous-wave, diode-end-pumped $Nd{:}GdVO_4$ laser system. This system not only serves as an effective method for

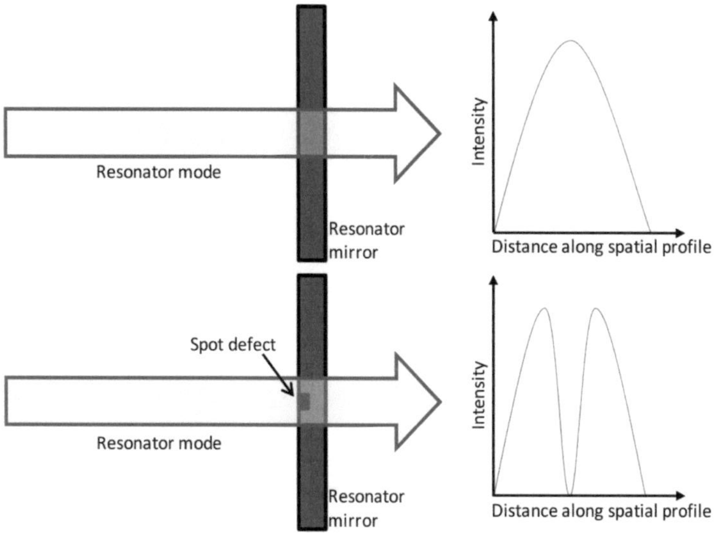

Figure 2. Schematic representation showing the principle of operation of a defect spot within a laser cavity to generate vortex laser emission. In a typical laser resonator, the laser beam output has a Gaussian intensity profile. By introducing an intra-cavity defect, we can break the symmetry of the resonator and force the oscillation of LG vortex modes.

generating vortex laser emission, but also as a laser which can be used to investigate non-linear wavelength conversion of vortex beams via the stimulated Raman scattering (SRS), and second harmonic (SHG) and sum-frequency generation (SFG) processes. There have been rather few studies of intra-cavity non-linear conversion of vortex laser beams, and the underlying dynamics of the transfer of topological charge under these processes have been little studied. The details of these non-linear conversion processes are discussed later in this chapter.

The end-pumped laser system investigated was comprised of an a-cut $Nd{:}GdVO_4$ crystal (0.3% Nd-doping), 4 × 4 × 20 mm^3 with a front surface coated with a high-reflecting coating ($R > 99.999\%$ at 1064 nm/$R < 0.5\%$ at 880 nm) and an end-mirror with reflectivity $R = 99.91\%$ at 1064 nm and radius of curvature, 250 mm. Key here is that the end-mirror had an array of circular damage spots (3 × 3 array of spots with increasing diameter from 40–160 μm with a spacing of 400 μm) laser micro-machined (to the level of the glass substrate) onto its surface which acted as defect spots [21]. The laser cavity was end-pumped with a fibre coupled (100 μm diameter, 0.22 NA) laser diode producing up to 30 W at 879 nm. The experimental layout of the system is shown schematically in **Figure 3**, along with a schematic of the array of damage spots on the resonator mirror.

Alignment and optimisation of the laser system required accurate positioning of the end-mirror to ensure that the laser cavity mode was aligned with each defect spot. The operation of the system aligned on each defect spot was analysed, and it was found that complete suppression of the TEM_{00} mode, and oscillation of the LG_{01} mode could only be achieved with the 40 μm diameter defect spot. Given that the cavity mode had a diameter of ~260 μm, this corresponded to a cavity to defect ratio of 0.15, which was consistent with reports in the literature.

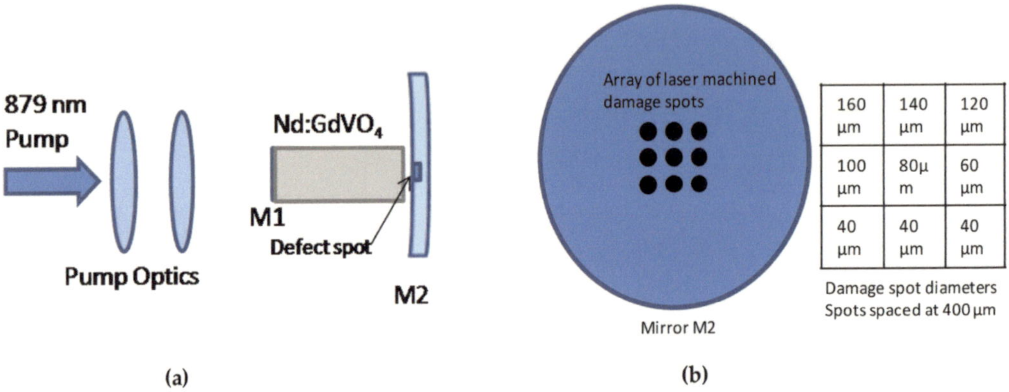

Figure 3. (a) Experimental layout of the diode-end pumped vortex laser utilising defect spots laser micro-machined onto the end-mirror [21]; and (b) the layout of the array of defect spots on mirror M2.

The threshold for oscillation of the LG_{01} mode from this system was achieved for an incident pump power of 0.1 W; maximum output power at 1064 nm was 0.4 W. The spatial profile of the laser emission was annular with a central dark spot, consistent with an LG_{01} mode. This spatial profile was retained across the pumping range. Beam quality factor (M^2) measurements of the output were made for incident pump powers of 0.1 W (just above laser threshold) and 2 W, and values of 2.2 and 3.1 were determined, respectively. An M^2 value of 2 is consistent with a vortex laser beam with topological charge of $\ell = 1$ [19]. The higher M^2 value at higher incident pump power is indicative of the onset of higher-order modes oscillating within the laser cavity, which is a consequence of the high laser gain and high cavity Q-factor.

The wavefront properties of the laser output were examined using an interferometric technique [4]. This technique yields a set of straight fringes, within which the position and order of the vortex beam phase-singularity can be determined through break-down in fringe structure- manifesting in a fork-like pattern. Both linear and spiral fringe patterns may be produced; in the case of a spiral fringe pattern, a spherical wavefront is formed using a short focal length lens [4]. Determination of the sign of the topological charge of the vortex beam using these methods relies on knowledge of the projection of the wavefront in the reference arm relative to the direction of propagation (in the case of straight fringe patterns); or knowledge of the radius of curvature of the reference beam in the case of spiral fringe patterns.

The spatial profile, along with the fork and spiral interferometer patterns for the laser output is shown in **Figure 4(a)–(c)**, respectively.

From **Figure 4(a)** it can be seen that an annular spatial profile is produced from this laser resonator, and from the interference patterns shown in **Figure 4(b)** and **(c)**, it can be seen that a single fork/spiral is observed, which is indicative of a phase singularity with a topological charge of 1. As the laser output was power scaled, the order of the topological charge and direction of rotation did not change. It was possible to alter the direction of the wavefront by altering the laser alignment through slight adjustment of the output coupler angle.

This result demonstrates the relative ease with which a vortex laser beam with a well-defined topological charge may be directly generated from an end-pumped solid-state laser system.

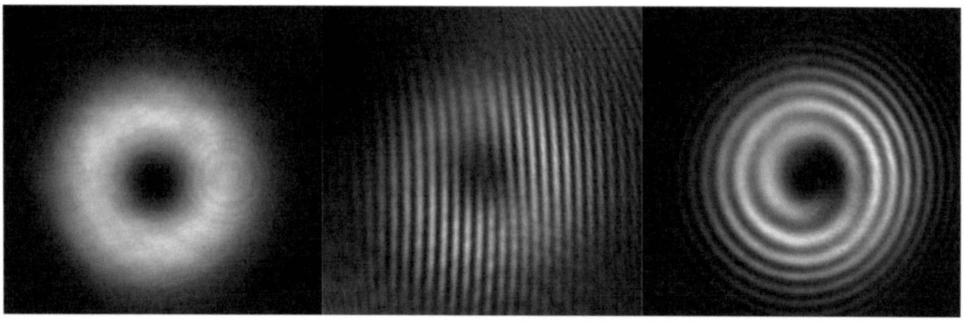

(a) (b) (c)

Figure 4. (a) Spatial profile of the 1063 nm emission from the laser resonator; and interferometer patterns showing (b) linear; and (c) spiral fringes [21].

This is significant as it precludes the requirement for additional vortex beam-shaping/forming components. As will be discussed further in this chapter, this simple resonator design facilitates investigation of intra-cavity non-linear wavelength conversion of vortex laser beams with relatively high efficiency.

2.2.2. Side pumped solid-state vortex laser

In a side-pumped laser resonator, the gain medium is pumped through a side facet as opposed to an end-face as described in the previous section. This has the effect of increasing the overall pump power that can be utilised in the system by distributing the thermal load throughout a larger area of the crystal [26]. In the side-pumped laser discussed in this section, the oscillating laser mode also undergoes reflection (total internal reflection) at the surface through which the pump radiation is delivered [24, 25]. In this configuration, oscillation of the vortex mode (LG_{01}) over the TEM_{00} (Gaussian mode) is achieved due to the thermal lens induced within the crystal. As the laser power is increased, the induced thermal lens affects the Gaussian mode faster (due to mode overlap considerations) and it is forced to be unstable (effectively being suppressed), while the LG_{01} vortex mode is still supported within the resonator. Therefore, this system operates with vortex output for a window of input pump powers, that is, enough pump power to induce a thermal lens sufficient to drive the system unstable for the Gaussian mode, while ensuring oscillation of the vortex mode. This allows for significantly high power vortex modes to be generated from this system.

The laser system comprises a laser diode side-pumped a-cut $Nd:GdVO_4$ slab (1% Nd-doping) with dimensions $2 \times 5 \times 20$ mm^3, around which a bounce resonator is formed. The y–z faces of the slab were wedged and anti-reflection coated for 1063 nm to prevent self-lasing [26]. The laser mode oscillates between two flat mirrors, denoted E.M (high reflecting at 1063 nm) and O.C. (coated R = 40% at 1063 nm). To maintain good spatial overlap between the resonator mode and the pumped volume, two cylindrical lenses were used intra-cavity, denoted $CL_{1,2}$ both with focal length f = 50 mm. The slab is side pumped by a laser diode bar array producing 55 W into an elliptical pump spot generated with a cylindrical lens (f = 12.7 mm). The system layout is shown in **Figure 5**.

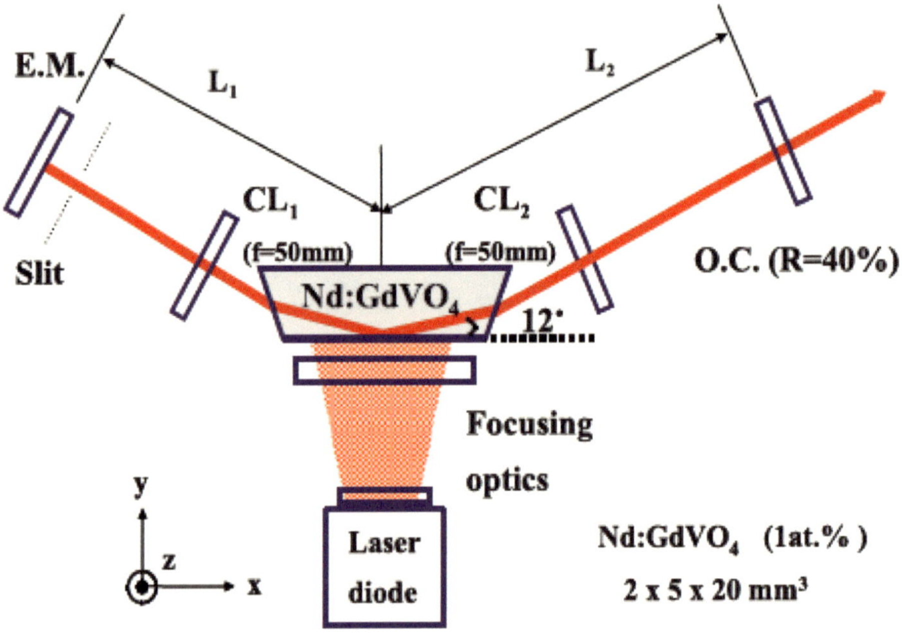

Figure 5. Layout of the side-pumped solid-state laser system [26].

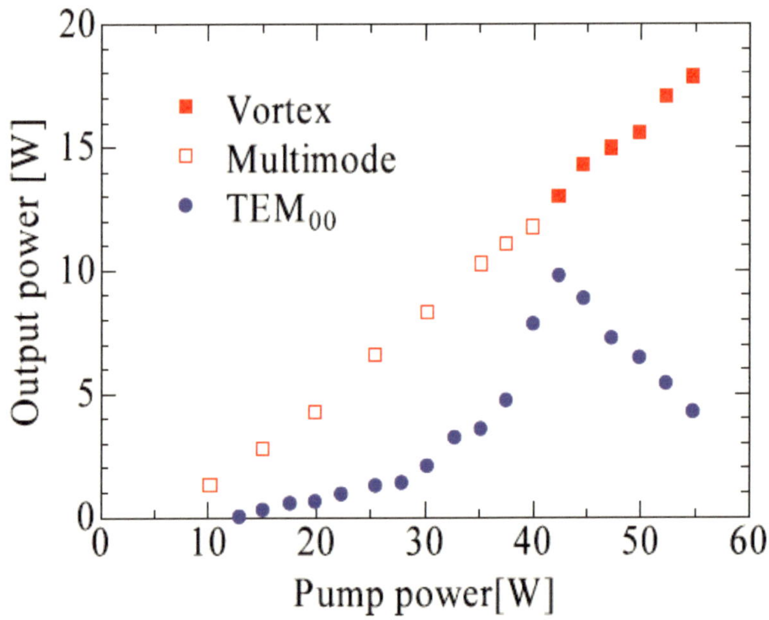

Figure 6. Power transfer curve for the side-pumped bounce laser. Vortex beam emission is observed for incident pump powers in the range 43–55W [26].

The threshold for lasing action was ~10 W incident pump power. Above this threshold, the laser oscillates with a Gaussian-like output profile. As the incident pump power is increased, oscillation of an LG_{01} mode dominates and the TEM_{00} mode is suppressed. The power transfer curve for this system is shown in **Figure 6**. Spatial profiles of the output laser mode are shown in **Figure 7(a)–(c)** for a range of incident pump powers.

From this system, a vortex beam with 17.8 W of power at 1064 nm is generated for 55 W incident diode pump power. **Figure 7(d)** shows the spiral interference pattern of the vortex beam generated at an incident pump power of 55 W.

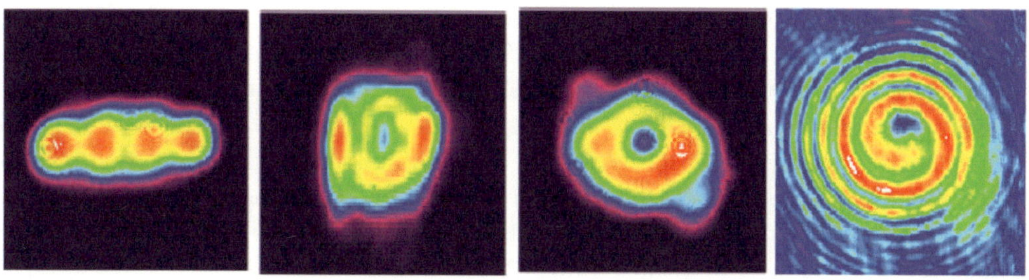

(a) (b) (c) (d)

Figure 7. Spatial profile of the emission from the laser system operating at powers of (a) less than 43 W; (b) 43 W; (c) 55 W. The spiral interference pattern of the output generated at a pump power of 55 W is shown in (d) [26].

It was observed that the topological charge of the vortex beam did not change as the system was power scaled, maintaining a value of $l = 1$.

This work demonstrates that very high-power vortex laser beams can be generated directly from a laser system without the need for additional beam shaping components. By exploiting thermal lensing (a property often considered a problem in laser systems), vortex beams with very high output power can be generated directly from a solid-state laser.

3. Non-linear wavelength conversion of vortex laser beams

Non-linear wavelength conversion of laser beams is an efficient and effective method of altering the wavelength of laser beams, and these methods have been employed extensively in the case of Gaussian beams [27, 28]. Non-linear wavelength conversion within a non-linear crystal can generally be achieved through reaching a threshold power/intensity, in the case of $\chi^{(3)}$ processes; or phase matching in the case of $\chi^{(2)}$ processes. As will be discussed in this section, in these non-linear conversion processes, the conservation of energy and momentum offers a unique, interesting and powerful method of controlling not only the wavelength of the vortex beam, but also to control/manipulate the topological charge and resultant spatial profile of the beam.

In this section, we cover non-linear wavelength conversion of vortex laser beams under the processes of stimulated Raman scattering (SRS); sum-frequency and second-harmonic generation (SFG, SHG); optical-parametric oscillation (OPO), and difference-frequency generation (DFG). Through this analysis, we investigate the dynamics of vortex laser beams and how each of these non-linear optical processes, in conjunction with cavity design, allows us to achieve selective control of the topological charge of the vortex beam.

3.1. Stimulated Raman scattering (SRS)

SRS is a third-order non-linear process $\chi^{(3)}$ which can be used to convert the wavelength of a laser beam through the Raman scattering process [29–31]. It can be performed both for intra-cavity and extra-cavity configurations, and is most commonly achieved with the use of crystalline Raman-active media. In the SRS process, a fundamental laser field excites a Raman-active resonant mode (phonon) of the crystalline material. This leads to a coupling of energy and as a result, a photon with energy difference between the incident photon and the resonant mode is scattered; this is referred to as the Stokes field. In the case of stimulated Raman scattering, a resonant cavity is used to oscillate the Stokes field. In this process, the scattered Stokes photons stimulate the generation of more Stokes photons at the same wavelength/frequency, akin to stimulated gain within a laser medium [26]. This leads to rapid build-up of the Stokes photons from noise, with a well-defined intensity threshold for the process. The wavelength-shifts which can be achieved are therefore dependant on the Raman-active modes which exist within a material. There are a plethora of Raman-active crystalline materials which have been demonstrated effective at wavelength shifting via SRS; some of most commonly applied crystals (and their primary Raman shift) include: $GdVO_4$ (885 cm^{-1}), YVO_4 (892 cm^{-1}), $KGdWO_4$ (768 and 901 cm^{-1}) and $BaWO_4$ (926 cm^{-1}) [31].

In the case of conventional laser fields with Gaussian spatial distributions, SRS is a well-studied process both for intra-cavity and extra-cavity configurations. However, relatively few investigations of SRS conversion of vortex laser beams have been undertaken. In this section, we investigate the generation of vortex laser beams within the context of end-pumped solid-state lasers, as a continuation of work presented in the previous section on direct-generation of vortex laser beams using defect spots on end-mirrors.

The SRS process requires the generation of very intense fields to achieve threshold [31]. Therefore, in the case of Gaussian fields, the SRS process is often restricted to the very centre of the oscillating Gaussian mode, where the intensity of the beam is highest. This generally leads to the generation of Stokes beams which are of higher beam quality, owing to the Raman-beam-clean-up effect [32]. This is an important consideration as the spatial profile of the 'fundamental' beam has a direct impact on the resultant Stokes beam. In the case where a non-Gaussian field, for example, a vortex field, oscillates within the laser cavity, achieving SRS threshold is more difficult due to the annular spatial profile of the oscillating mode.

We examine here, wavelength-shifting of a fundamental field at 1063–1173 nm via intra-cavity SRS within a self-Raman $Nd:GdVO_4$ laser crystal. The experimental setup is identical to that shown in **Figure 3**; however, key here is that the mirrors M1 and M2 are coated to oscillate both the fundamental and Stokes fields, M1 with $R > 99.999\%$ at 1064 and 1173 nm, and M2 with $R = 99.91\%$ at 1064 nm and $R = 99.4\%$ at 1173 nm. The same array of damage spots were used as detailed previously [21].

The threshold for oscillation of the vortex Stokes field at 1176 nm was ~ 2 W absorbed diode pump power, and a maximum Stokes power of 380 mW was generated from this laser for an absorbed pump power of 6.8 W. It was found that the Stokes beam retained an annular profile throughout the input pump power range. The spatial profile of the Stokes beam along with both linear and spiral interference patterns are shown in **Figure 8(a)–(c)**, respectively.

The annular spatial profile of the Stokes beam is a result of the SRS gain profile being dictated by that of the fundamental field mode. As the fundamental field is an LG_{01} mode, the SRS gain profile will also be annular in profile. Furthermore, the Stokes field is also affected by the damage spot laser machined onto the output mirror (M2). These two factors ensure that the Stokes field will retain an annular profile similar to that of the fundamental field.

The magnitude and sign of topological charge was always found to be the same for both the Stokes and fundamental fields. Here, it is interesting to consider what also happens to the phonon field which is also excited in the SRS process. Due to conservation of momentum rules, it is also possible for this field to receive topological charge from the fundamental field. To investigate this, we can consider conservation of orbital angular momentum under the process of SRS.

Let us denote the topological charge of the exciting fields as l_P, l_S and l_R for the fundamental, Stokes and material, respectively; and l'_F, l'_S, l'_R as the stimulated fields for the fundamental, Stokes and material, respectively. In a stimulated scattering event, the stimulated field must take on the same energy state as the stimulating field, so in this case, $l_S = l'_S$. Then in the stimulated scattering process, $l_F + l_S + l_R = 2l'_S + l'_R$.

We have observed that $l_F = l_S$ and so, there is no change in the topological charge state of the medium, that is, $l_R = l'_R$, that is, the material phonon does not receive orbital angular

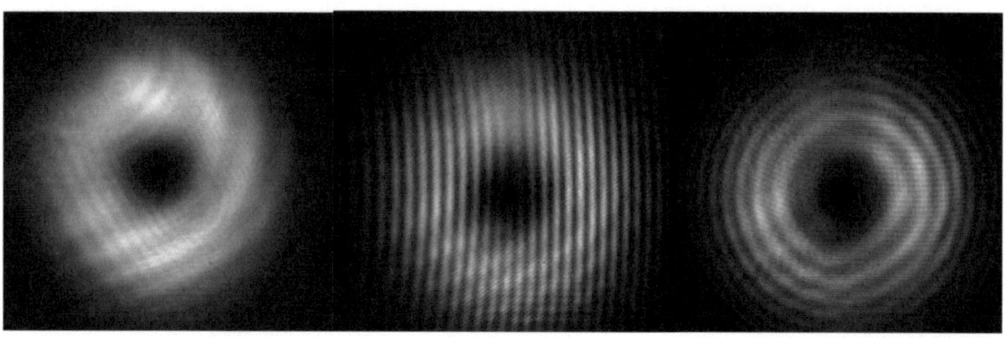

(a) (b) (c)

Figure 8. Images of (a) the spatial profile of the output Stokes beam; (b) linear interference pattern; and (c) spiral interference pattern [21].

momentum in the SRS process. It is interesting to note that while transfer of topological charge from photons to phonons was not observed in this work under the process of SRS, it has been reported in the literature through non-linear Brillouin interactions [33].

This work demonstrates how readily the wavelength of a vortex laser beam can be converted using the SRS process; in this case, a fundamental wavelength at 1063 nm is converted to 1173 nm in $GdVO_4$. Given the broad range of Raman-active crystalline materials that are available, the ability to generate new vortex laser wavelengths through the SRS process with the retainment of topological charge state is very powerful.

3.2. Sum-frequency generation and second harmonic generation

Sum-frequency generation and second harmonic generation (an example of sum-frequency generation where two photons of the same frequency are combined) are effective non-linear methods of decreasing the wavelength of laser fields [28]. It is most commonly used to convert wavelengths in the near-infrared to the visible region. For example, SHG of 1064 nm (Nd:YAG emission) is commonly used to generate green emission (532 nm). This is the non-linear process employed in the now ubiquitous green laser pointer.

SFG is a phase-matched, non-linear process and requires the conservation of both momentum and energy. As already discussed in the context of SRS, it is this conservation of momentum which imparts special relevance to vortex laser fields in that orbital angular momentum (OAM) must be conserved. In the case where two vortex beams with topological charge $l_{\omega 1}$ and $l_{\omega 2}$ are transformed under SFG, the resultant field must have topological charge which is the sum of the two initial fields, i.e. $l_{\omega 1+\omega 2} = l_{\omega 1} + l_{\omega 2}$.

Through this process, we have a powerful means of manipulating both the wavelength and the topological charge of the vortex laser beam. The process of second harmonic generation of vortex beams has been studied comprehensively in the context of extra-cavity conversion [34, 35], however, this process in an intra-cavity configuration is less well understood. In this work, we investigate this process in the context of an intra-cavity, end-pumped Raman laser, as detailed in the above section for SRS. In this case, the SFG/SHG crystal is incorporated into the laser cavity, and the process of SFG or SHG is achieved through appropriate phase matching. The resonator layout is shown in **Figure 9**.

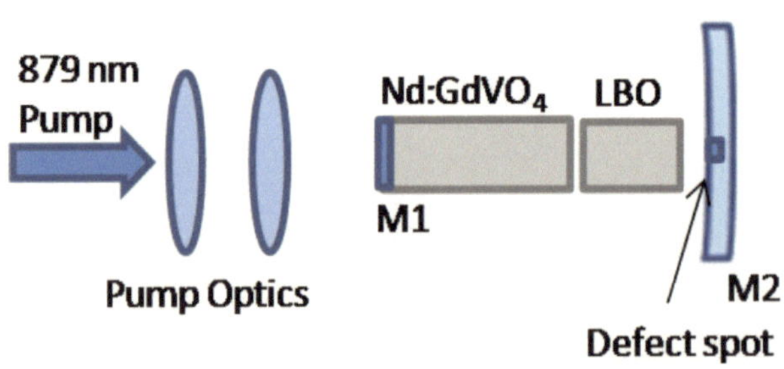

Figure 9. Layout of the intra-cavity vortex laser incorporating both SRS and SFG/SHG non-linear processes [22, 23].

The non-linear crystal used for SFG/SHG is a lithium triborate (LBO) crystal, cut for non-critical phase matching (NCPM), with dimensions, 4 × 4 × 10 mm^3. The LBO crystal was placed in a copper mount which could be temperature-controlled, to enable selective phase matching via temperature tuning. With the laser generating a fundamental wavelength of 1063 nm and a Stokes wavelength of 1173 nm, phase matching for SHG of the 1173 nm wavelength to 586 nm is achieved with the LBO crystal set to a temperature of 43.5°C, and SFG of the 1063 nm and 1173 nm fields is achieved at 93°C to generate output at 559 nm.

With this system, a maximum power of 727 mW at 586 nm, and 850 mW at 559 nm was achieved [22, 23]. The spatial profile of the SHG and SFG fields were similar to the annular profiles that were produced in the near-field, however these evolved to a spot with a central bright core in the far-field. The spatial profiles of the 586 nm and 559 nm fields in the near- and far-field are shown in **Figure 10**.

The beam quality-factor of the resultant beams were also measured. In the case of the near-field profiles, beam quality factors of M^2 ~ 3–3.6 were determined in both cases. A beam quality factor of $M^2 = 3$ is consistent with a vortex beam having topological charge $l = 2$.

The wavefront properties of the SFG and SHG beams were also analysed using the interferometric setup already described. Observations of the topological charge of the resultant SFG and SHG fields were consistent with conservation of orbital angular momentum, where the topological charge of the resultant field was always the sum of the topological charge of the initial fields, that is, $l_{SFG} = l_{\omega 1} + l_{\omega 2}$, wherein the degenerate case of SHG, $l_{2\omega} = 2l_{\omega}$. A comparison of the linear and spiral interference patterns of the Stokes (1173 nm) field and its SHG (586 nm) are shown in **Figure 11**.

The interference patterns clearly show that the Stokes field has a topological charge of $l = 1$, and the SHG field has a topological charge of $l = 2$ (three-pronged fork and two spirals). This shows that doubling of the topological charge takes place under SHG, further supporting the near-field spatial profile with a large central dark region as shown in **Figure 10(a)**.

In the case of SFG of the 1063 nm and 1173 nm fields, two different interference patterns were observed, one consistent with the generation of an SFG vortex field with topological charge

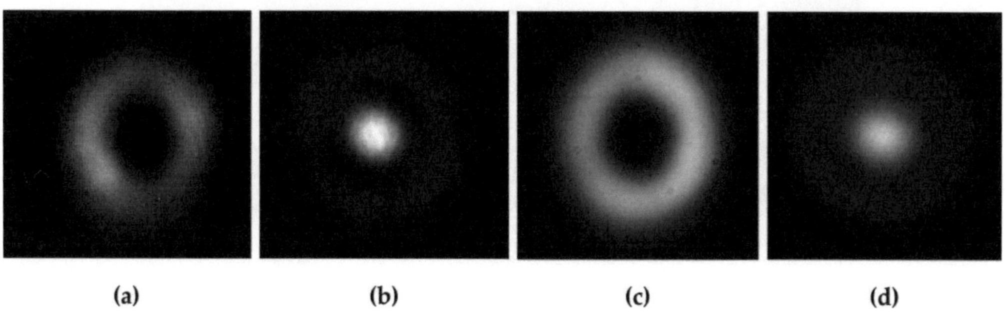

Figure 10. Spatial profiles of the 586 nm (SHG) field in the (a) near-field; and (b) far-field. Spatial profile of the 559 nm (SFG) field in the (c) near-field; and (d) far-field [22, 23].

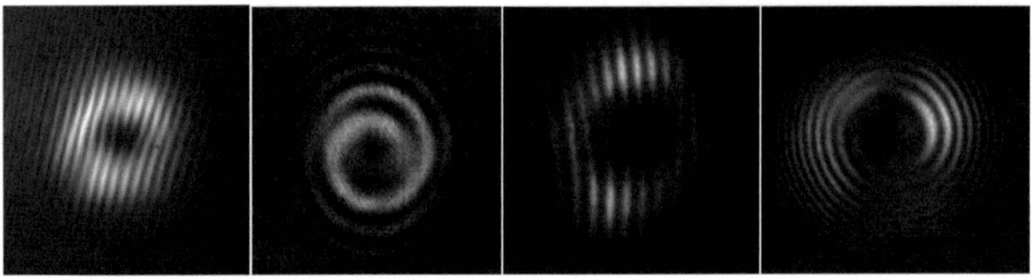

(a) (b) (c) (d)

Figure 11. Comparison of the interference patterns of the Stokes (1173 nm) field (a) linear; and (b) spiral interference patterns, with that of the SHG field (586 nm) (c) linear; and (d) spiral interference patterns [22].

$l = 2$, and the other consistent with an SFG field with topological charge of $l = 0$. It should be noted that in both cases, the same annular, near-field profile is generated. The linear interference patterns showing these two conditions are shown in **Figure 12**.

The two different topological charge states of the SFG field results from two different topological charge states of the 1063 nm and 1173 nm fields. In the case where the SFG field has a topological charge $l = 2$, both the 1063 nm and 1173 nm fields have topological charge $l = 1$. In the case where the SFG field has a topological charge of $l = 0$, the 1063 nm and 1173 nm fields have opposite sign of topological charge, that is, $l = +1$ and $l = -1$ or vice-versa. It is interesting to note that in this resonator configuration, where an LBO crystal is incorporated cf. that where only SRS takes place, it is possible to achieve a situation in which the 1063 nm and 1173 nm fields have opposite topological charge. The ability of the Stokes field to take on opposite topological charge to that of the fundamental field is likely due to the additional non-linear process of SFG competing with the SRS process in depleting the fundamental field.

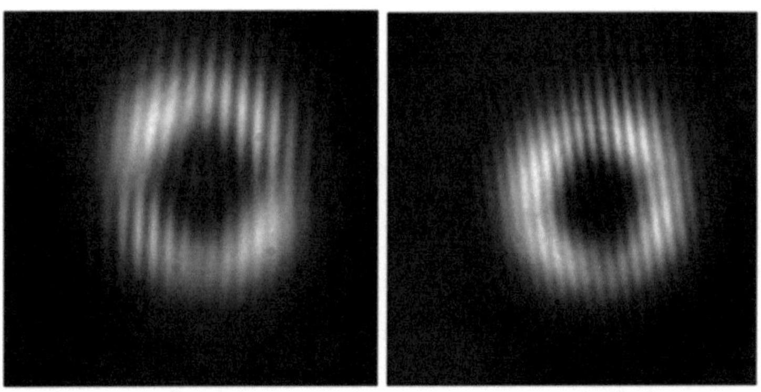

(a) (b)

Figure 12. Interference patterns generated from the SFG of 1163 nm and 1173 nm fields, showing (a) topological charge of 2; and (b) topological charge of 0 [23].

This generates instability within the system which affects mode symmetry, and enables oscillation of opposite topological charge.

What is significant in these results is that an SFG field can be generated with topological charge 0 and yet retain an annular spatial profile in the near-field. By definition, a vortex beam with zero topological charge should have a non-zero central core. Also of significance is that the spatial profile of the SHG and SFG fields is annular in the near-field, and develop a central bright spot in the far-field. This very interesting beam dynamics can be understood through a decomposition of the oscillating LG_{01} modes at 1063 nm and 1173 nm into their constituent Hermite-Gaussian (HG) modes, and examining how these modes evolve under the process of SFG/SHG and under free-space propagation. This process has been computationally modelled for comparison with experimental results.

In this model we first assume that the LG resonator modes for the fundamental and Stokes fields comprise Hermite-Gaussian (HG) modes, $HG_{1,0}$ and $HG_{0,1}$ where,

$$HG_{1,0} = x\,e^{-(x^2+y^2)} \tag{1}$$

$$HG_{0,1} = y\,e^{-(x^2+y^2)} \tag{2}$$

The expressions for the Lagurre-Gaussian vortex beams of the fundamental field with frequency (ω_1) and the Stokes field with frequency (ω_2) are given by:

$$LG_1^{\omega_1} = x\,e^{-(x^2+y^2)} + iy\,e^{-(x^2+y^2)}\,e^{i\Delta_1} \tag{3}$$

$$LG_{\pm 1}^{\omega_2} = x\,e^{-(x^2+y^2)} \pm iy\,e^{-(x^2+y^2)}\,e^{i\Delta_2} \tag{4}$$

where Δ_1 and Δ_2 are small phase mismatch terms which can manifest within the resonator [21]. In the expression for the Stokes field, ±1 denotes that the Stokes field can take a topological charge value of +1 or −1.

Now in the case of second harmonic generation, let us simply consider the case of SHG of ω_1. Here, the field strength of the SHG field is given by:

$$\begin{aligned} E_{SHG} &= LG_1^{\omega_1}\cdot LG_1^{\omega_1} = (x^2 - y^2 e^{2i\Delta_1})e^{-2(x^2+y^2)} + 2ixy\,e^{i\Delta_1}\,e^{-2(x^2+y^2)} \\ &= HG_{2,0} - HG_{0,2}\,e^{2i\Delta_1} + \tfrac{1}{2}(1 - e^{2i\Delta_1})HG_{0,0} + 2i\,e^{i\Delta_1}\,HG_{1,1} \end{aligned} \tag{5}$$

where $HG_{2,0} = \left(x^2 - \frac{1}{2}\right)e^{-2(x^2+y^2)}$, $HG_{0,2} = \left(y^2 - \frac{1}{2}\right)e^{-2(x^2+y^2)}$, $HG_{1,1} = x\,.\,y\,.\,e^{-2(x^2+y^2)}$

Note that in the above equation, the Gaussian term, in the form $HG_{0,0}$ manifests. This term does not have a central dark core in its spatial profile. Under propagation along an axis z, the contribution of Gouy phase on the modes as they propagate must also be considered. For HG modes, the Gouy phase term is represented as:

$$\delta_{m,n} = (m + n + 1)\tan^{-1}\frac{z}{z_R} \tag{6}$$

where m and n are the indices of the HG mode with the form $HG_{m,n}$, z is the propagation distance, and z_R is the Rayleigh length. For the HG modes under consideration here,

$$\delta(z) = \delta_{2,0} - \delta_{0,0} = \delta_{0,2} - \delta_{0,0} = \delta_{1,1} - \delta_{0,0} = 2\tan^{-1}\frac{z}{z_R}$$

The intensity of the SHG field is given by,

$$I_{SHG}(z) = \left|E_{SHG}\right|^2$$

$$\begin{aligned} I_{SHG}(z) = & \left|HG_{2,0}\right|^2 + \left|HG_{0,2}\right|^2 + 4\left|HG_{1,1}\right|^2 + \Delta_1{}^2\left|HG_{0,0}\right|^2 \\ & -2(1-2\Delta_1{}^2)HG_{2,0}HG_{0,2} + 2\Delta_1{}^2\cos(\Delta\delta) \\ & (HG_{2,0} + HG_{0,2})HG_{0,0} \end{aligned} \tag{7}$$

where for small Δ_1 $\langle\sin\Delta_1\rangle \simeq 0$, $\langle\cos\ \Delta_1\rangle \simeq 1$, $\langle\sin^2\Delta_1\rangle \simeq \Delta_1{}^2$

$$\delta(z) = \delta_{2,0} - \delta_{0,0} = \delta_{0,2} - \delta_{0,0} = \delta_{1,1} - \delta_{0,0} = 2\tan^{-1}\frac{z}{z_R}$$

From Eq. (7) we can see that Gouy phase impacts the spatial intensity profile of the SFG field in the far-field as its finite value (as Z increases) means that there is contribution to the overall spatial profile by the $HG_{0,0}$ mode. Following this derivation, we can also form expressions for the SFG field under conditions where the fundamental ω_1 and Stokes ω_2 fields have the same and opposite topological charge. The expressions for the intensity of the SFG fields in each case are given by:

$$\begin{aligned} I_{SFG}^{l=2}(z) = & \left|HG_{2,0}\right|^2 + \left|HG_{0,2}\right|^2 + (\Delta_1+\Delta_2)^2\left|HG_{0,0}\right|^2 + 4\left|HG_{1,1}\right|^2 - 2HG_{2,0}HG_{0,2} \\ & + \frac{(\Delta_1+\Delta_2)^2}{2}\cos\delta(HG_{2,0}+HG_{0,2})HG_{0,0} \end{aligned} \tag{8}$$

$$\begin{aligned} I_{SFG}^{l=0}(z) = & \left|HG_{2,0}\right|^2 + \left|HG_{0,2}\right|^2 + \left|HG_{0,0}\right|^2 - (\Delta_1+\Delta_2)^2\left|HG_{1,1}\right|^2 + 2HG_{2,0}HG_{0,2} \\ & + 2\cos\delta(HG_{2,0}+HG_{0,2})HG_{0,0} \end{aligned} \tag{9}$$

The spatial profile for the SHG field and the SFG field under the two different conditions where the fundamental and Stokes fields have the same, and opposite topological charge have been simulated, and are shown in **Figure 13**.

The simulated spatial intensity profiles of the SHG and SFG beams in the near- and far-fields very well replicate the spatial profiles observed experimentally in **Figure 10**. The interference patterns of the SFG fields under the conditions where the fundamental and Stokes have the same topological charge, and the opposite topological charge were simulated, and these are shown in **Figure 14(a)** and **(b)**, respectively.

The simulated interference patterns clearly replicate the experimental results and show that while the SFG field can have an annular spatial intensity profile in the near-field, it is also possible that the beam does not have a finite topological charge, and exhibits a topological charge $l = 0$ (in this case, occurring where fundamental and Stokes beams have opposite topological charge).

http://dx.doi.org/10.5772/66425

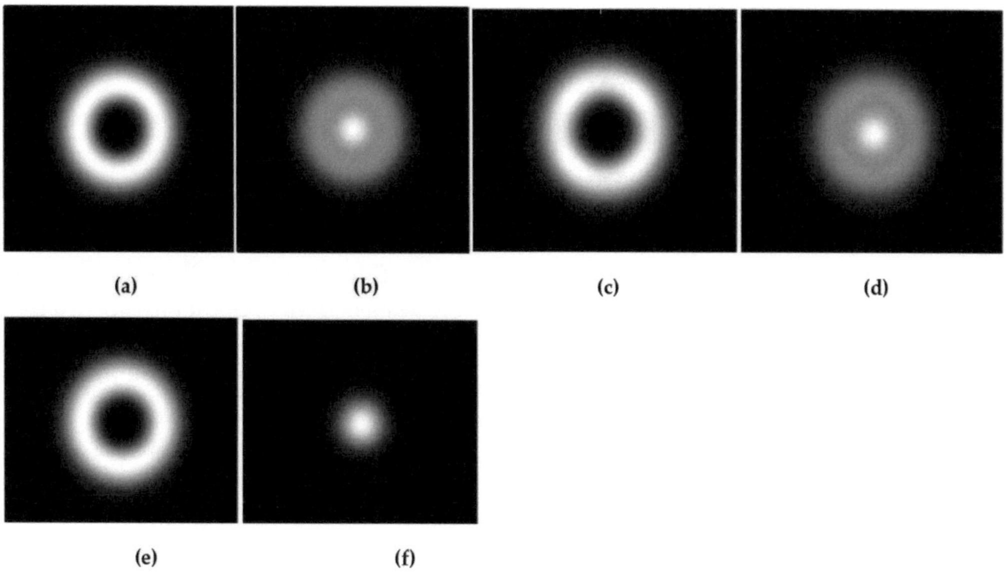

Figure 13. Simulated spatial intensity profiles of the SHG beam in (a) near-field; and (b) far-field. Simulated spatial intensity profile of the SFG field in (c) near-field; and (d) far-field, where the fundamental and Stokes fields have the same topological charge. Simulated spatial intensity profile of the SFG field in (e) near-field; and (f) far-field, where the fundamental and Stokes fields have opposite topological charge [22, 23].

This study of vortex beam dynamics under SHG and SFG within a laser cavity has yielded interesting results. Under the process of SHG or SFG, the wavelength of the vortex beam is converted as expected. The spatial intensity profile in the near-field is that of a vortex beam with a central dark spot with a size consistent with a vortex beam with the magnitude of topological charge equal to the sum of the magnitude of topological charge of the initial fields. What was unexpected was the evolution of the spatial intensity profile of the SHG/SFG vortex field as it propagates out of the laser cavity. While in the near-field, a perfect annular spatial

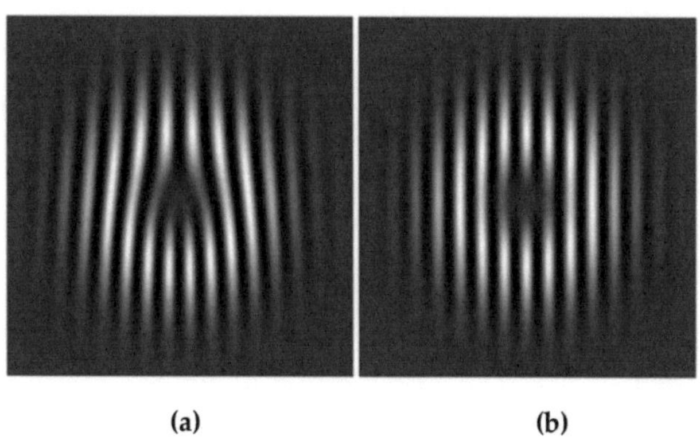

Figure 14. Simulated interference patterns for the SFG field where (a) both fundamental and Stokes fields have the same topological charge; and (b) fundamental and Stokes fields have opposite topological charge [23].

intensity profile is produced, as it propagates into the far-field, the profile evolves to resemble that of an annular beam with a central bright spot. Through decomposition of the vortex modes constituting the fundamental and Stokes modes, into HG modes, it is revealed that upon propagation out of the laser resonator, the contribution of Gouy phase shift must also be considered. This Gouy phase directly impacts the spatial profile of the vortex beam in the far field, resulting in the presence of a non-zero central region.

The variation in the spatial profile of the vortex beam in the near- and far-fields is interesting as this adds to the flexibility and range of applications for this type of laser beam. In one instance, the annular profile in the near-field may be re-imaged for applications where a central null is required, however in other applications, the far-field profile can be utilised where a Gaussian-like beam is required. Furthermore, it has been demonstrated that under SFG, it is possible to produce an annular beam without a net topological charge. This may also be of use in certain applications where an annular spatial profile is desired, but orbital angular momentum is not.

3.3. Optical parametric oscillation

OPO is an effective means of significantly increasing the wavelength of a laser beam. Using OPOs, laser beam wavelengths can be extended from the visible out to the infra-red [28, 36, 37]. The efficiency with which this can take place is dependent on the effective non-linearity of the non-linear crystal being used for conversion, and its transparency in the wavelength range being generated. In this non-linear process, an initial laser field, designated the 'pump' is split into two laser fields of lower energy (longer wavelength). The field with the longer wavelength is designated the 'idler', while that of the shorter wavelength, the 'signal'. The sum of the frequency of the signal and idler equal that of the pump, that is, $\omega_I + \omega_S = \omega_P$.

The dynamics of vortex laser beam transformation under optical parametric oscillation has been studied in the context of continuous-wave OPOs [38, 39]; however in these studies, the wavelength tuning diversity and power scaling properties of vortex beams produced in this non-linear process has not been examined. In this section, we examine the process of topological charge transfer and wavelength extension of vortex laser beams via the OPO process, with pump beams initially at wavelengths of 1064 and 532 nm. Conventional solid-state lasers producing nanosecond pulsed radiation at these wavelengths are used, along with spiral phase plates (see Section 2.1) to transform these Gaussian laser beams into vortex laser beams with topological charge $l = 1$. This vortex beam (pump field) is then injected into an optical parametric oscillator and converted to the signal and idler fields. By altering the resonator geometry and the crystal used for the OPO process, the dynamics by which the pump vortex beam is transformed, to generate signal and idler fields with different topological charge can be controlled. By tuning the phase matching conditions within the OPO, significant wavelength tuning of the generated signal and idler fields can be achieved.

Four different OPO setups are discussed here. The general OPO system layout is depicted in **Figure 15**, and the properties of each element used in the four configurations are summarised in **Table 1**.

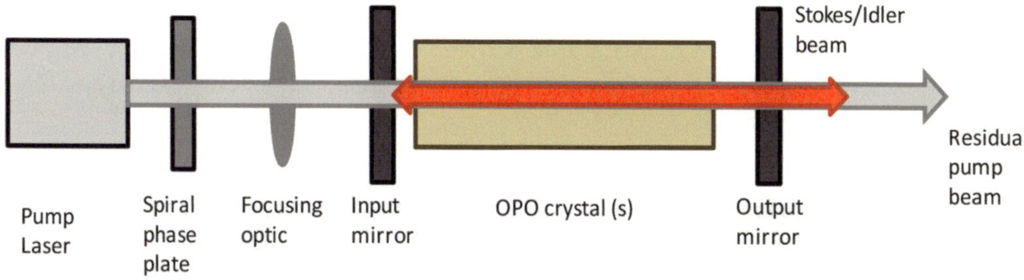

Figure 15. General vortex OPO system layout.

For each of the resonator configurations listed in Table 1, the topological charge transfer process from the pump to the signal and idler fields is different, and this is detailed in the following sections.

3.3.1. Plane-plane OPO resonator using KTP (Setup A)

In this system, a KTP crystal is used to phase match for degenerate OPO operation, in which the wavelength of the signal and idler fields are the same. The incident pump field is a vortex laser beam with topological charge $l = 1$ at a wavelength of 1064 nm, and the resultant signal

Setup	A (plane-plane OPO using KTP)	B (concave-concave OPO using KTP)	C (concave-plane OPO using cascaded KTP)	D (plane-concave OPO using cascaded LBO)
Pump laser	1064 nm, 45 ns, 50 Hz	1064 nm, 25 ns, 50 Hz	1064 nm, 25 ns, 50 Hz	532 nm, 25 ns, 50 Hz
Spiral phase plate	Designed to produce $\ell = 1$ vortex at 1064 nm	Designed to produce $\ell = 1$ vortex at 1064 nm	Designed to produce $\ell = 1$ vortex at 1064 nm	Designed to produce $\ell = 1$ vortex at 532 nm
Focusing lens/spot diameter on OPO crystal	$f = 700$ mm lens	Focal spot = 520 μm diameter	Focal spot = 450 μm diameter	Focal spot = 750 μm diameter
Input mirror	Flat, $R = 98\%$ at 2 μm; T = 90% at 1 μm	Radius of curvature (ROC) = 2000 mm, R = 98% at 2 μm; T = 90% at 1 μm	Radius of curvature (ROC) = 2000 mm, R = 98% at 2 μm; HT at 1 μm	Flat, HR at 980 nm, HT at 532 nm
OPO crystal	KTP, 5 × 5 × 30 mm^3, $\theta = 51.4^\circ$	KTP, 5 × 5 × 30 mm^3, $\theta = 51.4^\circ$	Two cascaded KTP crystals, 12 × 9 × 27 mm^3, $\theta = 53^\circ$	Two cascaded LBO crystals, 30 × 33 mm^3, $\theta = 90$, $\phi = 0$
Output mirror	Flat, R = 80% at 2 μm; HT at 1 μm	ROC = 100 mm, R = 80% at 2 μm; T = 80% at 1 μm	Flat, R = 50% at 2 μm; HT at 1 μm	Flat folding mirror with HR at < 980 nm and HT at > 1180 nm; combined with either: concave mirror, R = 80% at 980 nm (resonate signal) **OR** concave mirror, R = 60% at 1180 nm (resonate idler)

Table 1. Details of each resonator element used in the different OPO configurations examined in Refs. [40–43].

and idler fields have the same wavelength of 2128 nm. In this system, the topological charge of the pump beam was observed to evenly split between the signal and idler fields, in this case, they both receive a topological charge of $l = \frac{1}{2}$. Spatial profiles of the pump, signal and idler fields are shown in **Figure 16**.

It can be clearly seen that the spatial profile of the signal beam (**Figure 16(b)**) resembles a half-crescent shape, while that of the idler (**Figure 16(c)**) appears somewhat Gaussian. The topological charge state of the signal and idler fields is determined through frequency doubling of the fields. Here, the spatial profiles of the frequency-doubled signal and idler are shown in **Figure 17(a)** and **(b)**, respectively; and their corresponding interference patterns in **Figure 17(c)** and **(d)**, respectively.

The spatial profile and interference pattern (**Figure 17(a)** and **(c)**) of the frequency-doubled signal beam resembles that of a vortex beam with a topological charge of 1,confirming that the signal field has a topological charge of ½, knowing that the topological charge of the frequency-doubled field is double that of the original field (see Section 3.2). The spatial profile of the frequency-doubled idler shows an offset null region (**Figure 17(b)**); this offset is due to the effect of beam-walk-off for the polarisation of the idler beam (o-wave) in this OPO phase matching geometry. From the interference pattern (**Figure 17(d)**), it is clear that this beam also shows a topological charge of 1, also showing that the idler field also has a topological charge of ½. In this system, isotropic sharing of the topological charge of the incident pump field between the signal and idler takes place.

3.3.2. Concave-concave OPO resonator using KTP (Setup B)

In this system, in contrast to the plane-plane resonator, the resonator mirrors are both concave. The same KTP crystal was used to phase match the OPO process. In this system, wavelength tuning of the signal and idler fields was also investigated by changing the phase matching conditions.

It was found that the signal field could be tuned through a wavelength range of 1953–2158 nm [41]. Interestingly, in this system, isotropic topological charge sharing from the pump to signal and idler fields was not observed. Instead, anisotropic charge transfer occurred

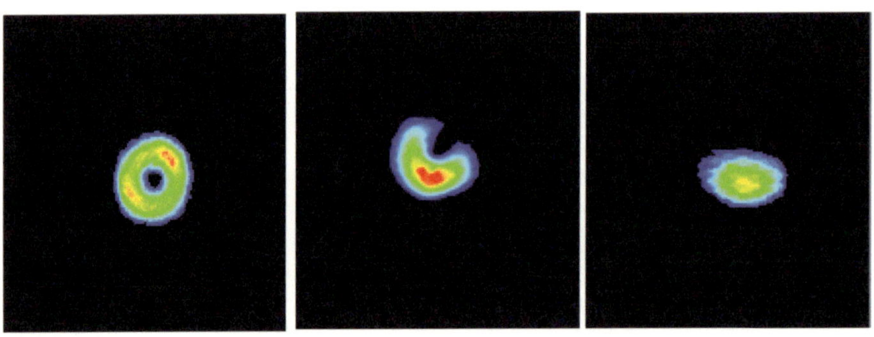

(a) **(b)** **(c)**

Figure 16. Spatial intensity profiles of (a) pump; (b) signal; (c) idler fields generated from the plane-plane OPO resonator using KTP in a degenerate phase matching configuration [40].

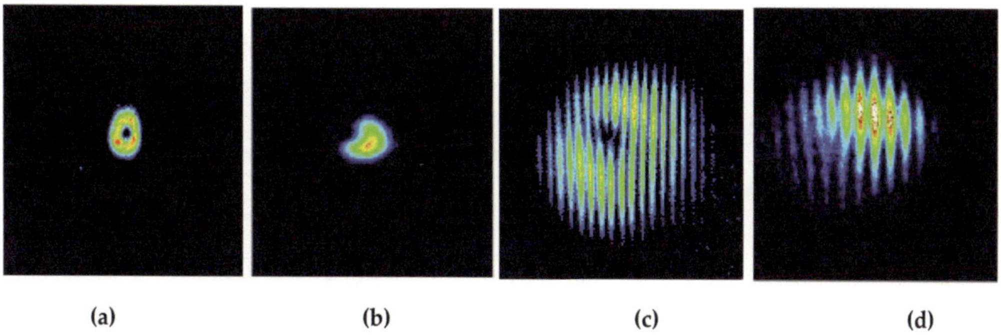

Figure 17. Spatial intensity profiles of (a) frequency-doubled signal, (b) frequency-doubled idler field, and interference patterns of (c) the frequency-doubled signal field, and (d) the frequency-doubled idler field [40].

between the pump and signal fields. In this case, the signal field always received the topological charge of the pump field, whilst the idler field remained Gaussian with topological charge $\ell = 0$. Images of the spatial profile of the pump, signal and idler fields are shown in **Figure 18(a)–(c)**, respectively. The topological charge state of the signal and idler beams was directly determined (not frequency-doubled as detailed in Section 3.3.1) through the

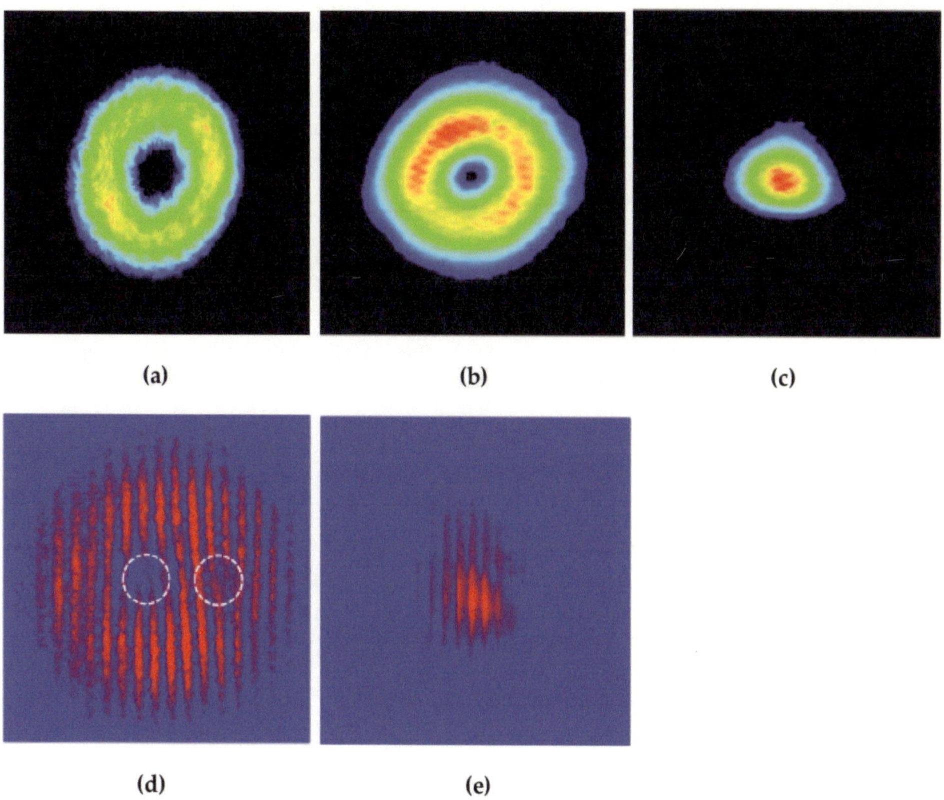

Figure 18. Spatial intensity patterns for (a) pump, (b) signal and (c) idler fields. The interference pattern generated through a self-referencing technique is shown for (d) the signal and (e) the idler fields [41].

use of a self-referencing interference technique [41]. In this technique, a set of two fork-like patterns are generated instead of one as generated through the Mach-Zehnder interferometer detailed previously. The intereference patterns for the signal and idler fields are shown in **Figure 18(d)** and **(e)**, respectively. The fork pattern seen in **Figure 18(d)** shows that the signal field has a topological charge $\ell = 1$, the same as the input pump beam, and the absence of any fork pattern in **Figure 18(e)** shows that the idler field has a topological charge $\ell = 0$.

The anisotropic topological charge transfer which occurs in this system can be understood through examination of how the signal and idler beams are produced within the OPO resonator. In contrast to the plane-plane resonator, the plane-concave resonator has a finite Rayleigh range of $Z_R = 50mm$. Due to this property, the effect of Gouy phase on the signal and idler fields must be considered, similar to that already discussed in the context of intra-cavity SFG/SHG. It is also important to realise that the pump and signal fields have the same polarisation (both e-waves), while the idler has an orthogonal polarisation (o-wave) due to the OPO phase matching condition used in the KTP crystal (type-II phase matching). Due to the finite Rayleigh length, there exists a phase shift between a Gaussian mode and the $\ell = 1$ vortex mode of ~1.2 rad. This effectively means that the overlap between a Gaussian mode and the vortex mode is reduced. Furthermore, the idler field also exhibits beam walk-off in the KTP crystal due to its polarisation state. The net effect of this walk-off is that the overlap between the idler mode and the vortex pump mode is impacted, significantly reducing the possibility of the idler mode taking on the profile of the vortex pump field. It is these two factors, Gouy phase shift and beam walk-off, within this OPO resonator configuration, which enforces topological charge transfer from the pump field to the signal field, while the idler field remains Gaussian.

With this concave-concave resonator design, the use of two KTP crystals simultaneously in a cascaded configuration was also investigated (**Setup C**) [42]. This enabled the wavelength tuning of both the signal and idler waves, and compensation of walk-off of the idler beam. Using this configuration, wavelength tuning of the signal beam across the range 1820 - 1954 nm and tuning of the idler beam across the range 2561–2335 nm could be achieved.

3.3.3. Plane-concave resonator using LBO (Setup D)

In this system a plano-concave OPO resonator was used, however, the KTP crystal was replaced for lithium niobate (LBO) cut for non-critical (type I) phase matching. In this phase matching scheme, an e-wave is converted into two o-waves $e \rightarrow o + o$. In this setup, two cascaded LBO crystals were used to enable wavelength tuning of both the signal and idler fields. Also, in this system, the pump field was generated at 532 nm instead of 1064 nm as used in the previous setups.

Initially, this system was setup to resonate both the signal and idler fields in a doubly resonant configuration. Due to a lack of walk-off of the signal or idler fields (due to the LBO crystal and phase matching geometry), no preferential topological charge transfer was observed from the pump to either of these fields. As a result, the signal and idler fields were both observed to exit the OPO resonator with mode structure which is an incoherent superposition of both

Gaussian and vortex mode profiles. To achieve preferential topological charge transfer from the pump field to either the signal or the idler field, it was necessary to resonate only one of these fields. In the case where only the signal field was oscillated, the topological charge of the pump was transferred to the signal, and similarly if only the idler field was resonated, the topological charge of the pump was transferred to the idler field. Hence, anisotropic topological charge transfer from the pump to the signal or idler fields could be selectively achieved. Wavelength tuning of the signal could be achieved across a range 850–990 nm, and that of the idler from 1130–1300 nm [43].

3.4. Difference-frequency generation of vortex beams

In addition to wavelength extension of vortex beams using an OPO, further wavelength extension of these vortex beams was explored by using DFG of the generated signal and idler fields. In this work, the signal and idler output from the plano-concave resonator incorporating cascaded KTP crystals (Setup C) was focussed into a zinc germanium phosphide (ZGP) crystal for DFG conversion under Type-I phase matching. It was found that through tuning of the signal and idler wavelengths, the resultant DFG field could be wavelength tuned through a range 6.3–12 μm, significantly extending the wavelength reach of vortex laser beams into the mid/far infra-red [42].

In this configuration, it was also found that conservation of topological charge was maintained in this process. In the DFG process, the conservation law of topological charge follows the relationship:

$$l_{\mathrm{DFG}} = \frac{\omega_S - \omega_I}{|\omega_S - \omega_I|}(l_S - l_I) \tag{10}$$

where ω_S and ω_I are the frequencies of the signal and idler fields, respectively, and l_{DFG}, l_S, and l_I are the topological charge of the DFG, signal and idler fields, respectively. From Eq. (10), it can be seen that the sign of the topological charge of the DFG field is dependent on the frequencies of the signal and idler fields. When the frequency of the signal is greater than that of the idler, a positive topological charge is produced, and when the frequency of the idler is greater than the signal, a negative topological charge is generated. Therefore, by swapping the frequency of the signal and idler fields, it is possible to swap the sign of the topological charge of the resultant DFG field, thereby imparting an additional method of controlling the topological charge state of the generated vortex field.

From these studies, it is clear that OPOs offer an effective method of significantly extending the wavelength diversity of vortex laser beams. Interestingly, the dynamics by which the beams are converted can be readily controlled through careful consideration of the OPO resonator design and the crystals which are used in the OPO process. Here, not only can control of the wavelength of the vortex beam be achieved, but also, the transfer of topological charge of the pump beam to the signal and idler beams can be manipulated. Significantly, the generation of a vortex laser beam with $l = \frac{1}{2}$ can be achieved through this non-linear conversion process. In contrast to conventional vortex laser beams with an annular spatial profile, this crescent spatial profile lends itself to numerous new applications including fabrication of novel devices such as meta-materials.

3. Summary

In this chapter, we have presented results detailing both the direct generation of vortex laser beams from solid-state laser systems, and their wavelength conversion via non-linear optical methods. The results show the great diversity of vortex outputs that can be generated, both in terms of spatial profile, topological charge and wavelength. It is through this continual examination of the dynamics of vortex laser beams, that the characteristics of these beams can be manipulated, and thus expanding the range of applications to which they can be applied.

Acknowledgements

The authors would like to acknowledge Prof. Helen Pask (Macquarie University) for her contributions to the work presented in this chapter.

Author details

Andrew J. Lee[1*] and Takashige Omatsu[2]

*Address all correspondence to: andrew.lee@mq.edu.au

1 Department of Physics and Astronomy, MQ Photonics Research Centre, Macquarie University, Australia

2 Graduate School of Advanced Integration Science, Chiba University, Chiba, Japan

References

[1] Allen A, Barnett SM, Padgett M. Optical Angular Momentum. Bristol, UK: Institute of Physics; 2003.

[2] Franke-Arnold S, Allen L, Padgett M. Advances in optical angular momentum. Las. Photon. Rev. 2008;**2**:299–313.

[3] Grier DG. A revolution in optical manipulation. Nature. 2003;**424**:810–816.

[4] Basistiy IV, Soskin MS, Vasnetsov MV. Optical wavefront dislocations and their properties. Opt. Commun. 1995;**119**:604–612.

[5] Jeffries GDM, Edgar JS, Zhao Y, Shelby JP, Fong C, Chiu TD. Using polarization-shaped optical vortex traps for single-cell nanosurgery. Nano Lett. 2007;**7**:415–420.

[6] Toyoda K, Miyamoto K, Aoki N, Morita R, Omatsu T. Using optical vortex to control the chirality of twisted metal nanostructures. Nano Lett. 2012;**12**:3645–3649.

[7] Omatsu T, Chujo K, Miyamoto K, Okida M, Nakamura K, Aoki N, Morita R. Metal microneedle fabrication using twisted light with spin. Opt. Express. 2010;**18**:17967–17973.

[8] Molina-Terriza G, Torres JP, Torner L. Twisted photons. Nat. Phys. 2007;**3**:305–310.

[9] Hell SW, Wichmann J. Breaking the diffraction resolution limit by stimulated emission: stimulated-emission-depletion fluorescence microscopy. Opt. Lett. 1994;**19**:780–782.

[10] Wildanger D, Medda R, Kastrup L, Hell SW. A compact STD microscope providing 3D nanoscale resolution. J. Microsc. 2009;**236**:35–43.

[11] Kotlyar VV, Almazov AA, Khonina SN, Soifer VA, Elfstrom H, Turnen J. Generation of phase singularity through diffracting a plane or Gaussian beam by a spiral phase plate. J. Opt. Soc. A. 2005;**22**:849–861.

[12] Heckenberg NR, McDuff R, Smith CP, White AG. Generation of optical phase singularities by computer generated holograms. Opt. Lett. 1992;**17**:221–223.

[13] Matsumoto N, Ando T, Inoue T, Ohtake Y, Fukuchi N, Hara T. Generation of high-quality higher-order Lagurre-Gaussian beams using liquid-crystal-on-silicon spatial light modulators. J. Opt. Soc. A. 2008;**25**:1642–1651.

[14] Beijersbergen MW, Allen L, van der Veen H, Woerdman JP. Astigmatic laser mode converters and transfer of orbital angular momentum. Opt. Commun. 1993;**96**:123–132.

[15] Allen L, Beijersbergen MW, Spreeuw RJC, Woerdman JP. Orbital angular momentum of light and the transformation of Laguerre-Gaussian laser modes. Phys. Rev. A. 1992;**45**:8185–8189.

[16] Zhan Q. Cylindrical vector beams: from mathematical concepts to applications. Adv. Opt. Photon. 2009;**1**:1–57.

[17] Bisson JF, Senatsky Y, Ueda KI. Generation of Laguerre-Gaussian modes in Nd:YAG laser using diffractive optical pumping. Las. Phys. Lett. 2005;**2**:327–333.

[18] Kim JW, High-power laser operation of the first-order Laguerre-Gaussian (LG_{01}) mode in a diode-laser-pumped Nd: YAG laser. J. Korean Phys. Soc. 2012;**61**:739–743.

[19] Ito A, Kozawa Y, Sato S. Generation of hollow scalar and vector beams using a spot-defect mirror. J. Opt. Soc. A. 2010;**27**:2072–2077.

[20] Kano K, Kozawa Y, Sato S. Generation of purely single transverse mode vortex beam from a He-Ne laser cavity with a spot-defect mirror. Int. J. Opt. 2012;**2012**:359141.

[21] Lee AJ, Omatsu T, Pask HM. Direct generation of a first-Stokes vortex laser beam from a self-Raman laser. Opt. Express. 2013;**21**:12401–12408.

[22] Lee AJ, Zhang C, Omatsu T, Pask HM. An intracavity, frequency-doubled self-Raman vortex laser. Opt. Express. 2014;**22**:5400–5409.

[23] Lee AJ, Pask HM, Omatsu T. A continuous-wave vortex Raman laser with sum frequency generation. Appl. Phys. B. 2016;**122**:64.

[24] Okida M, Omatsu T, Itoh M, Yatagai T. Direct generation of high power Laguerre-Gaussian output from a diode-pumped Nd:YVO_4 1.3 μm bounce laser. Opt. Express. 2007;**15**:7616–7622.

[25] Okida M, Hayashi Y, Omatsu T, Hamazaki J, Morita R. Characterisation of 1.06 μm optical vortex laser based on a side-pumped Nd:$GdVO_4$ bounce laser. Appl. Phys. B. 2009;**95**:69–73.

[26] Koechner W, Bass M. Solid-State Lasers. New York: Springer-Verlag; 2003.

[27] Shen YR. Principles of Nonlinear Optics. New York, NY, USA: Wiley-Interscience; 1984.

[28] Boyd R. Nonlinear Optics. Oxford, UK: Academic Press; 2008.

[29] Eckhardt G, Hellwarth RW, McClung FJ, Schwarz SE, Weiner D, Woodbury EJ. Stimulated Raman scattering from organic liquids. Phys. Rev. Lett. 1962;**9**:455–457.

[30] Eckhardt G, Bortfeld DP, Geller M. Stimulated emission of Stokes and anti-Stokes Raman lines from diamond, calcite and alpha-sulfur single crystals. Appl. Phys. Lett. 1963;**3**:137–138.

[31] Piper JA, Pask HM. Crystalline Raman lasers. IEEE J. Sel. Top. Quant. Electron. 2007;**13**:692–704.

[32] Murray JT, Austin WL, Powell RC. Intracavity Raman conversion and Raman beam cleanup. Opt. Mater. 1999;**11**:353–371.

[33] Zhu Z, Gao W, Mu C, Li H. Reversible orbital angular momentum photon-phonon conversion. Optica 2016;**3**:212–217.

[34] Padgett M, Allen L. Light with a twist in its tail. Contemp. Phys. 2000;**41**:275–285.

[35] Dholakia K, Simpson NB, Padgett MJ, Allen L. Second-harmonic generation and the orbital angular momentum of light. Phys. Rev. A. 1996;**54**:R3742–R3745.

[36] Cheng LK, Bosenberg WR, Tang CL. Broadly tunable optical parametric oscillation in β-BaB_2O_4. Appl. Phys. Lett. 1998;**53**:175–177.

[37] Bayer RL. Optical parametric oscillators. Quant. Electron. A. Treatise. Pt. B. 1975.

[38] Smith AV, Armstrong DJ. Generation of vortex beams by an image-rotating optical parametric oscillator. Opt. Express. 2003;**11**:868–873.

[39] Martinelli M, Huguenin JAO, Nussenzveig P, Khoury AZ. Orbital angular momentum exchange in an optical parametric oscillator. Phys. Rev. A. 2004;**70**:013812.

[40] Miyamoto K, Miyagi S, Yamada M, Furuki K, Aoki N, Okida M, Omatsu T. Optical vortex pumped mid-infrared optical parametric oscillator Opt. Express. 2011;**19**:12220–12226.

[41] Yusufu T, Tokizane Y, Yamada M, Miyamoto K, Omatsu T. Tunable 2-μm optical vortex parametric oscillator. Opt. Express. 2012;**20**:23666–23675.

[42] Furuki K, Horikawa MT, Ogawa A, Miyamoto K, Omatsu T. Tunable mid-infrared (6.3–12 μm) optical vortex pulse generation. Opt. Express. 2014;**21:**26351–26357.

[43] Abulikemu A, Yusufu T, Mamuti R, Miyamoto K, Omatsu T. Widely-tunable vortex output from a singly resonant optical parametric oscillator. Opt. Express. 2015;**23**:18338–18344.

Chapter 3

Superconducting Vortex-Antivortex Pairs: Nucleation and Confinement in Magnetically Coupled Superconductor-Ferromagnet Hybrids

Cinzia Di Giorgio, Domenico D'Agostino,
Anna Maria Cucolo, Maria Iavarone,
Alessandro Scarfato, Goran Karapetrov,
Steven Alan Moore, Massimiliano Polichetti,
Davide Mancusi, Sandro Pace, Valentyn Novosad,
Volodymir Yefremenko and Fabrizio Bobba

Additional information is available at the end of the chapter

http://dx.doi.org/10.5772/65954

Abstract

Superconducting vortices are a well known class of vortices, each of them carrying a single magnetic flux quantum. In this chapter the authors present the results of low temperature Magnetic Force Microscopy experiments to investigate the nucleation and dynamics of superconducting vortices in magnetically coupled Superconductor/Ferromagnet (S/F) heterostructures made by Nb/Py. It is here shown that by controlling the thicknesses of both S and F layer, the formation of spontaneous vortex-antivortex pairs (V-AV) can be favored and their confinement and mobility can be tuned. The experimental results are compared with two theoretical models dealing with the spontaneous nucleation of V/AV pairs in the limits of S thickness respectively greater and smaller than the London penetration depth. It is shown that vortex nucleation and confinement is regulated by the intensity of the out-of-plane component of the magnetization with respect to a critical magnetization set by the thickness of both S and F layers. Additionally, external field cooling processes were used to probe in-field vortex nucleation and V-AV unbalancing, whereas the sweeping of an external magnetic field when below the superconducting critical temperature was used to force the vortex into motion, probing the vortex mobility/rigidity and the vortex avalanche events.

Keywords: superconductivity, ferromagnetism, vortex-antivortex

1. Introduction

Superconductivity has a great potential to play a significant role in the development of new clean energy technologies by minimizing the losses in electrical current transport. However, the widespread use of superconducting materials is still limited by a few critical parameters, such as critical current, critical magnetic field, and critical temperature and, therefore, the active research on new superconducting technologies is underway.

Conventional superconductors are well described in the framework of the Ginzburg-Landau (G-L) theory [1]. Within G-L theory, the introduction of the G-L parameter $k = \frac{\xi}{\lambda}$ leads to differentiation between type I and type II superconductors for cases when $k > \frac{1}{\sqrt{2}}$ and $k < \frac{1}{\sqrt{2}}$, respectively. The two characteristic superconducting length scales, the coherence length (ξ) and penetration depth (λ), are intimately related to the superconducting material. The superconducting order parameter describing the local density of the superconducting Cooper pairs varies at length scale defined by the coherence length, while the penetration depth characterizes the distance at which the external magnetic field is exponentially screened from the interior of a superconducting sample (*Meissner effect*). In type II superconductors, the nucleation of quantized magnetic flux tubes, Abrikosov vortices, enables the persistence of the superconducting state in high applied magnetic fields up to the upper critical field H_{c2} at which the superconductivity is destroyed. The mixed state (or Shubnikov state) of type II superconductors persists in applied magnetic fields H between $H_{c1} < H < H_{c2}$, where $H_{c1}(T)$ and $H_{c2}(T)$ are the lower and upper critical fields, respectively, making these materials very technologically relevant. It has been theoretically predicted by Abrikosov [2] and later experimentally shown by Trauble and Essman [3] that the mixed state is a macroscopic quantum fluid having vortices each carrying a flux quantum $\Phi_0 = \frac{h}{2e}$, where h is the Planck constant and e is the electron charge. The vortex causes a local suppression of the superconducting order parameter on the length scale of ξ while λ measures the exponential decay of magnetic field and currents, when moving away from the vortex core. In 1957, Abrikosov [2] predicted a lattice arrangement of vortices (**Figure 1**) in order to minimize the energy of the system. The vortex lattice period d is set by the intensity of the external magnetic field H according to the relation $d = \sqrt{\frac{2\Phi_0}{\sqrt{3}H}}$. When an external current, exceeding the critical value, is applied to a superconductor in the mixed state, vortices are forced to move under the action of the Lorentz force causing energy dissipation. For this reason, control of the vortex dynamics is one of the main challenges for technological applications and fundamental science. In order to restore a dissipation-free regime, the driving Lorentz force has to be counterbalanced by a pinning force. In this scenario, the technological applications of type II superconductors deal with the capability to create and control pinning centers that locally induce a suppression or reduction of the superconducting order parameter.

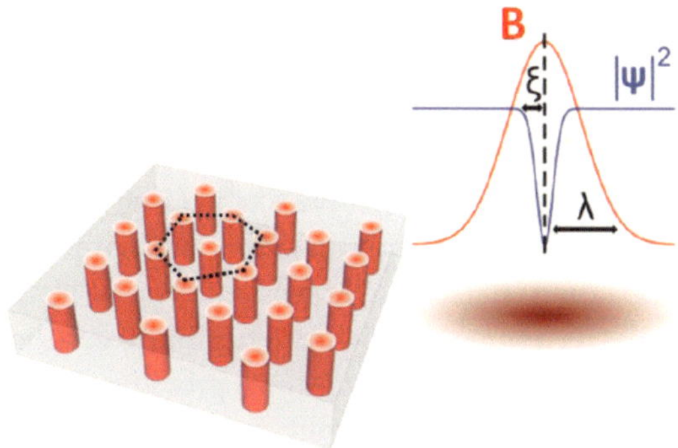

Figure 1. Periodic arrangement of vortices in a type-II superconductor in an external applied magnetic field. Each vortex has a normal core, where the superconducting order parameter $|\psi|^2$(blue line) drops to zero on the scale of ξ while the magnetic field profile (red line) exponentially decays on the scale of λ.

Lattice defects, dopant inclusions or peculiar sample geometry, have been proposed in order to impose a pinning potential for the superconducting vortices [4]. An enhancement of the critical current has been reported by bulk processing of the superconductors to create pinning centers and by lithographic patterning of arrays of pinning centers. Magnetic pinning centers have also been widely used for enhancing vortex pinning properties since they locally suppress the superconducting order parameter (pair-breaking effect of the local magnetic moment) and magnetically attract vortex lines. Several methods of introducing magnetic pinning centers have been employed from deposition of magnetic nanoparticles to lithographically defining magnetic nanotextures on the superconducting layer [5–30]. Magnetically coupled superconductor/ferromagnet thin film heterostructures in which the magnetic domains in the ferromagnet act as pinning centers have been of great interest due to ease of fabrication, scalability for future applications, and due to basic fundamental physics governing the superconductivity in these hybrid systems [31–52].

In the past decades a lot of effort has been focused on developing experimental techniques for studying vortex matter at the nanoscale. Since a collective behavior of vortex dynamics can be extracted, for example, from electronic and heat transport and magnetic measurements [53–61], the real challenge lies in the capability to investigate single vortex, lattice arrangements, and local mechanism of motion with a high spatial resolution. An overall view on the vortex lattice, and its structural characteristics, can be provided by small-angle neutron scattering in the reciprocal space [62, 63], and by Bitter decoration [3, 64], time-resolved magneto-optic techniques [65–68], and holography electron microscopy [69] in real space. The first observation of isolated vortices was pioneered by Essman and Trauble [3] in 1967. In a low magnetic field, they used small magnetic particles to decorate the surfaces of different superconductors in order to get information on the arrangement of vortices in the vortex lattice. By using this technique, large areas hundreds of microns square of the sample surface can be investigated by taking a snapshot of the lattice. More recently, real space imaging of superconducting vortices has been obtained by using scanning probe microscopy and spectroscopy (SPM/S) techniques. Among all of them, scanning SQUID microscopy [70] and scanning Hall probe microscopy [71, 72], with a submicron spatial resolution, have been successfully used to study the geometries, dynamics, and interactions of vortices in different systems. On the other hand, scanning tunneling microscopy (STM), with a subnanometric resolution, is the only technique able to image individual vortex cores by spatially mapping the amplitude of the order parameter [73–75]. The STM method is sensitive to the electronic properties of the sample surface and thus requires clean and flat surfaces. Although it provides a unique opportunity to image vortices at high magnetic fields (due to sensitivity to the order parameter rather than the magnetic profiling), STM technique cannot distinguish between the polarity of the vortices. On the other hand, magnetic force microscopy (MFM) provides information about the vortex polarity and requires less stringent surface quality, albeit the method is constrained to low enough magnetic field as to distinguish the magnetic profiles of individual flux quantum [22, 23, 47–50, 76–79]. MFM measures the force between a magnetic tip and the local magnetic moment on the surface of the investigated sample. In vacuum, the MFM operates in the so-called noncontact regime in which nonmagnetic short range tip-sample interactions are undetected. Being directly sensitive to the strength and direction of the stray field, MFM provides information that is not easily available elsewhere.

Recently, low temperature magnetic force microscopy experiments (MFM) have been performed by the authors to observe "spontaneous" vortex-antivortex pairs (V-AV), appearing in the absence of an external magnetic field, in magnetically coupled superconductor/ferromagnet (S/F) heterostructures made by niobium and permalloy (Py, $Ni_{80}Fe_{20}$) [47–50]. A thin film of SiO_2 separates S and F layers in order to prevent proximity effect [80]. Since the Curie temperature T_C of Py is much higher than the superconducting critical temperature T_s, a field cooling of Nb in the spatially non-uniform Py stray field occurs. The formation of quantum fluxes with opposite polarities, vortices and antivortices (V-AV), is a consequence of the peculiar stripe-like magnetic configuration of the Py. The magnetization vector in the Py film is slightly canted with respect to the film's plane so that the stray field, coming out from the Py surface due to the small and alternating out-of-plane components, causes the nucleation of V-AV pairs and guides their motion along such magnetic channels. Here, we detail MFM experimental results on spontaneous V-AV formation in Nb/Py bilayers as well as on vortex dynamics.

2. Theoretical models

In this chapter, the MFM results will be discussed in quantitative comparison with two different theoretical models dealing with the two opposite limits of superconductor film thickness greater [51] and smaller [52] than the penetration depth, $\frac{d_s}{\lambda} > 1$ and $\frac{d_s}{\lambda} < 1$, respectively. In the framework of S/F bilayers in which the ferromagnet exhibits alternating up-and-down out-of-plane magnetization vectors $\pm M_0$, the magnetization values required for spontaneous vortex nucleation were deduced, for given values of stripe domain width w of F layer, superconducting penetration depth λ and thickness d_s of S film. In agreement with the models, in the magnetically coupled Nb/Py system the vortex formation is due to the $\pm M_0$ out-of-plane components of Py magnetization. Hereinafter, we will define V or AV as the quantum fluxes formed on the top of $-M_0$ or $+M_0$ domains, respectively. Within the considered models [51, 52], by minimizing the total free energy of the S/F system, the critical magnetizations needed to nucleate the first V-AV pair can be deduced. In the limit of $\frac{w}{\lambda} > 1$ and $\frac{d_s}{\lambda} > 1$, the model of Laiho et al. has been taken into account [51]. The threshold magnetization values required to nucleate the first pair of spontaneous straight vortices M_{cs} (**Figure 2a**), which pierces through the superconducting film, or the first vortex semiloop M_{cl} (**Figure 2b**), which is bent within the superconducting film, result in $M_{cs} = 0.2\frac{d_s}{w}H_{c1}$ and $M_{cl} = \frac{H_{c1}}{8\ln(4w/\pi\lambda)}$, respectively. If $M_{cl} > M_{cs}$, the formation of straight vortices is energetically favorable, and vice versa. The energy profiles of a straight (red) and semiloop (blue) vortex are shown in **Figure 2(c)** indicating that the energy minimum of the straight vortex is achieved in the middle of the magnetic stripe domain, whereas the semiloop vortex crosses over the stripe domain wall.

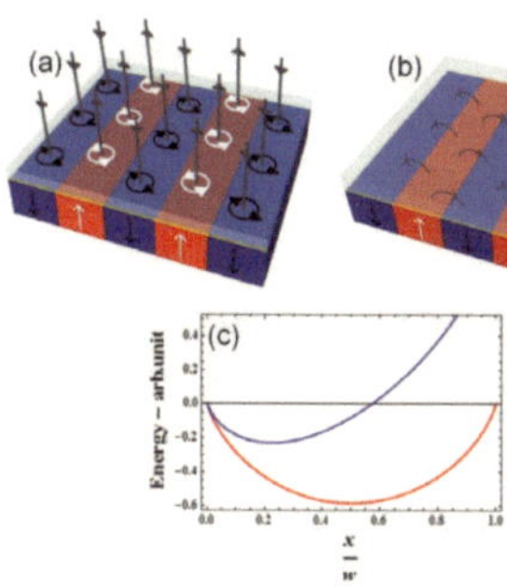

Figure 2. 3D sketch of magnetically coupled S/F heterostructure with the F layer in the stripe regime. (a) In black and white the screening supercurrents are shown, straight vortices and antivortices piercing the superconducting layer are represented by straight arrows. (b) Semiloop vortices close inside the superconducting layer. (c) Plot of the position-dependent energy of straight vortex (red) and semiloop vortex (blue) across a stripe.

On the other hand, in the opposite limit of $\frac{w}{\lambda} > 1$ and $\frac{d_s}{\lambda} < 1$, the model of Genkin et al. has to be taken into account [52], where the threshold magnetization for spontaneous straight vortex nucleation results in $M_c = \ln\left(\frac{\lambda_{\text{eff}}}{\xi}\right)\left(\frac{\Phi_0}{(4\pi)^2}\right)\frac{1}{w\lambda_{\text{eff}}}$, with $\lambda_{eff}(T) = \lambda(T)\coth\left(\frac{d_s}{\lambda(T)}\right)$ [81].

Spontaneous vortex formation will thus be energetically regulated by the threshold condition $M_0 > M_{c(s,l)}$.

Close to the superconducting critical temperature T_s, the superconducting lower critical field is almost zero, and the critical magnetizations for the nucleation of spontaneous vortices are lower than M_0. As a consequence, at $T \sim T_s$, the threshold condition $M_0 > M_{c(s,l)}$ is always satisfied and spontaneous vortices could be formed in the Nb layer but, being λ divergent, $\frac{w}{\lambda} < 1$ and there is no net magnetization within the vortex area. By further decreasing the temperature,$H_{c1}(T)$ increases with a corresponding increase in $M_{c(s,l)}$. For this reason, $M_0 < M_{c(s,l)}$ can occur and vortices can move out from the superconducting layer. Vortices escape from the S layer when $M_0 < M_{c(s,l)}$ and $U_{SV} < < U_{BL}$, i.e., the energy required to pin a vortex $U_{SV} = \frac{1}{4\pi} H_{c1} \Phi_0 d_s$ is much lower than the Bean-Livingstone barrier $U_{BL} = \frac{(\pi + 2)}{4\pi} H_{c1} \Phi_0 \lambda$ [82]. This "escape condition" $U_{SV} < < U_{BL}$ is always verified for the semiloops while it is satisfied by the straight vortices only when $d_s < < (\pi + 2)\lambda$ [51].

3. S/F heterostructures

In this chapter, we focus on Nb/SiO_2/Py heterostructures with 1 and 2 μm thick Py layers and Nb thickness in the range of 50–360 nm. In all cases, a 10 nm thin insulating SiO_2 was placed between the S and F layers in order to have only a magnetic coupling between Nb and Py. The choice of the insulating material is not as crucial as the choice of its thickness. It should be thick enough to prevent electrical proximity effects, which are, in general, short-range (Å to few nanometers) but not as much to reduce the magnetic coupling between F and S layers, which is a long-range interaction. The experiments brought us to the conclusion that 10 nm of SiO_2 is sufficient to reach such a goal. All the heterostructures were made by sputtering deposition as described in [47–50].

Nb films were characterized by both transport and magnetic measurements, showing a superconducting critical temperature of $T_s = (8.8 \pm 0.1)K$. From transport measurements [47] and by using the dirty limit expression as derived by Gorkov [83] and Kes and Tsuei [84], $\xi(0K) = 12\text{nm}$ and $\lambda(0K) = 61\text{nm}$ were inferred. As a consequence, the superconducting lower critical field was calculated to be $H_{c1}(0K) = 720Oe$. At the MFM measurement temperature of 6 K $\xi(6K) = 21$ nm, $\lambda(6K) = 68$ nm, and $H_{c1}(6K) = 418$ Oe have been derived.

Py is a ferromagnetic material where competing magnetic energies (magnetostatic, exchange, magneto-elastic, domain wall, and anisotropy) determine the domain configurations. In thin films, periodic stripe-like domains occur above a critical thickness of $t_c = 2\pi \frac{A}{K_u}$ [85], where A is the exchange constant and K_u is the perpendicular anisotropy constant [86, 87]. We remark that K_u and consequently the critical thickness t_c can be strongly affected by the deposition parameters [88]. In our case, by considering the typical value $A = 1 \times 10^{-6} \text{erg}/\text{cm}$, $t_c = 100 \div 300nm$ is calculated. Stripe domains appear as a consequence of a slight magnetization canting with respect to the overall in-plane orientation. The small out-of-plane components ($\pm M_0$) point alternatively in upward and downward directions across adjacent stripes. The width w of the stripes can be controlled by the Py thickness d_m following the phenomenological relation $w = \alpha\sqrt{d_m}$ [87].

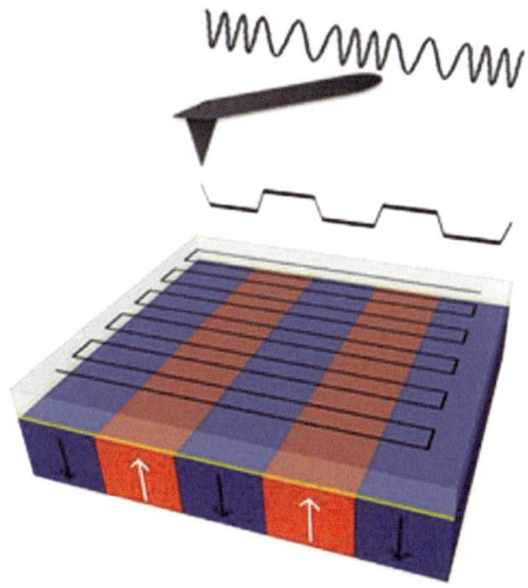

Figure 3. In frequency modulation-MFM mode, a resonating cantilever scans the sample surface line by line. The stray field coming out from it shifts the resonance frequency of the cantilever. The frequency increases (decreases) in an attractive (repulsive) regime.

In our experiments, the choice of Py as ferromagnetic material is thus due to the high control we can exert on w by changing the ferromagnetic thickness d_m. Indeed, w affects the value of the critical magnetization M_{cs}. For the same reason, several values of the superconducting thickness d_s have been considered in order to tune M_{cs}, by keeping the same superconducting material. Indeed, once the S layer is set, the superconducting critical field $H_{c1} = \frac{\Phi_0}{4\pi\lambda^2}\ln\left(\frac{\lambda}{\xi}\right)$, which depends on the London penetration depth λ and on the coherence length ξ, is automatically fixed.

The magnetic properties of the Nb/Py hybrids were analyzed by means of a vibrating sample magnetometer insert of a Quantum Design PPMS and a cryogenic ultrahigh vacuum scanning force microscope equipped with a magnetic tip and operating in frequency modulation-magnetic force microscopy (FM-MFM) mode. **Figure 3** illustrates the working principle of the FM-MFM technique: the frequency shift $df = f - f_0$ of the resonating cantilever, due to the stray field coming out from the sample, is acquired line by line (f is the oscillation frequency measured during tip-sample interaction and f_0 is the free frequency). MFM maps were obtained by scanning at constant tip-sample heights and the attractive/repulsive tip-sample interaction is mapped using color contrast. For each sample, we scanned several different areas of the surface to get good statistics. During the measurements the temperature stability was within 0.01 K. We used a commercial Si cantilever, covered with ferromagnetic Co/Cr film, with resonance frequency $f_0 \approx 75kHz$, elastic constant $k \approx 2.8 N/_m$, and nominal low magnetic moment $\mu \approx 0.3 \times 10^{-13} emu$.

The tip coercivity $H_{c,tip} = 550 \div 600Oe$ was deduced from the inversion of the magnetic contrast by applying an external magnetic field. Before measuring, the tips were magnetized in downward direction along their longitudinal axis. As a consequence, an attractive (repulsive) force appears as a dark (bright) contrast region in the MFM maps. In the presence of vortices, V (AV) appears as darker (brighter) spots on the magnetic background. We remark that the tip-sample distance plays a key role during the MFM experiments. When crossing T_s, the cantilever was pulled away from the sample surface (about 10 μm), to minimize the influence of the tip's stray field on V-AV nucleation and spatial configuration.

Typical magnetic hysteresis loop of Py-1 μm and Py-1 μm, in perpendicular (**Figure 4a** and **b**) and parallel (**Figure 4c** and **d**) configurations with respect to the film plane are reported in **Figure 4**.

By measuring the saturation fields $H_{s\parallel} = 130 \div 160Oe$ and $H_{s\perp} = 11 \div 12kOe$, the saturation magnetization M_s as well as the uniaxial anisotropy constant K_u can be estimated, resulting in $M_s = \frac{H_{s\perp} + H_{s\parallel}}{4\pi} \approx 900G$ and $K_u = \frac{H_{s\parallel} M_s}{2} \approx 6.3 \times 10^4 erg/_{cm^3}$ [89]. The comparison between parallel and perpendicular saturation field values confirms the presence of an easy axis mainly oriented in the film's plane, whereas the ratio $\frac{K_u}{K_d} < < 1$ (here, $K_d = 2\pi M_s^2$ is the stray field energy density) indicates a weak perpendicular anisotropy.

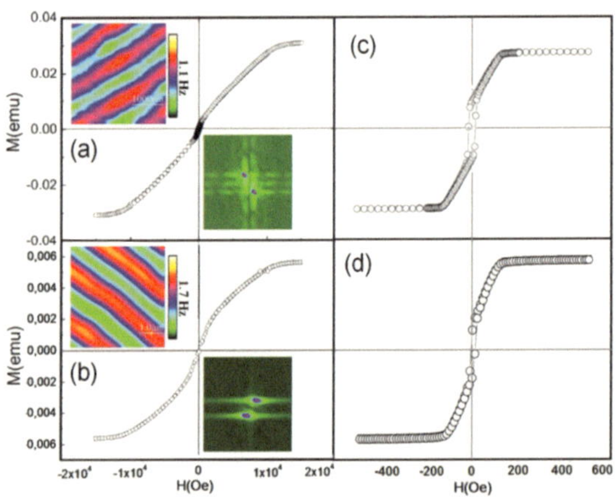

Figure 4. (a) Magnetic hysteresis loop of 1 μm-Py in perpendicular applied magnetic field. Top left corner: MFM map of Nb(100 nm)/Py(1 μm) at $T = 12$ K and tip-Py separation of $h = 140$ nm. Bottom right corner: FFT of the MFM map. (b) Magnetic hysteresis loop of 2 μm-Py in perpendicular applied magnetic field. Top corner: MFM map of Nb(200 nm)/Py(2 μm) at $T = 12$ K and $h = 180$ nm. Bottom right corner: FFT of the MFM map. (c) and (d) Magnetic hysteresis loop of 1 and 2 μm-Py in parallel applied magnetic field.

The MFM maps shown in the insets of **Figure 4(a)** and **(b)**, respectively, on Nb/Py(1 μm) and Nb/Py(2 μm) samples, were taken at $T = 12$ K with relative separation of the tip from the Py surface of 140 and 380 nm, respectively. Before the MFM experiments, the magnetic stripes were oriented along a preferred direction by applying an in-plane external magnetic field greater than $H_{S\,\|}$. Frequency spans of the MFM images of 1.1 Hz in Nb/Py(1 μm) and 1.7 Hz in Nb/Py(2 μm), even though the tip-Py(2 μm) separation is higher, indicates that the magnetic signal coming out from 2 μm-Py sample surface is definitely stronger than 2 μm-Py. In addition to this, in the 2 μm ferromagnetic layer, not only the stripe conformation is much more straight and regular, but also the magnetic roughness along the single stripe is significantly lower, as measured by a frequency shift of around 0.16 Hz in 2 μm-Py and 0.4 Hz in 1 μm-Py layer. From the fast Fourier transform (FFT) analysis of the MFM maps, acquired in different areas of the sample surface, an average stripe width of 490 nm ± 2% and 790 nm ± 4%, for the 1 μm-Py and 2 μm-Py layers, respectively, can be inferred, confirming theoretical expectations $w = \alpha\sqrt{d_m}$ [87].

In order to quantitatively compare the MFM results with the theoretical threshold conditions for vortex nucleation, the thickness d_s of Nb films and the magnetic domain width w were tuned by changing thin film deposition rate and time and measured by statistical analyses of MFM maps by FFT. Moreover, the knowledge of $\xi(6K)$ and $\lambda(6K)$, derived from transport and magnetic measurements, allows the estimate of $H_{c1}(6K)$.

Nb (nm)	Py (μm)	w[nm]+ Δw	w/λ	d_s/λ	M_{cs}(6K) (G)	M_{cl}(6K) (G)	M_c(6K) (G)
50	1.0	490 ± 2%	7	0.74	–	–	15.9
100				1.5	15.1	32.6	–
150				2.2	24.9	33.9	–
200				2.9	33.9	34.2	–
360				5.3	61.5	34.3	–
120	2.0	790 ± 4%	12	1.8	11.9	26.1	–
200				2.9	21.1	25.2	–

By increasing Py thickness, the stripe width increases following a square root dependence. Note that M_{cs} is always lower than M_{cl}, except for Nb(360nm)/Py(1 μm).

Table 1. Characteristic parameters of measured Nb/Py bilayers and relative critical magnetization values.

In **Table 1**, the thickness of the superconducting films and the magnetic domain width are compared to the magnetic size of the vortex and the strength of the critical magnetizations is derived. For all the analyzed hybrids, the ratio $\frac{w}{\lambda}$ is always greater than 1. On the other hand, we point out that the Nb(50 nm)/Py(1 μm) sample, being within the limit $\frac{d_s}{\lambda} < 1$, satisfies the conditions of model [52], whereas all other samples are in agreement with model [51], having $\frac{d_s}{\lambda} > 1$. From **Table 1**, one can note that the formation of spontaneous straight vortices is energetically favored in most cases, since it results $M_{cs} < M_{cl}$. Only in the case of Nb(360 nm)/Py(1 μm) the semiloop vortices are expected. At the MFM imaging temperature, vortices will stay in the superconducting layer if the intensity of the out-of-plane components M_0 of Py magnetization is enough to sustain them (the temperature-dependent threshold conditions are satisfied). If under threshold, they can still be confined in the superconductor when the energy of the single vortex U_{sv} is higher than the energy of the Bean-Livingston barrier U_{BL}, so that the escape condition is not satisfied.

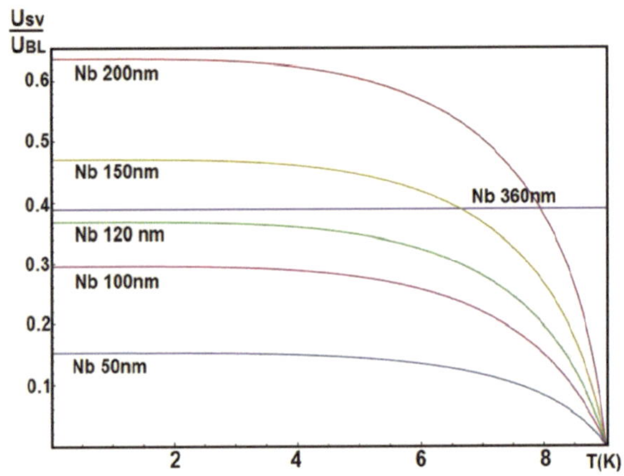

Figure 5. Temperature-dependence of the ratio $\frac{U_{SV}}{U_{BL}}$ in the studied Nb/Py heterostructures.

In **Figure 5**, we report the plot of the ratio $\frac{U_{sv}}{U_{BL}}$ as a function of the temperature for all of our samples, showing that spontaneous V-AV formation is only regulated by M_0 value, thus ruling out Bean-Livingston confinement. By comparing the calculated values of the critical magnetizations of 15.9 and 15.1 G for Nb(50 nm)/Py(1 μm) and Nb(100 nm)/Py(1 μm), respectively, with the measured value $M_0 = 16G$ [47, 90, 91], we expect a spontaneous formation of V-AV in both samples, even though the threshold values are very close to the measured M_0.

4. Superconducting vortex nucleation

In **Figure 6(a)** and **(b)**, the MFM images of Nb(100 nm)/Py(1 um) and Nb(50 nm)/Py(1 um) below the superconducting critical temperature are shown. As expected, Nb(100 nm)/Py(1 μm) (**Figure 6a**) forms spontaneous vortices and antivortices in the center of the oppositely polarized stripes, with a vortex polarity collinear with the magnetization of the underlying stripe domain [50]. In a scan area of 3.8 μm × 3.8 μm, we observe a small vortex density, with unequal number of vortices and antivortices, with "up" polarity vortices dominating. To gain further insight into the imbalanced vortex–antivortex phenomenon, field cooling (FC) measurements in both positive and negative magnetic fields were performed. In general, a change in the relative density of vortices and antivortices is always expected after a field cooling process. Indeed, the effect of the external magnetic field on the magnetization vectors is to enhance the collinear magnetization components and compensate (partially or totally) the anticollinear ones. When the external magnetic field totally compensates the stray field of the ferromagnet, vortices will not nucleate on the top of them, as it happens in **Figure 6(c)**. If the stripe stray field is only partially compensated, vortices might be still induced. However, their density will result lower with respect to the vortex family nucleated on stripes collinear to the external field.

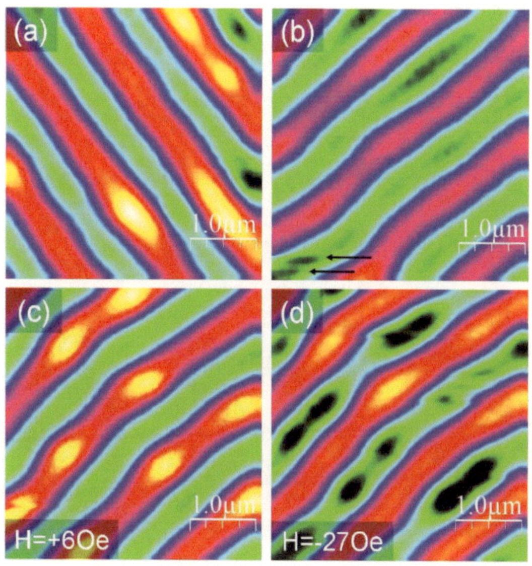

Figure 6. MFM maps in zero field cooling of (a) Nb(100 nm)/Py(1 μm), T = 6 K, h = 130 nm; (b) Nb(50 nm)/Py(1 μm), T = 6 K, h = 200 nm. MFM of Nb(100 nm)/Py(1 μm) at T = 6 K, and h = 180 nm field cooled in (c) H = +6 Oe, (d) H = −27 Oe.

Figure 6(c) shows a MFM image acquired after a FC in H = +6 Oe. Antivortices appear above the proper stripes whereas no vortices are present above oppositely polarized magnetic stripes. On the other hand, the map acquired after a FC in higher negative field H = -27 Oe (**Figure 6d**) still shows the presence of both V and AV, even though the density of vortices with the same polarity as the external applied field becomes higher. The absence of vortices after a field cooling in H = 6 G (**Figure 6c**) and the presence of antivortices after a field cooling in H = −27G (**Figure 6d**) suggest the local unbalancing of Py out-of-plane magnetization components. Indeed, while a positive field of 6 G is enough to compensate the "negative stripe," preventing vortex formation, a negative field of 27 G has only a partial effect, still letting some antivortices to pierce the Nb film. This result points out a local residual out-of-plane magnetization of −16.5 G.

No clear evidence of spontaneous V-AV formation was shown by Nb(50 nm)/Py(1 μm) but instabilities or jumps in the MFM image (marked with arrows in **Figure 6b**), and contrast modulation along the stripes were measured. In reference [50], it was shown that these jumps, which always appear in the direction of the fast-scan axis, are due to the interaction of the vortex with the magnetic tip itself. Jumps due to the vortex motion are also visible in **Figure 6(c)** and **(d)** and their geometrical confinement inside the stripes is proof of the role of the Py out-of-plane component as a strong magnetic pinning source acting against the possibility for the vortices to move perpendicularly to the stripe domains, by crossing the domain wall barrier. The behavior below the superconducting critical temperature of the thickest superconductor samples, Nb(360 nm)/Py(1 μm), Nb(200 nm)/Py(1 μm), and Nb(150 nm)/Py(1 μm) is presented in **Figure 7** [50].

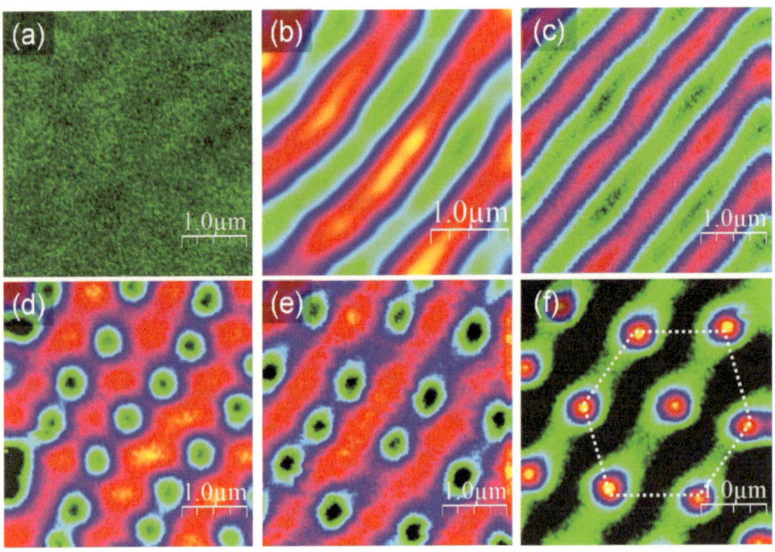

Figure 7. MFM maps in zero field cooling of (a) Nb(360 nm)/Py(1 μm), $T = 6$ K, $h = 110$ nm; (b) Nb(200 nm)/Py(1 μm), $T = 6$ K, $h = 60$ nm; (c) Nb(150 nm)/Py(1 μm), $T = 6$ K, $h = 110$ nm. MFM maps of (d) Nb(360 nm)/Py(1 μm), (e) Nb(200 nm)/Py(1 μm), and (f) Nb(150 nm)/Py(1 μm) field cooled respectively in H = −16 Oe, H = −11.5 Oe, and H = +10 Oe.

The Nb diamagnetism causes the attenuation of the stripe contrast as the thickness of the superconducting layer grows. Keeping the tip-sample separation fixed at $h = 110$ nm above the Nb surface, a low magnetic contrast is observed in the thickest sample (Nb 360 nm, **Figure 7a**), whereas the magnetic stripes appear visible whenever the Nb thickness is at or below 150 nm (**Figure 7c**). Clearly, a more efficient screening of the Py out-of-plane magnetization component occurs in the thickest superconducting layer. In Nb(200 nm)/Py, **Figure 7(b)**, we reduced the tip-sample separation to $h = 60$ nm in order to gain sensitivity. All attempts to unveil spontaneous V-AV in Nb(360–200–150 nm)/Py(1 μm) failed, thus confirming the agreement between the theoretical model [51] and the measured value of local magnetization of 16 G, estimated from the transport measurements. Since the stray field from 1 um-Py film was not sufficient by itself to induce vortices, these Nb/Py samples were cooled down in an out-of-plane external applied magnetic field. In **Figure 7(d)–(f)**, three MFM maps acquired at T = 6 K on Nb(360 nm)/Py(1 μm), Nb(200 nm)/Py(1 μm), and Nb(150 nm)/Py(1 μm), respectively, field cooled in H = -16Oe, H= -11.5 Oe, and H = +10O e, are reported. As expected, only the vortices parallel to the external field direction are created. In **Figure 7(f)**, the intensity of the field was tuned in order to get a vortex-vortex distance matching the formation of a triangular (or hexagonal) vortex lattice, matching the stripe confinement.

Finally, the formation of spontaneous V-AV pairs due to thicker Py layer was demonstrated in Nb(200 nm)/Py(2 μm) (**Figure 8a**)and Nb(120 nm)/Py(2 μm) (**Figure 8b**). The experimental evidence of spontaneous V-AV nucleation and its comparison with the model [51], to which these samples within the limit $\frac{d_s}{\lambda} > 1$ refer, allow us to infer the lower limit of the 2 μm-Py out-of-plane component value resulting in $\left|M_{0} - _{Py2\mu m}\right| > 21.1G$.

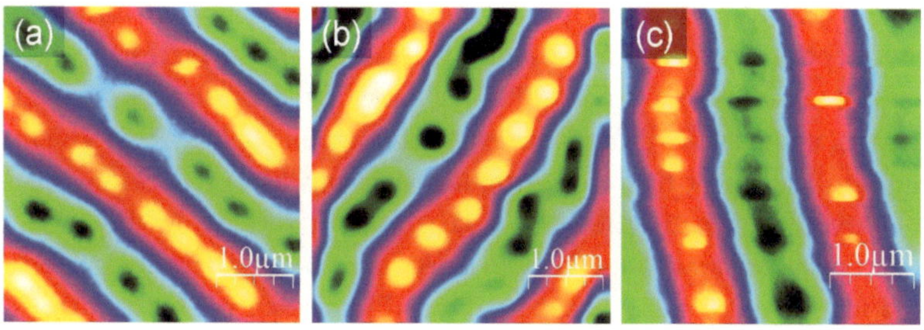

Figure 8. MFM maps in zero field cooling of (a) Nb(200 nm)/Py(2 μm), T = 6 K, h = 180 nm; (b) Nb(120 nm)/Py(2 μm), T = 6 K, h = 180 nm; (c) MFM map of Nb(200 nm)/Py(2 μm) field cooled in H = +60 Oe, h = 150 nm.

In these samples we observe a high and almost uniform vortex density along the stripes as well as the tendency for spontaneous vortices and antivortices to be paired with each other. We correlate these experimental results to the stronger magnetic template, together with wider magnetic stripe domains, and the thickest superconducting layer. As compared to the 1 μm-Py layer samples, the stripe conformation in 2 μm-Py samples is more straight and regular, the magnetic signal coming out from the surface is stronger and the magnetic roughness along the single stripe is smaller, thus highlighting a much more uniform canting of the ferromagnet's magnetization. The frequency signal of the vortex compared to the stripe's magnetic background is 0.97 mHz in Nb(200 nm)/Py(2 μm), 0.3 mHz in Nb(120 nm)/Py(2 μm), and 0.4 mHz in Nb(100 nm)/Py(1 μm), indicating that, as expected, superconducting leaks occur in the thinnest samples.

We speculate that the decoupling of V-AV pairs in Nb(100 nm)/Py(1 μm) may be affected by the tendency of the magnetic field lines coming out from a vortex to close inside the leak, instead of the paired antivortex. As well as by the presence of smaller magnetic stripe domains so that any inhomogeneity in the stripe width induces very inhomogeneous vortex density. In **Figure 8(c)**, the low temperature MFM map of Nb(200 nm)/Py(2 μm) after a field cooling in H = −60 Oe is shown. The strength of the field is not enough to completely compensate the $+M_0$

magnetic domain and, as a consequence, both the families of vortices and antivortices are still in the sample, albeit in different numbers. We notice a very high vortex mobility inferred from the frequent "vortex jumps" facilitated by the scanning magnetic tip.

5. Superconducting vortex dynamics

In **Figure 9**, the comparison between vortex motion in Nb(360 nm)/Py(1 μm) and Nb(200 nm)/Py(2 μm) under the sweeping of the magnetic field is reported. **Figure 9(a)–(d)** shows the behavior of vortices in Nb(360 nm)/Py(1 μm), after a field cooling in H = −21 Oe and by sweeping the field up to positive values. After an initial phase, where the vortex configuration appeared rigid, we noticed that few vortices start moving and, at H = +80 Oe (**Figure 9a**), a nonuniform spatial distribution of the vortex density takes place. As a consequence of jamming events or influenced by the intrinsic pinning, anomalous accumulations of vortices can occur. We speculate that the presence of few higher-energy pinning centers acts as an obstacle against the possibility for other vortices to move along a stripe. By further increasing the external field pressure, a switching event happened at H = +122 Oe, that was captured in **Figure 9(b)**, and an antivortex avalanche enters during the external magnetic field sweep. The regular vortex pattern present in the lower half of **Figure 9(b)** that was recorded before the avalanche is suddenly destroyed and a disordered flux distribution sets up in the upper half of the **Figure 9(b)**. From this point, we kept the field constant and we imaged the vortex arrangement that appeared not to match the Py stripe pattern (**Figure 9c**). We found that the antivortex disorder (with respect to the underlying magnetic background) remains present while the external field is reduced to zero (**Figure 9d**). To check if the disordered vortex pattern was not due to any modification of the Py stripes, the sample was consequently warmed up above the superconducting critical temperature of the Nb, and the stripe domains proved to remain unchanged from the original configuration.

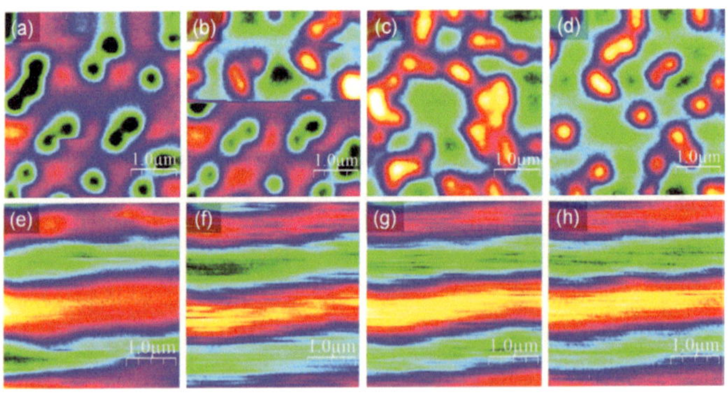

Figure 9. (a–d) MFM maps of Nb(360 nm)/Py(1 μm) at T = 6 K, h = 110 nm. After a field cooling in H = −21G, the field has been swept up +80 Oe (a), from +80 Oe to +122 G (b), kept constant at +122 Oe (c), brought down to 0 G (d). (e)–(h) MFM maps of Nb(200 nm)/Py(2 μm) at T = 6 K, h = 250 nm. The field has been applied below T_s and swept from 0 to −600 G.

The scenario of the vortex dynamic is completely different in Nb(200 nm)/Py(2 μm), where spontaneous vortices appear below the superconducting critical temperature. In this case, there is no need to cool down the sample in a negative (positive) magnetic field and then sweep it to the opposite polarity, since both vortices and antivortices are already in the sample. The extremely high mobility of the spontaneous vortices was imaged by keeping the fast-scan axis as parallel as possible to the stripes in order to follow the vortex motion. **Figure 9(e)–(h)** shows the MFM maps acquired while the field is sweeping respectively from −60 to −94 Oe, from −159 to −191 Oe, from −289 to −323 Oe, and from −483 to −516 Oe and, due to the continuous motion of the vortices under the tip apex, it was not possible to get a clear image of a single vortex. By sweeping the magnetic field down to -600 G, no occurrences of avalanches were recorded.

6. Magnetization measurements

Temperature-dependent low-field magnetization *M*(*T*) curves have been acquired on Nb/Py samples in zero-field-cooling (ZFC) and field-cooling (FC) processes as follows. The samples were first cooled down to 5 K in zero magnetic field, then a small field was applied and the ZFC curve was obtained by measuring the magnetization as a function of the temperature during the warming of the samples up to 10 K.

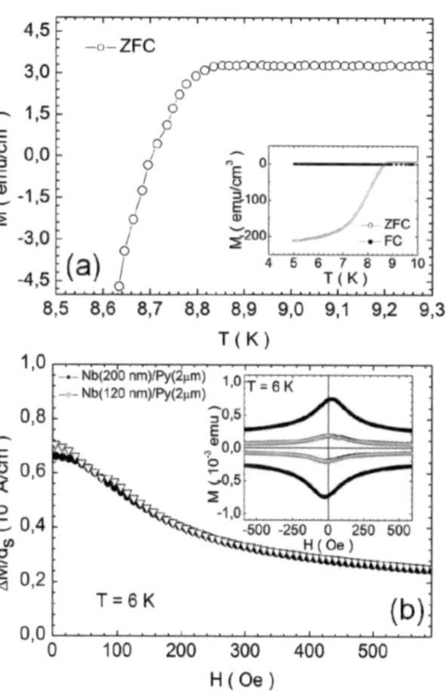

Figure 10. (a) ZFC magnetization measured as function of *T* during the warming-up of Nb(200 nm)/Py(2 μm) e in 20 Oe, perpendicular to the film plane. In the inset: ZFC and FC magnetization curves. (b) Field dependence of the difference ΔM between the upper demagnetization branch and the lower magnetization branch of the loops in Nb(200 nm)/ Py(2 μm) and Nb(120 nm)/Py(2 μm) normalized to the Nb layer thickness d_S. In the inset: hysteresis loops for both samples, at *T* = 6 K without the contribution of the Py film, in the perpendicular field configuration.

After that, the FC curve was measured while cooling the sample down to 5 K, in the presence of the applied magnetic field. In the inset of **Figure 10(a)**, we report both the ZFC and FC curves measured in applied magnetic field of 20 Oe perpendicular to the film plane for the sample Nb(200 nm)/Py (2 μm). The ZFC curve, in the main graph of **Figure 10(a)**, shows the characteristic behavior of a superconducting $M(T)$, with the shielding of the magnetic field starting just below the superconducting critical temperature. From the magnetic hysteresis loops measured below T_s in the external field perpendicular to the film plane, for samples with the same Py thickness (2 μm), but different Nb thickness (200 and 120 nm), the value of the critical current density j_c remains the same. This indicates that the vortex pinning is dominated by the underlying ferromagnetic layer rather than by an intrinsic pinning in Nb films.

The magnetic response of the Nb layer at $T = 6$ K, shown in the inset of **Figure 10(b)**, was determined by subtracting from the $M(H)$ measured at $T < T_s$ the same curve measured at $T > T_s$. From the hysteresis loops of the SC layers, one can evaluate the critical current density by calculating the ratio $\Delta M/d_s$, where ΔM is the difference between the upper demagnetization and the lower magnetization branches of the loops and d_s is the Nb layer thickness. As shown in **Figure 10(b)**, the $\Delta M/d_s$ curves are perfectly overlapping.

7. Conclusions

In this work, we studied vortex-antivortex formation in magnetically coupled Nb/Py bilayers, by varying both the superconducting and ferromagnetic thicknesses. By studying the magnetostatic interaction between S and F layers satisfying the constraint $\frac{w}{\lambda} > 1$, the threshold condition $M_0 > M_{c(s,l)}$ to form spontaneous V-AV (straight or semiloops) was experimentally shown and compared to theoretical predictions. In cases for both superconducting layer thickness greater or smaller than the penetration depth, $\frac{d_s}{\lambda} > 1$ or $\frac{d_s}{\lambda} < 1$, the temperature-dependent expressions of the critical magnetizations $M_{c(s,l)}(T)$ were found and experimentally analyzed. By analyzing the temperature behavior of $M_{c(s,l)}(T)$, one can deduce that vortices are always formed right below the superconducting critical temperature T_s, where $M_{c(s,l)}(T_s) = 0$. As the temperature decreases, $M_{c(s,l)}(T)$ increases and the threshold condition can result to be no longer satisfied, allowing the exit of the superconducting vortices from the S layer whenever the escape condition $d_s < < (\pi + 2)\lambda$ is respected. We proved that our samples always satisfy the escape condition, addressing the occurrences of spontaneous V-AV formation to a M_0 value higher than $M_{c(s,l)}(T)$. By referring to [52] for Nb(50 nm)/Py(1 μm) and to [51] for Nb(100 nm)/Py(1 μm), we should expect spontaneous V-AVs if $M_{0-Py(1\mu m)} > 15.9G$ and $M_{0-Py(1\mu m)} > 15.1G$, respectively. From transport measurements

[47, 90, 91], $M_{0-Py(1\mu m)} \approx 16G$ was estimated and from the imaging of spontaneous V-AVs in Nb(100 nm)/Py(1 μm), $M_{0-Py(1\mu m)} > 15.1G$ was confirmed by MFM. On the other hand, the vortex nucleation in Nb(50 nm)/Py(1 μm) still leaves some open questions. A strong indication of the vortex nucleation in this sample is the presence of jumps appearing in the MFM map only below T_S. These jumps are the signatures of the interaction between the magnetic tip and the superconducting vortex. It is not surprising to find clearer evidence of spontaneous V-AVs in Nb(100 nm)/Py(1 μm) rather than in Nb(50 nm)/Py(1 μm). In fact, when the penetration depth $\lambda(T)$ starts being greater than the superconducting thickness, it has to be corrected into $\lambda_{eff}(T) = \lambda(T)\coth\left(\frac{d_S}{\lambda(T)}\right)$ [81]. As a consequence, as $\lambda_{eff}(T) > \lambda(T)$, the vortices swell and a greater M_0 value is required to accommodate them on the stripes. From a theoretical point of view, this results in using model [52] instead of [51] that fails to satisfy the validity condition $\frac{d_S}{\lambda} > 1$. In **Figure 11**, we report the behavior of $M_c(d_S)$ for the two models, together with the dependence of λ_{eff} on the superconducting thickness. In the framework of model [52], any further reduction in the superconducting thickness, due for example to the presence of few oxide layers, will favor V-AV formation, provided that the condition $\frac{w}{\lambda} < 1$ is satisfied. In the case of Py(1 μm), where $w \approx 490$ nm, the thinnest superconducting layer satisfying the model results to be ideally 10 nm thick.

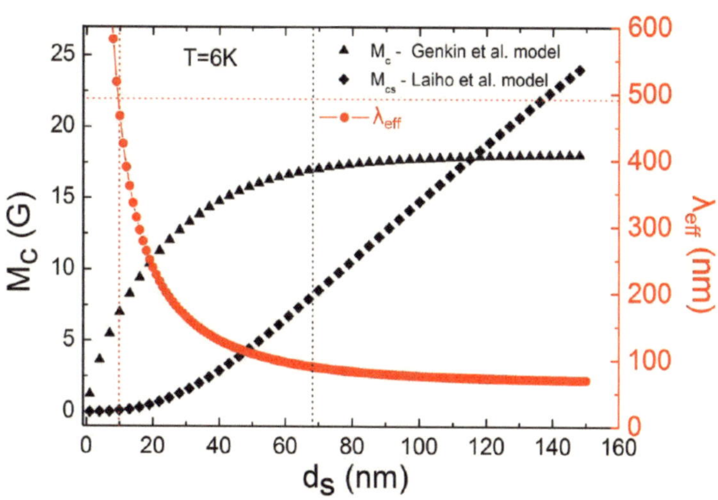

Figure 11. The behavior of $M_c(d_S)$, $M_{cs}(d_S)$, and $\lambda_{eff}(d_S)$ at $T = 6$ K is reported. The intersection points between black dashed line- M_c and black-dashed line- M_{cs} draw, respectively, the lower limit of the model [51] and the upper limit of [52]. The intersection point between the red dashed lines draws the lower limit of the model [52], resulting in $d_S \approx 10nm$ in the studied S/F system.

In summary, we were able to estimate the value of ferromagnet's spontaneous out-of-plane magnetization M_0 based on our MFM results. We have shown experimentally the robustness of the two theoretical models describing spontaneous vortex formation in the S/F bilayer. The field cooled experiments demonstrate that either vortices or antivortices, depending on the sign of the external field, can be formed in the samples that lack sufficient magnetization to form spontaneous V-AV pairs. The zero-field cooled experiments on samples fulfilling the condition for spontaneous V-AV formation show that the V-AV population density can be unbalanced. We also studied the dynamics of vortex and antivortex lattice under a changing applied magnetic field. Different behavior was observed in the case of spontaneous V-AVs compared to the case of Vs (or AVs) formed in external field cooling. After a field cooling in a negative static field, Nb(360 nm)/Py(1 μm), in the under-threshold regime, correctly shows Vs populating the proper stripes. This vortex configuration appears rigid when the field is swept from negative to positive values, up to 122 Oe, when an avalanche of antivortices penetrates the superconducting layer. In fact, once this critical field is reached, vortices are locally driven out and antivortices completely penetrate inside the sample, regardless of underlying magnetic template. We directly imaged already the vortex lattice before and after an avalanche in the Nb(200 nm)/Py(1 μm) sample [48]. In that case, even though the antivortices suddenly penetrated the Nb layer, the magnetic confinement imposed by the Py stripe domains was still visible. We attribute the antivortex dislocation occurring in Nb(360 nm)/Py(1 μm) to the stronger bulk pinning in the thicker Nb films: the influence of the magnetic template on the Nb surface decreases as much as the Nb thickness increases and, in a thick superconducting layer, during as abrupt a phenomenon such as the avalanche, the antivortices can assume a disordered configuration. On the other hand, if the magnetic field is swept in the presence of spontaneous V-AVs, a completely different vortex dynamic occurs. No avalanches were observed in the Nb(200 nm)/Py(2 μm) sample by ramping the external magnetic field from 0 to −600 Oe, but a continuous motion of V-AVs occurs, as revealed in the MFM data. The magnetic template guided the vortices along the magnetic channels, preventing them from crossing the stripes.

Acknowledgements

Work at Temple University was supported by the U.S. Department of Energy, Office of Basic Energy Sciences, Division of Materials Sciences and Engineering under Award DE-SC0004556. Work at Argonne National Laboratory was supported by UChicago Argonne, LLC, Operator of Argonne National Laboratory ('Argonne'). Argonne, a U.S. Department of Energy Office of Science laboratory, is operated under Contract No. DE-AC02-06CH11357. We also acknowledge the support of the MIUR (Italian Ministry for Higher Education and Research) under the project "Rientro dei cervelli."

http://dx.doi.org/10.5772/65954

Author details

Cinzia Di Giorgio[1], Domenico D'Agostino[1], Anna Maria Cucolo[1,2], Maria Iavarone[3], Alessandro Scarfato[1,6], Goran Karapetrov[4], Steven Alan Moore[3], Massimiliano Polichetti[1,2], Davide Mancusi[1], Sandro Pace[1,2], Valentyn Novosad[5], Volodymir Yefremenko[5] and Fabrizio Bobba[1,2*]

*Address all correspondence to: fbobba@unisa.it

1 Physics Department "E.R. Caianiello," University of Salerno, Fisciano, SA, Italy

2 CNR-SPIN Salerno, Fisciano, SA, Italy

3 Physics Department, Temple University, Philadelphia, PA, USA

4 Physics Department, Drexel University, Philadelphia, PA, USA

5 Materials Science Division, Argonne National Laboratory, Argonne, IL, USA

6 Department of Quantum Matter Physics, University of Geneva, Geneva, Switzerland

References

[1] V. L Ginzburg and L. D. Landau, J. Exptl. Theoret. Phys. 20, 1064 (1950).

[2] A.A. Abrikosov, Sov. Phys. JETP 5, 1174 (1957).

[3] U. Essmann and H. Trauble, Phys. Lett. 24, 526 (1967).

[4] A. M. Campbell and J.E. Evetts, Critical Currents in Superconductors - Monographs on Physics, Taylor & Francis Ltd., London (1972).

[5] Y. Otani, B. Pannetier, J. P. Nozieres, and D. Givord, J. Magn. Magn. Mater. 126, 622 (1993).

[6] M. Baert, V. V. Metlushko, R. Jonckheere, V. V. Moschalkov, and Y. Bruynseraede, Phys. Rev. Lett. 74, 3269 (1995).

[7] J. I. Martin, M. Velez, J. Nogues, and I. K. Schuller, Phys. Rev. Lett. 79, 1929 (1997).

[8] J. I. Martin, Y. Jaccard, A. Hoffman, J. Nogues, J. M. George, J. L. Vicent, and I. K. Schuller, J. Appl. Phys. 84, 411 (1998).

[9] J. I. Martin, M. Velez, J. Nogues, A. Hoffman, Y. Jaccard, and I. K. Schuller, J. Magn. Magn. Mater. 177, 915 (1998).

[10] J. I. Martin, M. Velez, A. Hoffman, I. K. Schuller, and J. L. Vicent, Phys. Rev. Lett. 83, 1022 (1999).

[11] M. V. Milošević and F. M. Peeters, Phys. Rev. B 68, 094510 (2003).

[12] D. B. Jan, J. Y. Coulter, M. E. Hawley, L. N. Bulaevskii, M. P. Maley, Q. X. Jia, B. B. Maranville, F. Hellman, and X. Q. Pan, Appl. Phys. Lett. 82, 778 (2003).

[13] M. Lange, M. J. Van Bael, Y. Bruynseraede, and V. V. Moshchalkov, Phys. Rev. Lett. 90, 19 (2003).

[14] M.V. Milošević and F. M. Peeters, Physica C 404, 281 (2004).

[15] M.V. Milošević and F. M. Peeters, Phys. Rev. Lett. 93, 267006 (2004).

[16] M. V. Milošević and F. M. Peeters, Phys. Rev. B 69, 104522 (2004).

[17] M.V. Milošević and F. M. Peeters, Phys. Rev. Lett. 94, 227001 (2005).

[18] A. V. Silhanek, W. Gillijns, M. V. Milošević, A. Volodin, V. V. Moshchalkov, and F. M. Peeters, Phys. Rev. B 76, 100502(R) (2007).

[19] A. Angrisani Armenio, C. Bell, J. Aarts, and C. Attanasio, Phys. Rev. B 76, 054502 (2007).

[20] T. G. Holesinger, L. Civale, B. Maiorov, D. M. Feldmann, J. Y. Coulter, D. J. Miller, V. A. Maroni, Z. Chen, D. C. Larbalestier, R. Feenstra, X. Li, Y. Huang, T. Kodenkandath, W. Zhang, M. W. Rupich, and A. P. Malozemoff, Adv. Mater. 20, 391 (2008).

[21] T. Aytug, M. Paranthaman, K. J. Leonard, K. Kim, A. O. Ijaduola, Y. Zhang, E. Tuncer, J. R. Thompson, and D.K. Christen, J. Appl. Phys. 104, 043906 (2008).

[22] T. Shapoval, V. Metlushko, M. Wolf, V. Neu, B. Holzapfel, and L. Schultz, Physica C 470, 867 (2010).

[23] T. Shapoval, V. Metlushko, M. Wolf, B. Holzapfel, V. Neu, and L. Schultz, Phys. Rev. B 81, 092505 (2010).

[24] A. V. Kapra, V. R. Misko, D. Y. Vodolazov, and F. M. Peeters, Supercond. Sci. Technol. 24, 024014 (2011).

[25] D.H. Tran, W.B.K. Putri, C.H. Wie, B. Kang, N.H. Lee, W.N. Kang, J.Y. Lee, W.K. Seong, Thin Solid Films 526, 241 (2012).

[26] O. Polat, M. Ertuŭgrul, J. R. Thompson, K. J. Leonard, J. W. Sinclair, M. P. Paranthaman, S. H. Wee, Y. L. Zuev, X. Xiong, V. Selvamanickam, D. K. Christen, and T. Aytuŭg, Supercond. Sci. Technol. 25, 025018 (2012).

[27] V. Selvamanickam, Y. Chen, Y. Zhang, A. Guevara, T. Shi, Y. Yao, G. Majkic, C. Lei, E. Galtsyan, and D. J. Miller, Supercond. Sci. Technol. 25, 045012 (2012).

[28] C.F. Tsai, J.H. Lee, and H. Wang, Supercond. Sci. Technol. 25, 075016 (2012).

[29] V. Selvamanickam, Y. Chen, T. Shi, Y. Liu, N. D. Khatri, J. Liu,Y. Yao, X. Xiong, C. Lei, S. Soloveichik, E. Galstyan, and G. Majkic, Supercond. Sci. Technol. 26, 035006 (2013).

[30] A.V. Khoryushin, P. B. Mozhaev, J. E. Mozhaeva, N. H. Andersen, J.C. Grivel, J. B. Hansen, and C. S. Jacobsen, Physica C 485, 39 (2013).

[31] L. N. Bulaevskii, and E. M. Chudnovsky, M. P. Maley, Appl. Phys. Lett. 76, 18 (2000).

[32] M. Lange, M. J. Van Bael, V. V. Moshchalkov, and Y. Bruynseraede, Appl. Phys. Lett. 81, 322 (2002).

[33] S. Erdin, I. F. Lyuksyutov, V. L. Pokrovsky, and V. M. Vinokur, Phys. Rev. Lett. 88, 1 (2002).

[34] I. F. Lyuksyutov and V. L. Pokrovsky, Adv. Phys. 54, 1 (2005).

[35] M.V. Milošević, G.R. Berdiyorov, and F. M. Peeters, Phys. Rev. Lett. 95, 147004 (2005).

[36] W. Gillijns, A. Yu. Aladyshkin, A. V. Silhanek, and V. V. Moshchalkov, Phys. Rev. B 76, 060503(R) (2007).

[37] A. Y. Aladyshkin, A. V. Silhanek, W. Gillijns, and V. V. Moshchalkov, Supercond. Sci. Technol. 22, 053001 (2009).

[38] G. Karapetrov, M. V. Milošević, M. Iavarone, J. Fedor, A. Belkin, V. Novosad, and F. M. Peeters, Phys. Rev. B 80, 180506(R) (2009).

[39] G. Carapella, P. Sabatino, and G. Costabile, Phys. Rev. B 81, 054503 (2010).

[40] H. Yamazaki, N. Shannon, and H. Takagi, Phys. Rev. B 81, 094503 (2010).

[41] L. Y. Zhu, M. Z. Cieplak, and C. L. Chien, Phys. Rev. B 82, 060503(R) (2010).

[42] V. Vlasko-Vlasov, U. Welp, W. Kwok, D. Rosenmann, H. Claus, A. A. Buzdin, and A. Melnikov, Phys. Rev. B 82, 100502(R) (2010).

[43] A. Yu. Aladyshkin, A. S. Mel'nikov, I. M. Nefedov, D. A. Savinov, M. A. Silaev, and I. A. Shereshevskii, Phys. Rev B 85, 184528 (2012).

[44] V. Vlasko-Vlasov, A. Buzdin, A. Melnikov, U. Welp, D. Rosenmann, L. Uspenskaya, V. Fratello, and W. Kwok, Phys. Rev. B 85, 064505 (2012).

[45] M. Z. Cieplak, Z. Adamus, M. Ko'nczykowski, L. Y. Zhu, X. M. Cheng, and C. L. Chien, Phys. Rev. B 87, 014519 (2013).

[46] M. Iavarone, S. A. Moore, J. Fedor, S. T. Ciocys, G. Karapetrov, J. Pearson, V. Novosad, and S. D. Bader, Nat. Comm. 5, 4766 (2014).

[47] M. Iavarone, A. Scarfato, F. Bobba, M. Longobardi, G. Karapetrov, V. Novosad, V. Yefremenko, F. Giubileo, and A. M. Cucolo, Phys. Rev B 84, 024506 (2011).

[48] M. Iavarone, A. Scarfato, F. Bobba, M. Longobardi, S. A. Moore, G. Karapetrov, V. Yefremenko, V. Novosad, and A. M. Cucolo, IEEE Trans. Magn. 48, 3275 (2012).

[49] A. M. Cucolo, A. Scarfato, M. Iavarone, M. Longobardi, F. Bobba, G. Karapetrov, V. Novosad, and V. Yefremenko, J. Supercond. Nov. Magn. 25, 2167 (2012).

[50] F. Bobba, C. Di Giorgio, A. Scarfato, M. Longobardi, M. Iavarone, S. A. Moore, G. Karapetrov, V. Novosad, V. Yefremenko, and A. M. Cucolo, Phys. Rev. B 89, 214502 (2014).

[51] R. Laiho, E. Lähderanta, E.B. Sonin, and K.B. Traito, Phys. Rev. B 67, 144522 (2003).

[52] G.M. Genkin, V.V. Skuzovaktin, I.D. Tokman, J. Magn. Magn. Mater. 130, 51 (1993).

[53] J. J. Krebs, B. T. Jonker, and G. A. Prinz, J. Appl. Phys. 61, 3744 (1987).

[54] A. Berger and M. R. Pufall, Appl. Phys. Lett. 71, 965 (1997).

[55] L. Louail, K. Ounadjela, and R. Stamps, J. Magn. Magn. Mater. 167, L189 (1997).

[56] W.S. Kim, W. Andra, and W. Kleeman, Phys. Rev. B 58, 6346 (1998).

[57] A. Berger and M. R. Pufall, J. Appl. Phys. 85, 4583 (1999).

[58] K. Ha and R. C. O'Handley, J. Appl. Phys. 87, 5944 (2000).

[59] G. Grimaldi, A. Leo, A. Nigro, S. Pace, and R. P. Huebener, Phys. Rev. B 80, 144521 (2009).

[60] G. Grimaldi, A. Leo, A. Nigro, A. V. Silhanek, N. Verellen, V. V. Moshchalkov, M.V. Milošević, A. Casaburi, R. Cristiano, and S. Pace, Appl. Phys. Lett. 100, 202601 (2012).

[61] G. Grimaldi, A. Leo, P. Sabatino, G. Carapella, A. Nigro, S. Pace, V. V. Moshchalkov, and A. V. Silhanek, Phys. Rev. B 92, 024513 (2015).

[62] A. Huxley, Springer Series in Solid-State Sciences Volume: 132 Pages: 301–320 Published:2002 Workshop on Vortices in Unconventional Superconductors and SuperFluids Location: Dresden, Germany, Date: Feb 28 – Mar 03, 2000.

[63] S. Mühlbauer, C. Pfleiderer, P. Böni, M. Laver, E. M. Forgan, D. Fort, U. Keiderling, and G. Behr, Phys. Rev. Lett. 102, 136408 (2009).

[64] P. L. Gammel, D. J. Bishop, G. J. Dolan, J. R. Kwo, C. A. Murray, L. F. Schneemeyer, and J. V. Waszczak, Phys. Rev. Lett. 59, 2592 (1987).

[65] P. Vavassori, Appl. Phys. Lett. 77, 1605 (2000).

[66] G. Gubbiotti, G. Carlotti, M. G. Pini, P. Politi, A. Rettori, P. Vavassori, M. Ciria, and R. C. O'Handley, Phys. Rev. B 65, 214420 (2002).

[67] P. E. Goa, H. Hauglin, A. A. F. Olsen, D. Shantsev, and T. H. Johansen, Appl. Phys. Lett. 82, 1 (2003).

[68] H. Stillrich, C. Menk, R. Fromter, and H. P. Oepen, J. Appl. Phys. 105, 07C308 (2009).

[69] A. Tonomura, J. Supercond. 7, 2 (1994).

[70] S. Okayasu, IEEE T. Appl. Supercond. 15, 2 (2005).

[71] A. Oral, S. J. Bending, R. G. Humphreys, and M. Henini, J. Low Temp. Phys. 105, 1135 (1996).

[72] R. B. G. Kramer, A.V. Silhanek, W. Gillijns, and V.V. Moshchalkov, Phys. Rev. X 1, 021004 (2011).

[73] A. Kohen, T. Cren, Th. Proslier, Y. Noat, W. Sacks, D. Roditchev, F. Giubileo, F. Bobba, A. M. Cucolo, N. Zhigadlo, S. M. Kazakov, and J. Karpinski, Appl. Phys. Lett. 86, 212503 (2005).

[74] G. Karapetrov, J. Fedor, M. Iavarone, M. T. Marshall, and R. Divan, Appl. Phys. Lett. 87, 162515 (2005).

[75] H. Suderow, I Guillamón, J G Rodrigo, and S Vieira, Supercond. Sci. Technol. 27, 063001 (2014).

[76] A.P. Volodin and M.V. Marchevsky, Ultramiroscopy 42–44, 757 (1992).

[77] A. Volodin, K. Temst, C. Van Haesendonck, and Y. Bruynseraede, Appl. Phys. Lett. 73, 8 (1998).

[78] E. W. J. Straver, J. E. Hoffman, O. M. Auslaender, D. Rugar, and K. A. Moler, Appl. Phys. Lett. 93, 172514 (2008).

[79] O. M. Auslaender, L. Luan, E. W. J. Straver, J. E. Hoffman, N. C. Koshnick, E. Zeldov, D. A. Bonn, R. Liang, W. N. Hardy, and K. A. Moler, Nat. Phys. 5, 35 (2009).

[80] A.I. Buzdin, Rev. Mod. Phys. 77, 935 (2005).

[81] A.I. Gubin, K.S. Il'in, S.A. Vitusevich, M. Siegel, and N. Klein, Phys. Rev. B 72, 064503 (2005).

[82] B.V. Petukhov, and V.R. Chechetkin, Zh. Eksp. Teor. Fiz. 65, 1653 (1973) [Sov. Phys. JETP].

[83] L.P. Gorkov, Zh. Eksp. Teor. Fiz. 36, 1918 (1959) [Sov. Phys. JEPT 9, 1364 (1959)].

[84] P.H. Kes and C.C. Tsuei, Phys. Rev. B 28, 5126 (1983).

[85] J. B. Youssef, N. Vukadinovic, D. Billet, and M. Labrune, Phys. Rev. B 69, 174402 (2004).

[86] Y. Murayama, J. Phys. Soc. Jpn. 21, 2253 (1966).

[87] S. Chikazumi, Physics of Ferromagnetism, Oxford University Press Inc., New York (1997).

[88] N. Amos, R. Fernandez, R. Ikkawi, B. Lee, A. Lavrenov, A. Krichevsky, D. Litvinov, and S. Khizroev, J. Appl. Phys. 103, 07E732 (2008).

[89] D. Mancusi, C. Di Giorgio, F. Bobba, A. Scarfato, A. M. Cucolo, M. Iavarone, S. A. Moore, G. Karapetrov, V. Novosad, and V. Yefremenko, Supercond. Sci. Technol. 27 (2014) 125002.

[90] A. Belkin, V. Novosad, M. Iavarone, J. Fedor, J.E. Pearson, A. Petrean-Troncalli, G. Karapetrov, Appl. Phys. Lett. 93, 0725110 (2008).

[91] A. Belkin, V. Novosad, M. Iavarone, R. Divan, J. Hiller, T. Proslier, J.E. Pearson, G. Karapetrov, Appl. Phys. Lett. 96, 092513 (2010).

Chapter 4

Optical Vortices Illumination Enables the Creation of Chiral Nanostructures

Takashige Omatsu, Katsuhiko Miyamoto and
Ryuji Morita

Additional information is available at the end of the chapter

http://dx.doi.org/10.5772/67073

Abstract

We discovered that optical vortices with an annular spatial form and an orbital angular momentum owing to a helical wave front enable us to twist materials, such as metal, silicon and azo-polymer, to form various structured matters including microneedles, chiral nanostructures and chiral surface reliefs. Such structured matters will potentially open the door to advanced devices, for instance, silicon photonic device, biomedical micro-electro-mechanical systems, ultrasensitive detector for chiral chemical composites and plasmonic metasurfaces for chiral chemical reactions.

Keywords: singular optics, optical vortices, orbital angular momentum, laser materials processing, chiral structured materials

1. Introduction

Optical vortices [1–3] carry an annular spatial form and an orbital angular momentum (l) owing to an associated helical wave front with phase singularity $l\phi$ (where l is an integer and ϕ is the azimuthal angle) (**Figure 1**) and they have widely received much attention in a variety of fields, such as optical trapping and manipulations [4–6], space division multiplexing optical telecommunications [7, 8], quantum physics [9] and "super-resolution" microscopes with a spatial resolution beyond the diffraction limit [10–12].

In recent years, we discovered that optical vortices enable us to twist various materials, for example, metal, silicon and azo-polymer, so as to form a variety of structured matters including microneedles, chiral nanostructures and chiral surface reliefs. Such structured matters created by optical vortex illumination will potentially open the door to various material sciences and for

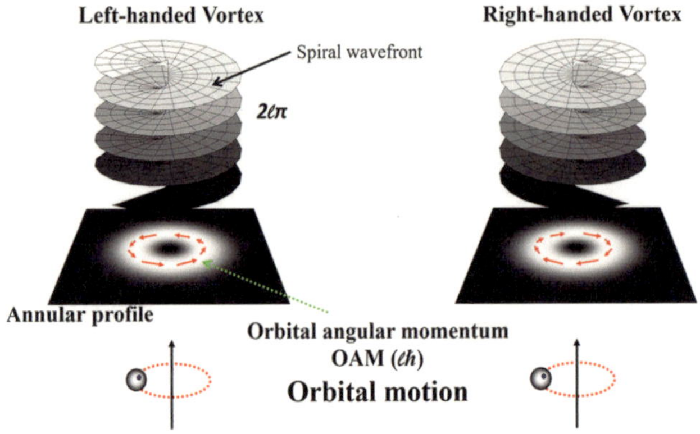

Figure 1. Optical vortex.

instance, they will enable advanced photonic devices [13], biomedical micro-electro-mechanical systems [14], ultrasensitive detector for chiral chemical composites [15] and plasmonic metasurfaces for chiral chemical reactions [16], etc.

2. Microneedle

2.1. Metal microneedle [17]

Metal microneedles, in particular, two-dimensional metal microneedle arrays should allow many applications, such as high-speed microscopic imaging [18], energy-saving field emission displays [19] and biomedical micro-electro-mechanical systems.

To date, several methods to fabricate metal microneedles have been proposed [20–22]; however, they are mostly based on a bottom-up technique including several chemical processes. Thus, their time- and cost-efficiencies are limited. Laser ablation, in which a target is broken down into its compositional elements, that is, ions and electrons, by the laser pulse illumination, has been widely studied for materials processing such as microdrilling [23], cutting [24] and scribing of metals [25], dielectric materials and semiconductors; however, it is ill-suited to collect compositional elements and create structured matters.

Laser ablation using optical vortex pulse (optical vortex laser ablation) [26], in which compositional elements created through laser ablation process receive orbital angular momentum from optical vortex pulse, enables us to fabricate structured materials. It allows us to form a metal microneedle with typically a 10 μm height and a <0.3 μm tip diameter merely by deposition of a few optical vortex pulses onto a metal, so as to fabricate two-dimensional metal microneedle arrays at high time- and cost-efficiencies [17].

A schematic diagram of an experimental setup for metal microneedle is shown in **Figure 2**. The circular polarization associated with a helical electric field also adds spin (*s*) angular momentum to the light [27]. The resulting circularly polarized optical vortex exhibits a total angular momentum (*J*) defined as the sum of the orbital (*l*) and spin (*s*) angular momenta [28]. The quarter-wave plate also suppressed the polarization dependence of the ablation efficiency.

http://dx.doi.org/10.5772/67073

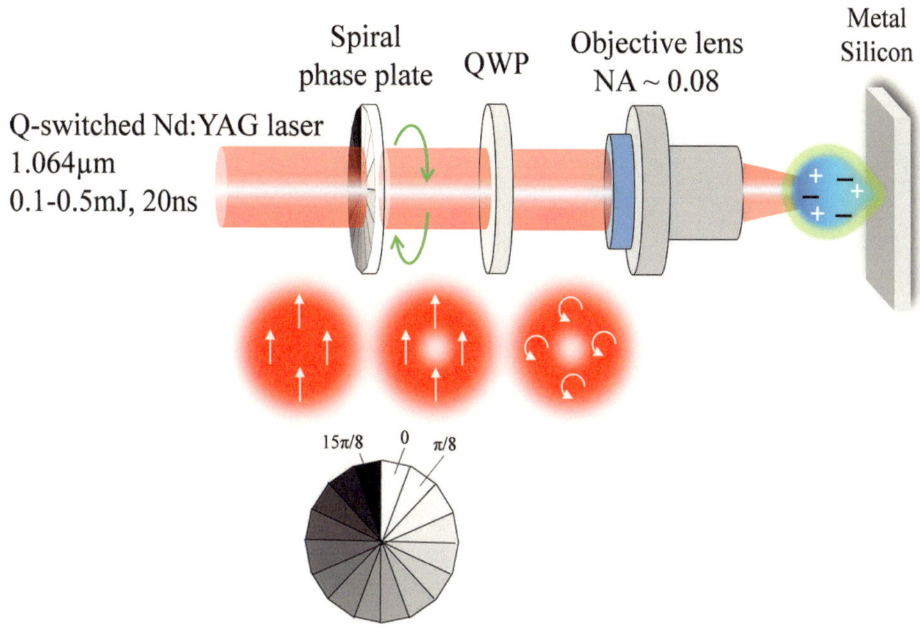

Figure 2. Schematic diagram of optical vortex laser ablation.

The target used was a polished tantalum plate with a 1-mm thickness. The pump laser used was a conventional Q-switched Nd:YAG laser with a wavelength of 1064 nm, a pulse duration of 30 ns and a Gaussian spatial form and its output was converted to the circularly polarized optical vortex by utilizing a spiral phase plate (SPP) [29], fabricated by electron beam etching, with azimuthal 2π phase and a quarter-wave plate. The optical vortex with $J = 2$ was focused to be ~130-μm annular spot onto the target output by an objective lens (M Plan Apo NIR, magnification factor 10, NA 0.26 from Mitutoyo Co.). The output energy on the sample surface was then fixed at 2 mJ. The ablated Ta plate was observed using a confocal laser-scanning microscope (Keyence VK-9700/VK9710GS) with a spatial resolution of 30 nm in both depth and transverse displacements. All experiments were performed at atmospheric pressure and room temperature.

Laser-scanning microscope images of processed Ta surfaces by optical vortex pumping are shown in **Figure 3**. After the single-shot deposition, a small bump with an approximately 4.4 μm height and 9.2 μm thickness appears at the center of the processed surface. When four optical vortex pulses were overlaid, the bump was shaped to be a needle with a height of ~10 μm and a tip diameter of less than 0.5 μm.

When $J = 0$ (the direction of the vortex is opposite to that of the circular polarization), in which the orbital and spin angular momenta work against with each other, the surface of Ta plate has no small bump and a lot of debris along the azimuthal direction around the outer circumference. Even when several pulses were overlaid onto the surface, only a small bump with an aspect ratio of <3 is structured.

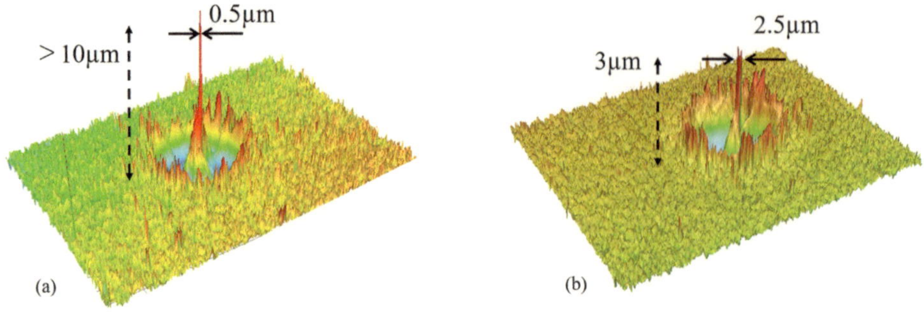

Figure 3. Metal microneedles fabricated by illumination of optical vortices with (a) $J = 2$ and (b) $J = 0$.

The following model will support the aforementioned experiments. The focused optical vortex forces the laser-induced melted or vaporized matters to revolve azimuthally along the annular intensity profile of the optical vortex. After that, the melted matter is confined in the dark core and accumulates at the center of the processed surface, resulting in a structured microneedle. We also successfully fabricated a two-dimensional and uniformly well-shaped 5 × 6 microneedle array with an average length and tip diameter of 11 and 0.5 μm by optical vortex laser ablation (**Figure 4**). Two optical vortex pulses are then overlaid on a target and after that, the target was translated.

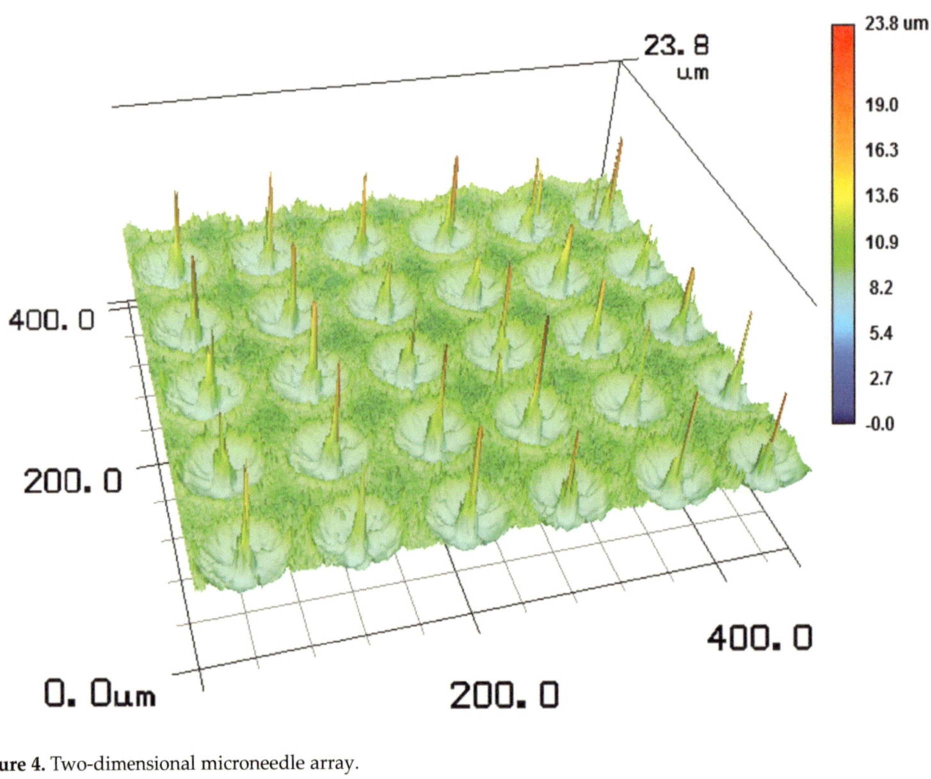

Figure 4. Two-dimensional microneedle array.

2.2. Silicon microneedle [30]

Monocrystalline silicon nano- or microstructures can improve significantly the performance of several photonic devices, such as photonic crystals [31], optical waveguides [32], photovoltaic devices [33], field emission arrays and metamaterials. Optical vortex laser ablation might provide us to fabricate silicon nanostructures at high cost- and time-efficiencies.

The experimental setup was almost identical with that in the metal microneedle fabrication. The pump laser used in this study was a picosecond Nd:YAG laser with a wavelength of 1064 nm, a pulse repetition rate of 10 Hz and a pulse duration of 20 ps and its output was converted to a circularly polarized optical vortex with a total angular momentum of 2. The optical vortex pulse was focused to be a 60-μm annular spot on the Si target by an object lens. A polished (100) monocrystalline silicon plate was used as the target. The ablated surface of the Si plate with a sputtered platinum coating was observed by a scanning electron microscope (JEOL, JSM-6010LA) with a spatial resolution of 8 nm at 3 kV. These experiments were also performed at atmospheric pressure and room temperature.

Picosecond optical vortex pulse with an energy of 0.6 mJ, which is sufficiently higher than the ablation threshold of ~0.03 mJ, was deposited on the target, so as to fabricate a debris-free needle with a height of 14 μm (the "length" defined as the length between the top and bottom ends of the needle was also measured to be 15 μm), a tip curvature of 160 nm and a thickness, defined as the full-width at 50% height of the needle, of approximately 2.9 μm (**Figure 5a**).

Worse heating effects arising from nanosecond pulse illumination (wavelength, 1064 nm; pulse energy, 0.6 mJ; pulse duration, 20 ns) suppress such needles formation, thereby yielding only a bump with a height of ~10μm (a length of ~12 μm) and a thickness of ~9.8 μm (>3 times that of the needle obtained by picosecond pulse illumination), even at the high-energy pumping (**Figure 5b**). These results indicate that the silicon needle formation requires picosecond pulse illumination with less heating effects.

Several overlaid picosecond vortex pulses enabled us to shape the needle with a height of ~40 μm. The experimental height, length and thickness of the needle as a function of the vortex pulse energy are shown in **Figure 6**.

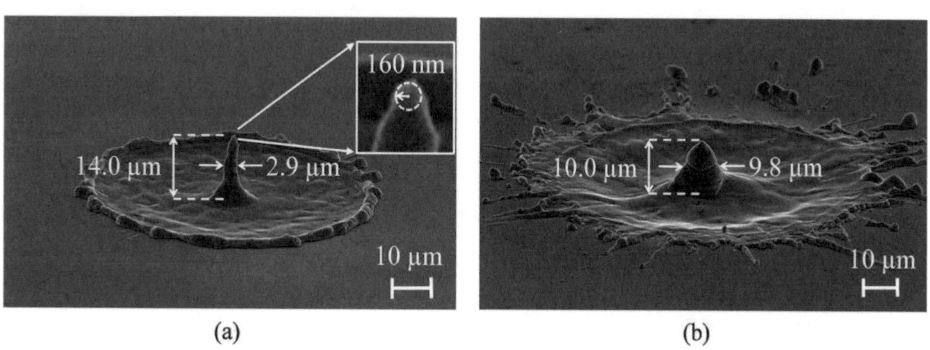

Figure 5. Monocrystalline silicon needles fabricated by (a) picosecond and (b) nanosecond optical vortex pulses.

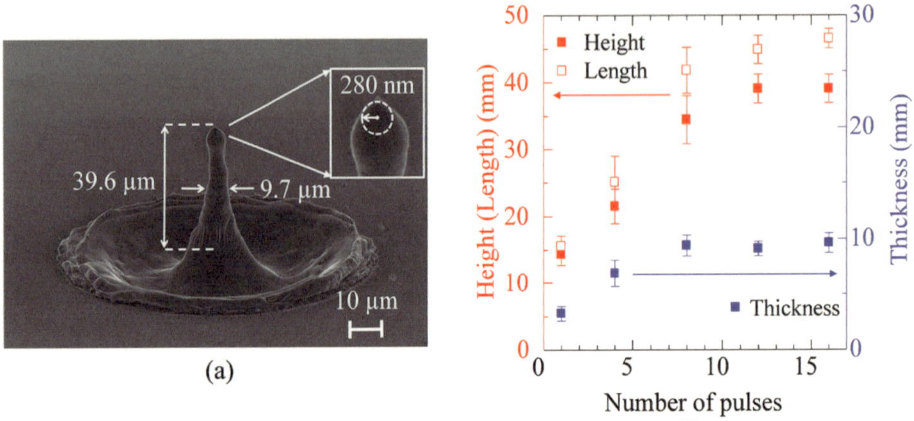

(a)

(b)

Figure 6. (a) Monocrystalline silicon needle by 10 overlaid vortex pulses and (b) height and thickness of the needle as a function of number of overlaid pulses.

The needle fabricated by picosecond vortex pulse irradiation exhibits the identical electron backscattering diffraction pattern to that of the silicon substrate with a lattice index of (1 0 0), although the needle had submicron-sized voids arising from thermal shock by illumination with optical vortex pulses (**Figure 7**). Namely, the monocrystalline needle was epitaxially grown on the silicon substrate.

In general, the silicon irradiated by ultrafast (femtosecond or picosecond) laser pulse was recrystallized mostly to form polycrystalline materials [34, 35]. How does the optical vortex illumination create such monocrystalline silicon structures?

The temporal dynamics of the silicon needle formation was investigated by utilizing an ultra-high-speed camera with a frame rate of 5 × 10⁶ frame/s. Optical vortex pulse deposition provides the melted silicon optical radiation forces such as an optical angular momentum and a forward scattering force, $F_s(r)$, written as follows:

$$F_S \propto r^2 e^{-2r^2/\omega_0^2} \quad (1)$$

where $|u(r)|^2$ is the intensity profile of the optical vortex, r is the radial coordinate of the optical field and ω_0 is the beam waist of the vortex pulse on the Si substrate, respectively. The resulting radial gradient $\Delta F(r)$ of the forward scattering force around the dark core ($r < \omega_0$) is then given by

$$\Delta F_S \propto -2r e^{-2r^2/\omega_0^2} \cdot \left(1 - 2\frac{r^2}{\omega_0^2}\right) \approx -2r. \quad (2)$$

The radial gradient $\Delta F(r)$ acts as a restoring force to collect the melted silicon within the dark core of the optical vortex. Also, the vaporization-induced recoil pressure [36] directs the melted silicon to the dark core. After the vortex pulse is gone (the recoil pressure is also gone), the melted silicon further transports toward the dark core by thermal diffusion effects. An

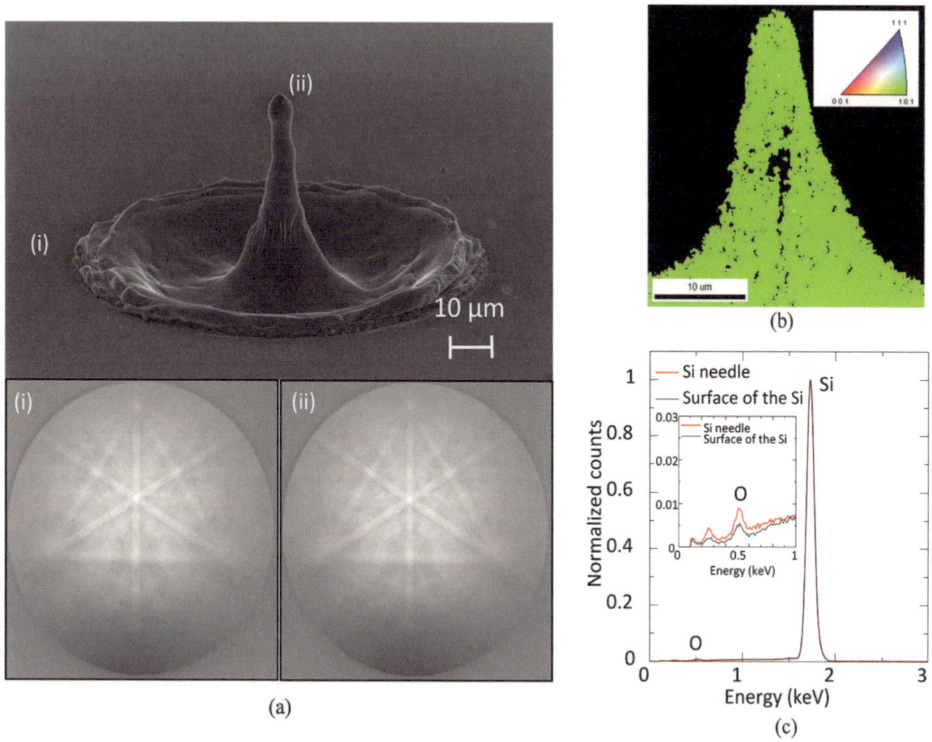

Figure 7. (a) Electron backscattering pattern of the silicon needle, (b) lattice index of the silicon needle and (c) energy dispersive X-ray spectrum of the silicon needle.

additional 200–600 ns later, the silicon is supercooled to recrystallize at the core (**Figure 8a**). A capillary wave [36] induced by the optical vortex illumination also ejects superfluous silicon droplets with a radius a given by the following formula,

$$a = \frac{3\pi\sigma}{4\rho v^2 \sqrt{1 - M^2}} \tag{3}$$

where σ is the surface tension coefficient (770 mN/m), v is the velocity of the silicon droplet (~50 m/s; the velocity was estimated from **Figure 8a**) and M is the Mach number (~0.3), respectively. The estimated a (~1.3 μm) is consistent with experimental one (~2.0 μm). Such slow solid-liquid hydrodynamics at a microsecond time scale enables us to establish the monocrystalline silicon needle.

Also, the silicon was pumped by a circularly polarized annular beam with the pulse energy of 0.8 mJ (without any orbital angular momentum) produced by a damaged mirror. The silicon cone-shaped structure with a height of ~9.0 μm and a thickness of ~3.6 μm then was formed. The silicon droplets flied away with a wide divergence angle θ (>6°) from the silicon substrate, so as to impact efficient accumulation of the silicon on the substrate (**Figure 8b**). We conclude that such straight flight straight flight of silicon droplets with spinning motion is induced by the optical vortices to establish the silicon needle.

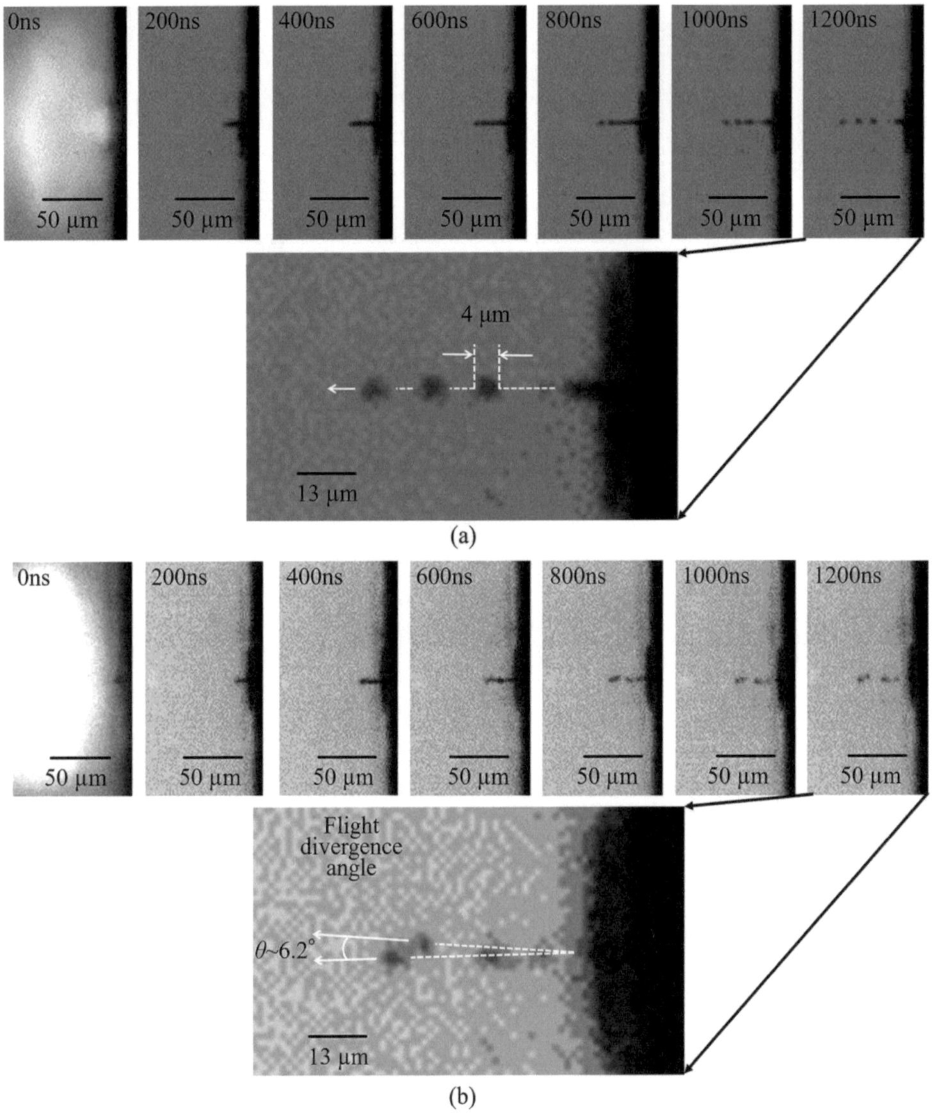

Figure 8. Temporal evolutions of silicon needle formation pumped by (a) optical vortex pulse and (b) annular beam without orbital angular momentum.

The spinning motion of the droplets cannot be directly observed, because of a relatively low spatial resolution (~1.2 μm) of the high-speed camera and it should be further investigated by utilizing nanosecond pump–probe analysis with high spatial and temporal resolutions [37].

Also, note that the spin–orbital angular momentum coupling effect is negligible in the present experiments based on optical vortex pumping with a low numerical aperture and a short pulse duration. The fabrication efficiency of the needles is almost 100%, though the individual needles to exhibit an individual length and height with a relatively large standard error of 3–4 μm.

http://dx.doi.org/10.5772/67073

3. Chiral structures

3.1. Chiral metal nanoneedle [38, 39]

Chiral metal nanoneedles will explore potentially various material sciences, for instance, selective assignment of the chirality and optical activity of molecules and chemical composites on a nanoscale and asymmetric chemical synthesis on plasmonic metasurfaces. However, no technique for twisting metal to form chiral metal nanoneedles has been established yet. Laser material processing has not been used to produce chiral metal nanoneedles, either.

Optical vortex with a helical wave front carries a handedness determined by the sign of the orbital angular momentum. If optical vortices can transfer their handedness to the melted metal through laser ablation processes, they will twist metal to form chiral metal nanoneedles.

Schematic diagram of experimental setup is almost identical with that as shown in **Figure 2**. A circularly polarized nanosecond optical vortex pulse (wavelength 1 μm, pulse width 30 ns, total angular momentum 2) was focused to be an annular spatial form with a diameter of <65 μm onto a metal. Four vortex pulses then were overlaid. The pulse energy was ~0.3 mJ, corresponding to <1/6 that used in Section 2. The handedness of the optical vortex pulse was reversed by inverting the SPP and QWP.

A needle with a tip curvature of <72 nm and a height of <10 μm, respectively, was formed at the center of the ablated zone with a smooth outline (**Figure 9a**) and its conical surface was also twisted azimuthally in the clockwise direction (**Figure 9b**). At higher energy pumping, the chiral nanoneedle formation was inhibited. When the handedness of the optical vortex (total angular momentum 2) was reversed, a needle was twisted azimuthally in the counter-clockwise direction (**Figure 9c** and **d**). These results, in which the handedness of the optical vortex can determine the twisting direction, for example, chirality of the nanoneedle, evidence that the metal melt by optical vortex illumination is forced to revolve azimuthally around the dark core by orbital angular momentum transfer effects.

The tip curvature of the twisted nanoneedle was further found to be inversely proportional to NA of the focusing optics (vortex pulse fluence was fixed to <9 J/cm^2) and the minimum value was measured to be 36 nm, which is <1/25 of the optical vortex wavelength (1064 nm) (**Figure 10**).

The electric resistance of the nanoneedle was measured to be <0.05 Ω by using two 50-μm-diameter tungsten probes. This value was identical to that of the substrate. Energy-dispersive X-ray (EDX) spectrum of the nanoneedle was also almost identical to that of the substrate, evidencing that the nanoneedle is perfectly metallic (**Figure 11**).

To make clear whether the wave front or polarization helicity primarily contributes to chiral nanoneedle fabrication or whether both helicities have similar contributions to chiral nanoneedle formation, the structures of chiral nanoneedles were further studied by using optical vortices with various values of the angular momenta J, l and s. The numerical aperture of the objective lens and the optical vortex pulse energy were then tuned in the range of 0.04–0.15 and 0.2–0.8 mJ, respectively, so as to maintain a constant beam waist and a constant fluence.

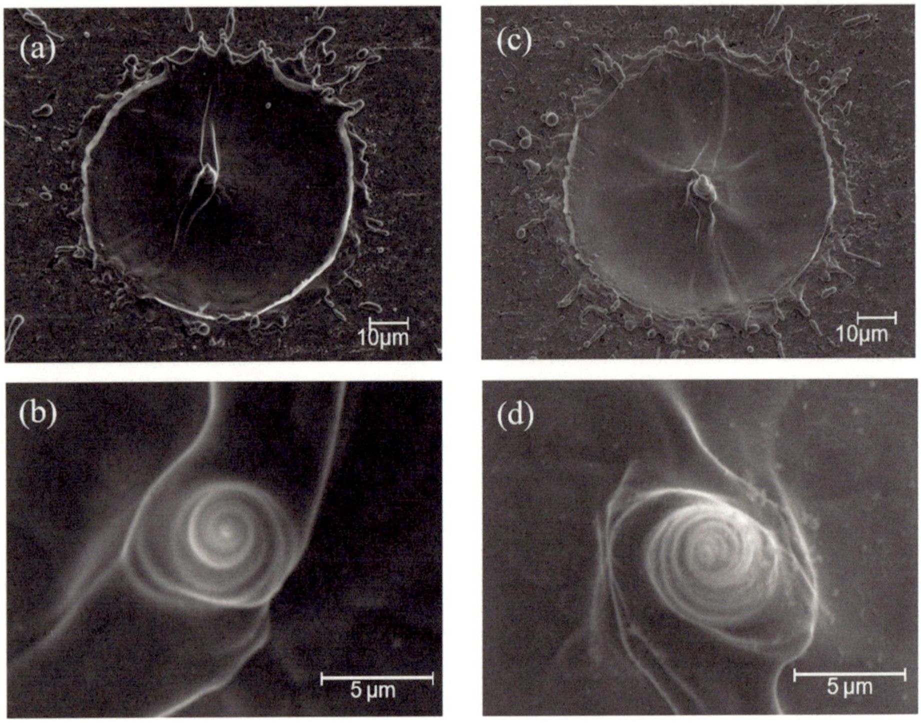

Figure 9. Chiral metal nanoneedles. (a) and (b) Left-handed needle, (c) and (d) right-handed needle.

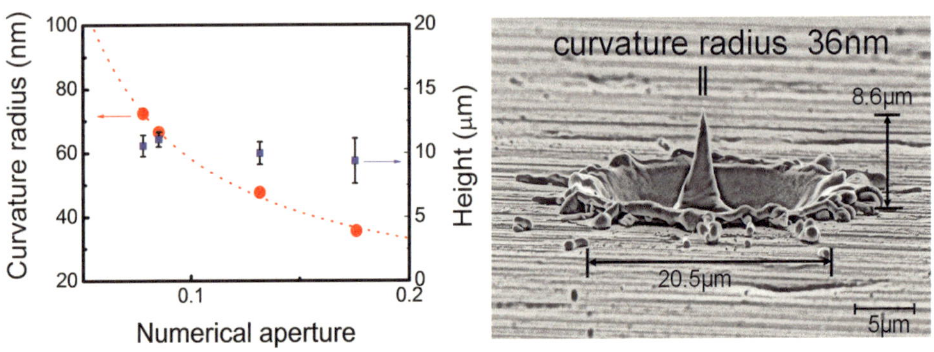

(a) (b)

Figure 10. (a) Tip curvature of the nanoneedle as a function of NA and (b) nanoneedle with a tip curvature of 36 nm.

The chirality of the nanoneedles is determined only by the handedness of orbital angular momentum *l* (the handedness of spin angular momentum *s* does not matter). The spiral density of the nanoneedle (defined as the number of turns divided by the length of the nanoneedle) was determined by the magnitude of *J* rather than *l* and it also increased as increasing

http://dx.doi.org/10.5772/67073

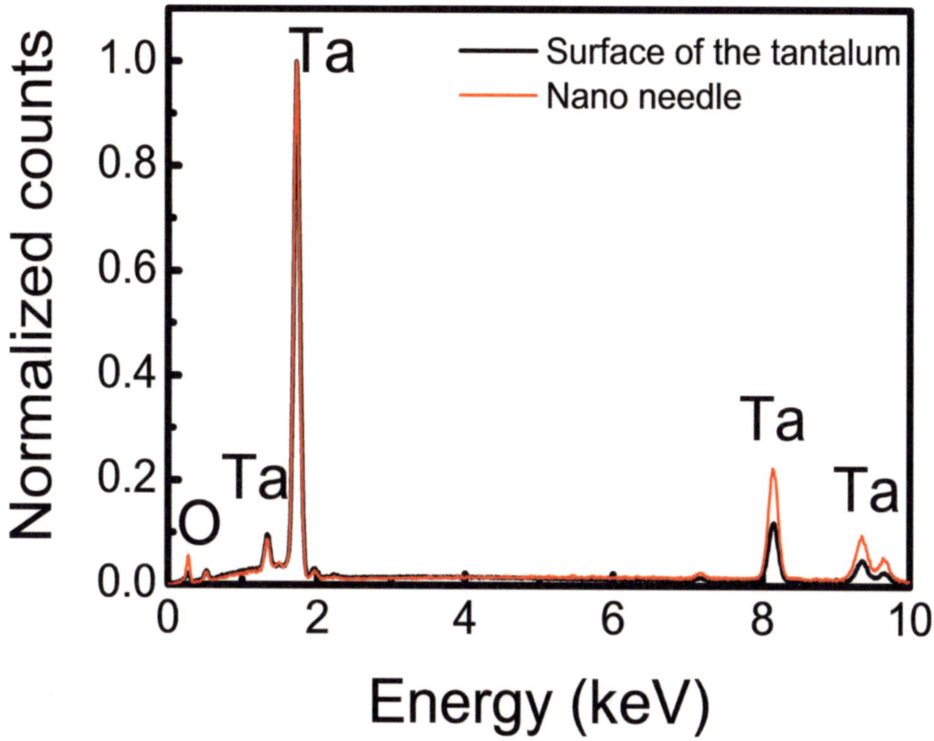

Figure 11. Energy dispersive X-ray spectrum of the 36-nm chiral metal nanoneedle.

Figure 12. Chiral nanoneedles at total angular momenta *J* of 0, 1 and 2.

J (**Figure 12**). Namely, it is found that nanoneedles created by linearly polarized second-order ($l = 2$, $s = 0$) and circularly polarized first-order ($l = 1$, $s = 1$) optical vortices exhibit the same chirality and spiral density (**Figure 13**).

3.2. Other chiral structures [40]

Chiral monocrystalline silicon nanostructures are difficult to fabricate even by utilizing advanced chemical technologies [41, 42].

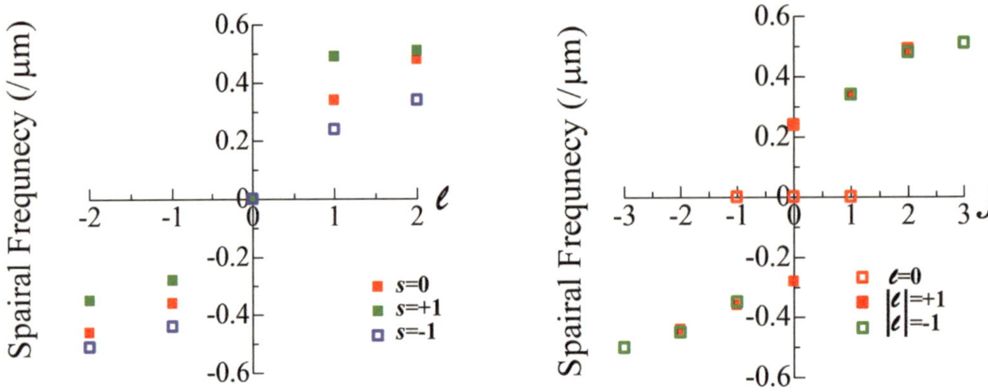

Figure 13. Spiral frequency of chiral metal nanoneedle at various *J*, *l* and *s*.

The target used was a polished (1 0 0) monocrystalline silicon plate. The circularly polarized nanosecond optical vortex pulse (total angular momentum 2) was focused to be a φ25-μm annular spot onto a silicon. With this system, the experimental ablation threshold of the silicon was measured to be 0.02 mJ.

At a pulse energy of 0.1 mJ, a silicon cone with a spiral conical surface (chiral silicon cone) was fabricated and its tip curvature and height were measured to be 110 nm and 1.0 μm, respectively (**Figure 14**).

The length of the chiral silicon cone was also measured to be 4.8 μm. The chirality of the silicon cone was also reversed by inverting the handedness of the optical vortex pulse.

Also, note that the silicon cone had the lattice index of (1 0 0) and its electron diffraction pattern and Raman spectrum were fully identical with those for the silicon substrate.

Even a copper, which has been widely investigated as a plasmonic metal, was twisted by optical vortex illumination, so as to form chiral needles (**Figure 15**).

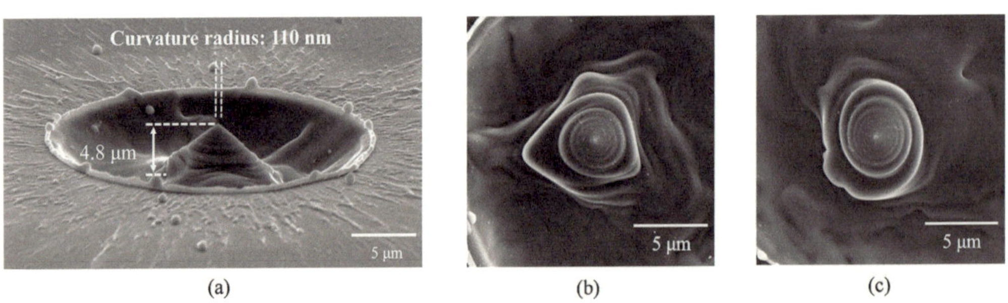

Figure 14. (a) Chiral silicon cone, (b) right-handed cone and (c) left-handed cone.

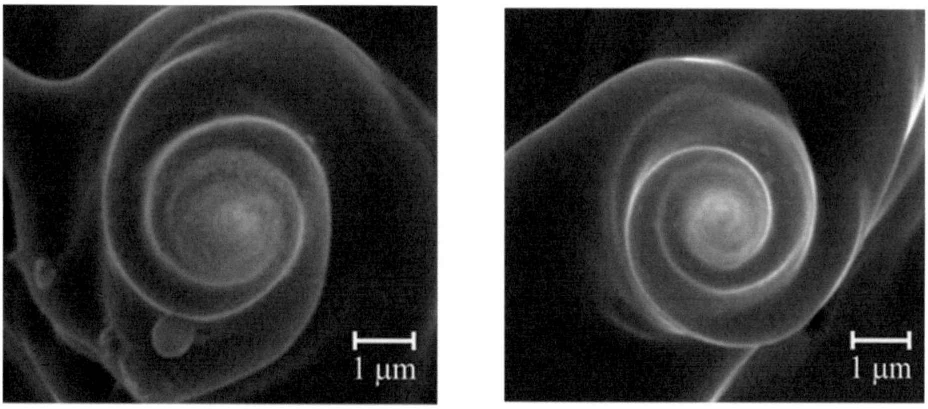

Figure 15. Left-handed and right-handed copper needles.

4. Surface relief

4.1. Chiral surface relief formation [43]

Surface relief formation has been established on azo-polymer films through mass transport owing to a driving force based on an optical gradient force, anisotropic photo-fluidity and cis-trans photoisomerization [44, 45]. Such surface relief provides many photonic devices, for instance, holographic memories [46], holographic waveguides [47] and photonic circuits.

In general, the mass transport driving force arises from the spatially inhomogenous intensity of the structured illumination, so as to direct the azo-polymer from a bright fringe toward a dark fringe along the polarization direction of the light. Thus, a spiral surface relief formation even by using circularly polarization is mostly inhibited.

Ambrosio et al. demonstrated the formation of spiral surface relief (termed "spiral relief" in their work), in which tightly focused higher-order optical vortices with a high numerical aperture (NA ~1.3) objective lens create a spiral surface relief with a shallow depth (10–20 nm) through azimuthal mass transport owing to interference between longitudinal and transverse optical fields [48].

Recently, we first demonstrated the formation of a single-arm chiral surface relief with a deep modulation depth of over 1 μm by using a lower-order optical vortex together with the spin angular momentum associated with the circular polarization.

A ~4 μm-thick-spin-coated azo-polymer (poly-orange tom-1, POT) film [49] used had absorption band in the wavelength range of 300–550 nm; thus, it exhibits a photo-isomerization behavior by green laser irradiation (**Figure 16**). A continuous-wave frequency-doubled

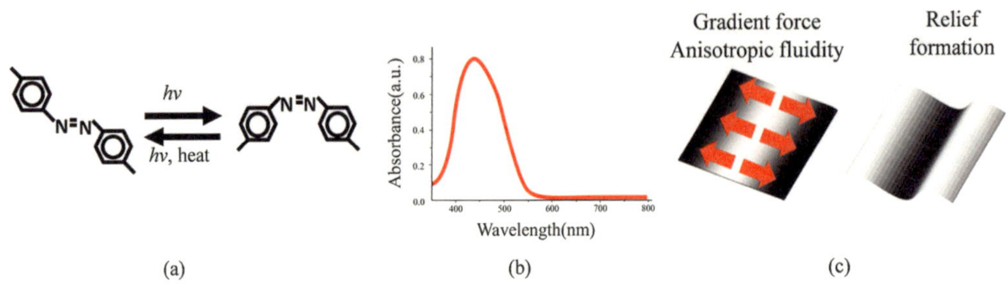

Figure 16. (a) Photoisomerization of azo-polymer, (b) absorption of poly orange-tom 1 and (c) mass transport on azo-polymer film to form surface relief.

Nd:YVO_4 laser output (a wavelength of 532 nm) was converted to be a circularly polarized first-order optical vortex by employing a computer-generated hologram displayed on a spatial light modulator (Hamamatsu photonics, X10468-03) and a quarter-wave plate. The sign of the orbital angular momentum was then made the same (or opposite) to that of the spin angular momentum, resulting in the total angular momentum J of 2 or 0. The optical vortex beam with a power of ~300 μW was focused to an annular spot with a diameter of 4 μm on the azo-polymer film by an objective lens with NA ~0.45, corresponding to a focused spot intensity of ~2.5 kW/cm^2. An exposure time on the film was fixed to be 8 s. All experiments were performed at room temperature and in the atmosphere.

In the case of a linearly polarized optical vortex, the mass transport occurs along the polarization direction to collect azo-polymer toward the dark core, so as to establish a cat-shaped (non-spiral) surface relief (**Figure 17a**). Loosely focusing in our experiments impacts to produce transverse optical field, thereby preventing the spiral surface relief.

A circularly polarized optical vortex with positive orbital and spin angular momenta forces the orbital motion of the azo-polymer in a clockwise direction to complete a single-arm spiral structure. The resulting chiral surface relief then had a height of ~1 μm and a diameter of ~5 μm, respectively (**Figure 17b**). In contrast, a circularly polarized optical vortex beam with negative spin angular momentum inhibited the spiral surface relief formation to establish only non-spiral bump relief (**Figure 17c**).

Only circularly polarized optical vortex with positive (or negative) orbital and positive (or negative) spin angular momenta can produce such clockwise (or counter-clockwise) spiral surface relief on the azo-polymer film for all topological charges in a range of 1 3. The experimental data are summarized in **Figure 18**.

In general, mass transport arising from the photo-isomerization occurs at a temperature below the glass transition temperature of the polymer, meaning that the polymer volume before and after the relief formation should be preserved [50]. In fact, the spiral surface relief can be easily erased by the spatially uniform green laser illumination (0.36 W/cm^2), indicating that photo-isomerization but rather heat-induced effects (melting, expansions, ablation, etc.) contributes predominantly to the spiral surface relief formation (**Figure 19**).

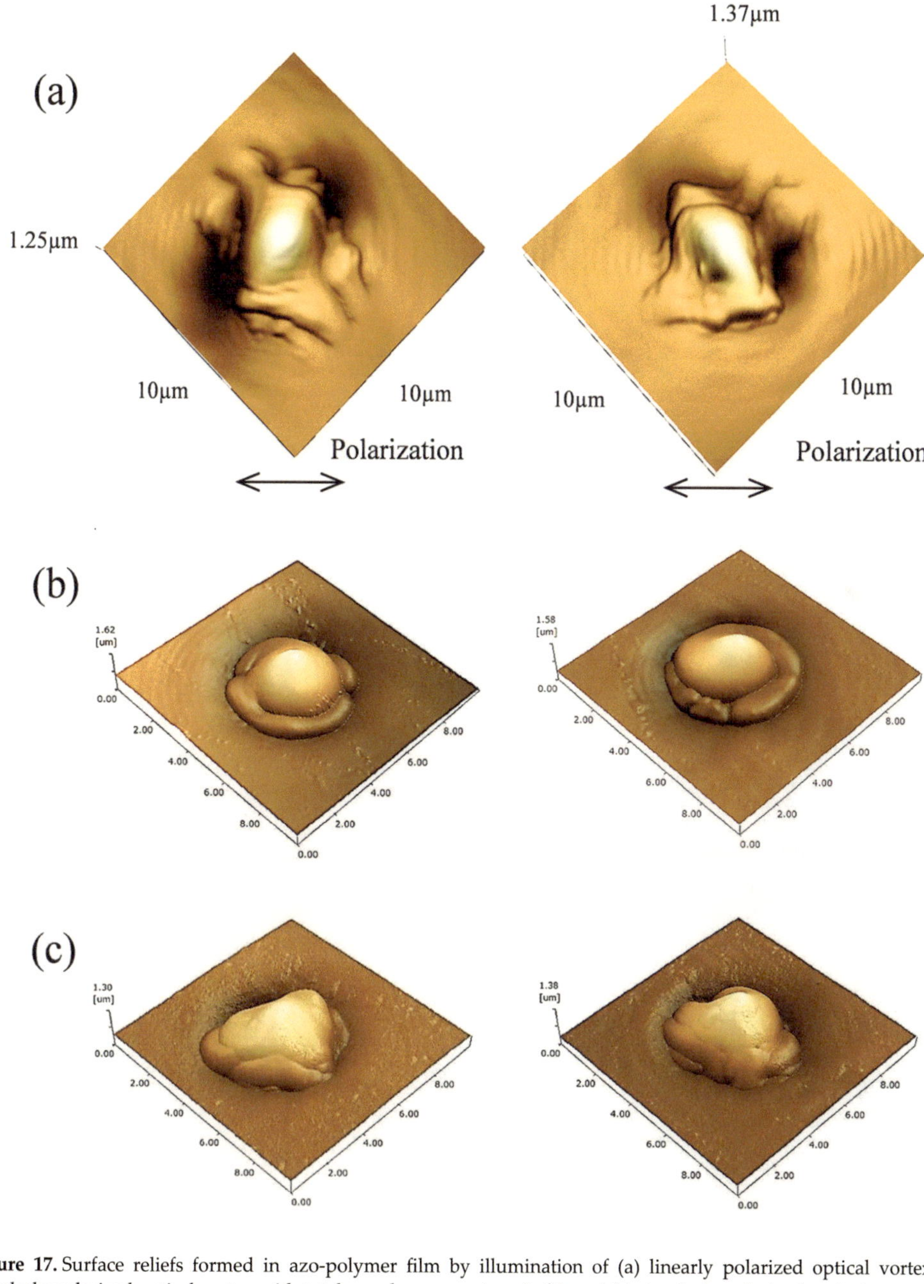

Figure 17. Surface reliefs formed in azo-polymer film by illumination of (a) linearly polarized optical vortex, (b) circularly polarized optical vortex with total angular momentum *J* of 2, and (c) circularly polarized optical vortex with total angular momentum *J* of 0.

4.2. Theoretical discussion [51]

Why can organic azo-polymers be deformed to create a chiral surface relief by circularly polarized optical vortices with positive (or negative) orbital and positive (or negative) spin angular momenta?

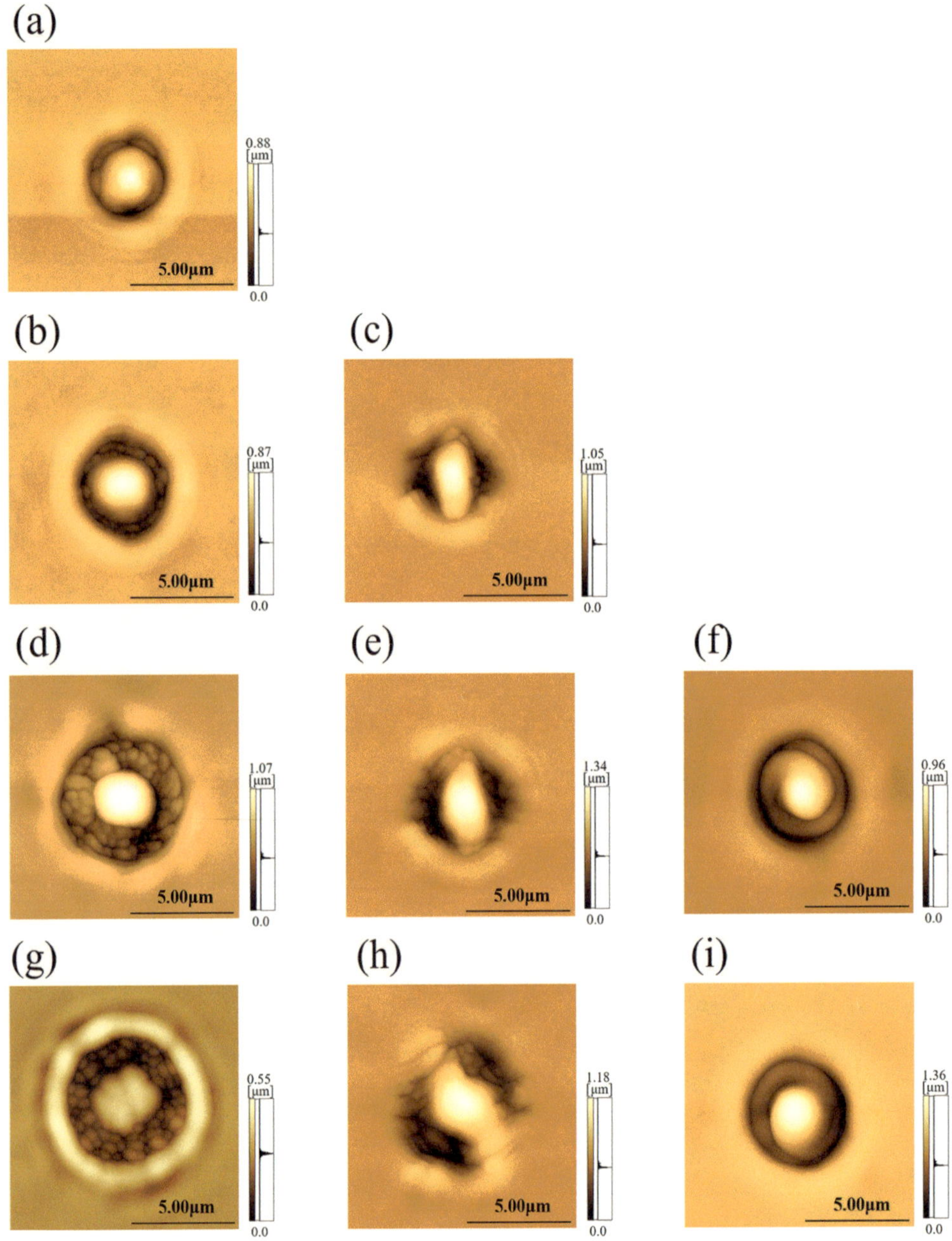

Figure 18. Surface reliefs formed in an azo-polymer thin film by irradiation with optical vortices with spin and orbital angular momenta of (a) $J = 0$ ($l = 1, s = -1$), (b) $J = 1$ ($l = 2, s = -1$), (c) $J = 1$ ($l = 1, s = 0$), (d) $J = 2$ ($l = 3, s = -1$), (e) $J = 2$ ($l = 2, s = 0$), (f) $J = 2$ ($l = 1, s = 1$), (g) $J = 3$ ($l = 4, s = -1$), (h) $J = 3$ ($l = 3, s = 0$), and (i) $J = 3$ ($l = 2, s = 1$).

Conventional optical manipulation, in which particles play a role as electric dipoles with optically induced surface charges at the interface between themselves and a solution, has been theoretically well established [52, 53]. However, there were few theoretical works concerning the formation of spiral structures in isotropic and homogeneous materials using the optical radiation force, in which the wavefront-sensitive light-induced mass transport occurs.

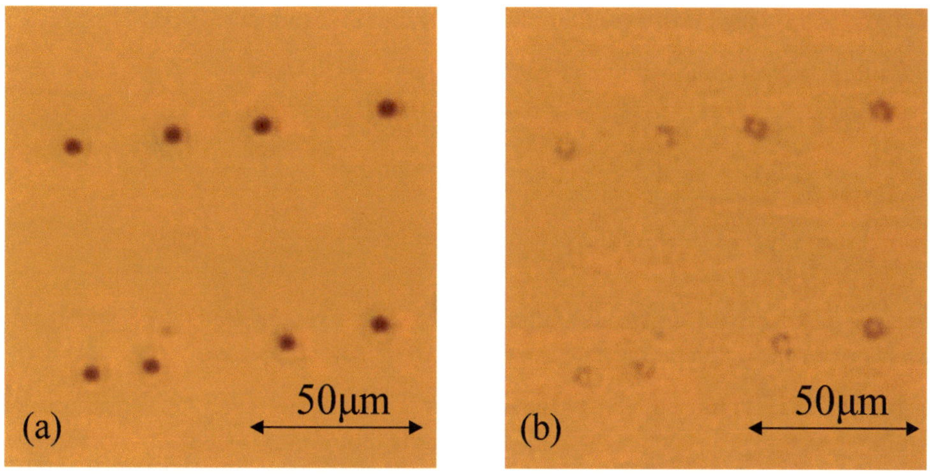

Figure 19. (a) Microscope image of surface relief before uniform green laser irradiation and (b) microscope image of surface relief after uniform green laser irradiation.

A few researchers have proposed theoretical models for such light-induced wavefront-sensitive mass transport in the azo-polymer, in which the azo-molecule is considered to be a microscopic dipole; however, their models based on a finite difference time-domain method are difficult to understand directly how the azo-polymer homogeneous materials deform by optical vortex illumination and how the spin and orbital angular momenta contribute to wavefront-sensitive structures formation? [54–56]

An analytical formula for the optical vortex induced optical radiation force in an isotropic and homogeneous material is proposed, so as to directly understand how the homogenous materials deform by the irradiation of optical vortices with arbitrary orbital and spin angular momenta.

The optically induced polarization charge density on an isotropic and homogeneous material surface is zero; thus, the time-averaged optical radiation force F arising from optically induced electric polarization is expressed by

$$F = \langle -i\omega\, \varepsilon_0 \chi E \times B \rangle = \frac{\omega\, \varepsilon_0}{2}\left[\chi_r \operatorname{Im}(E \times B^*) + \chi_i \operatorname{Re}(E \times B^*)\right] \tag{4}$$

where **E** is the electric field vector, **B** is the magnetic flux density vector, ε_0 is the dielectric constant in a vacuum, ω is the angular frequency of the optical field and χ ($= \chi_r + i\chi_i$) is the macroscopic complex electric susceptibility, respectively. The angle brackets < > then denote the time average.

Assuming a paraxial approximation, in which the continuous-wave optical vortex beam is loosely focused and it propagates along the z-axis without diffraction in the material, the scholar electric field, $E(r, \phi, z)$, of the right-handed optical vortex with a positive topological charge in cylindrical coordinates is given by,

$$E(r, \phi, z) = A_\ell(r)\, e^{ikz}\, e^{i\ell\phi}\, e^{-i\omega t} \tag{5}$$

where $A_l(r)$ is the axisymmetric amplitude and k is the wave number.

The resulting optical radiation force **F** can be expressed as follows:

$$F(r,\phi,z) = \frac{\varepsilon_0\chi_r}{4}\left\{\frac{\sqrt{1-s^2}}{2}\frac{\partial A_\ell^2}{\partial r}\left(-\cos\phi\cdot e_x+\sin\phi\cdot e_y\right)+\left(\frac{1}{2}\frac{\partial A_\ell^2}{\partial r}-\ell s\frac{A_\ell^2}{r}\right)e_r\right\}$$
$$+\frac{\varepsilon_0\chi_i}{4}\left\{\ell\frac{\sqrt{1-s^2}}{2}\frac{A_\ell^2}{r}\left(\sin\phi\cdot e_x+\cos\phi\cdot e_y\right)+\left(\ell\frac{A_\ell^2}{r}-\frac{s}{2}\frac{\partial A_\ell^2}{\partial r}\right)e_\phi+2k A_\ell^2 e_z\right\} \tag{6}$$

where $\mathbf{e}_r$, $\mathbf{e}_f$ and $\mathbf{e}_z$ are the unit vectors along the r, ϕ and z-axes, respectively; $\mathbf{e}_x$ and $\mathbf{e}_y$ are also unit vectors for the polarization state in a Cartesian coordinate system.

The optical radiation force, given by the first and second terms in Eq. (6), is proportional to the gradient of the optical intensity along the polarization direction and it plays a role as the mass transport driving force. The radial optical radiation force, expressed as the third and fourth terms in Eq. (6), acts as a restoring (or repulsive) force to direct the materials toward (or outwards from) the dark inner core of the optical vortex with a mode field radius of $x_{0.}$

The axisymmetric amplitude $A_l(r)$ is given as

$$A_\ell(r)=\frac{A_0}{\sqrt{\ell!}}\left(\sqrt{2}\frac{r}{\omega_0}\right)^{|\ell|}e^{-\frac{r^2}{\omega_0^2}}. \tag{7}$$

Also, note that the radial optical radiation force is proportional to the real part χ_r of the polarization susceptibility. The optical absorption force (fifth and sixth terms), which is proportional to the imaginary part χ_i of the electric susceptibility, also gives the torque to the materials. The optical absorption force given by the seventh and eighth also induces the orbital motion of materials around the dark core. The forward scattering force given by the last term also serves as a restorative force to confine the materials to the inner dark core, thereby yielding the radial force proportional to the gradient of the forward scattering force.

A linearly x-polarized optical vortex without spin angular momentum ($s = 0$) exhibits the optical radiation force **F** given by,

$$F(r,\phi,z) = \frac{\varepsilon_0}{2}\left(A_\ell^2\right)\cdot\left[\left\{\left(\frac{\ell}{r}-\frac{2r}{\omega_0^2}\right)\chi_r\sin\phi-\frac{\ell}{r}\chi_i\cos\phi\right\}e_y+k\chi_i e_z\right]. \tag{8}$$

The optical radiation force provides mostly the clockwise torque to the materials along the y direction (along the vertical direction to the polarization) and thereby, it prevents chiral structures formation. The complex electric susceptibility of the material was then assumed to be $2 + 2i$, corresponding to that of the azo-polymer. The gradient of forward scattering force will also drive the materials toward the dark core of the optical vortex, resulting in the mass transport along the x direction (along the direction of polarization).

The circularly polarized optical vortex ($s = \pm 0$) produces the optical radiation force written as follows:

$$F(r,\phi,z) = \frac{\varepsilon_0\chi_r}{2}\cdot\left(|\ell|-\ell s-\frac{2r^2}{\omega_0^2}\right)\left(\frac{A_\ell^2}{r}\right)e_r+\frac{\varepsilon_0\chi_i}{2}\cdot\left\{\left(\ell-|\ell|s+s\frac{2r^2}{\omega_0^2}\right)\left(\frac{A_\ell^2}{r}\right)e_\phi+2kA_\ell^2 e_z\right\}. \tag{9}$$

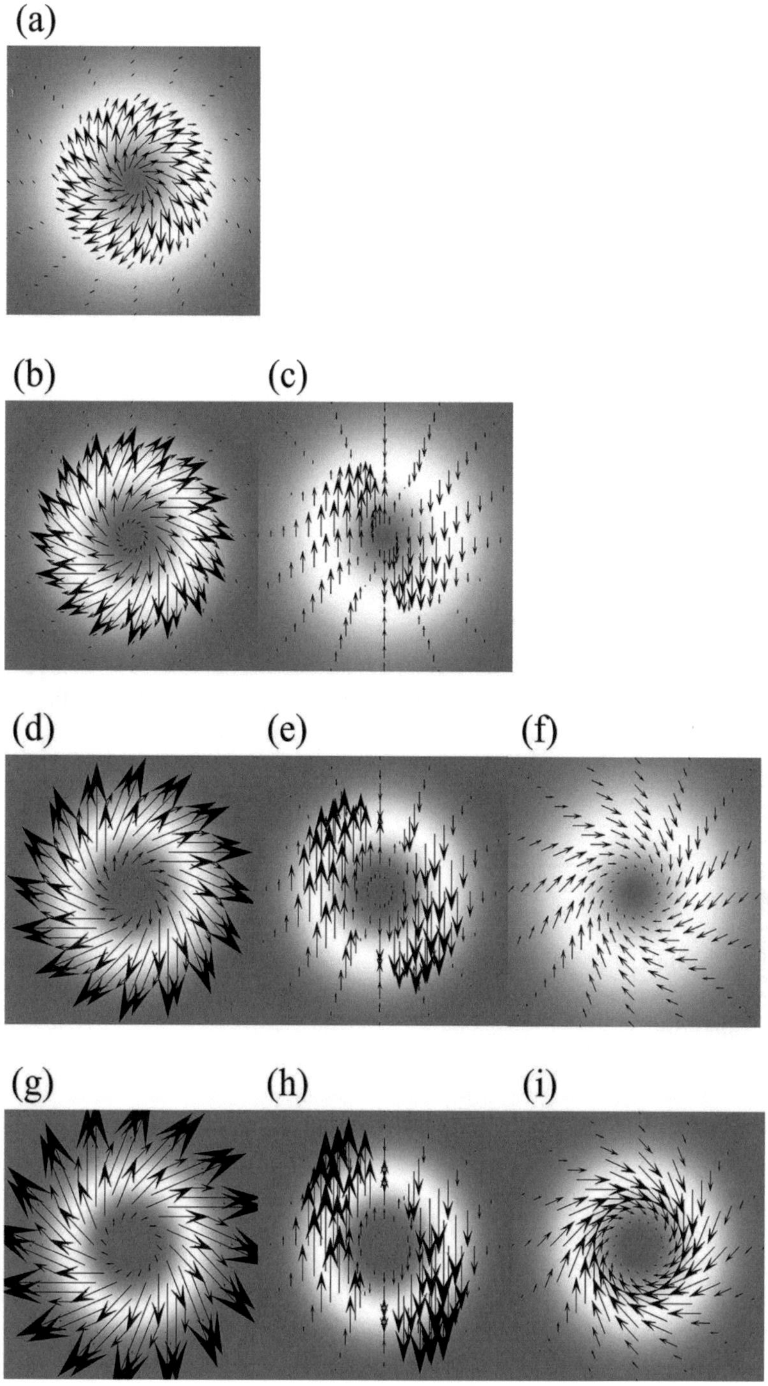

Figure 20. Spatial distributions of the radial and azimuthal optical radiation forces produced by optical vortices with total angular momenta of (a) $J = 0$ ($l = 1$, $s = -1$), (b) $J = 1$ ($l = 2$, $s = -1$), (c) $J = 1$ ($l = 1$, $s = 0$), (d) $J = 2$ ($l = 3$, $s = -1$), (e) $J = 2$ ($l = 2$, $s = 0$), (f) $J = 2$ ($l = 1$, $s = 1$), (g) $J = 3$ ($l = 4$, $s = -1$), (h) $J = 3$ ($l = 3$, $s = 0$), and (i) $J = 3$ ($l = 2$, $s = 1$). The optical field of the vortex, shown in white, underlies the optical radiation forces.

When the product of l and s is positive, the radial optical radiation force acts as a restoring force for the materials toward the dark core and the azimuthal optical absorption force drives the orbital motion of the materials around the dark core. The forward scattering force also plays a role as the restoring force to confine the materials inside the dark core, thereby establishing a chiral structure.

In the case of the negative product of l and s, the optical vortex provides the repulsive radial optical radiation force, which competes with the effective radial force produced by the forward scattering force, so as to prevent the efficient confinement of the materials in the dark core. Furthermore, the radial radiation force and azimuthal absorption force are reversed outside the dark core. The resulting non-twisted orbital motion of the materials will occur, thereby yielding non-chiral structures (**Figure 20**).

These results, in which only optical vortices with a positive product of orbital and spin angular momenta (constructive spin-orbital coupling) enable the formation of chiral structures, can support well the experiments of the chiral surface relief formation in the azo-polymer film. The degeneracy among optical vortices with the same total angular momentum J is then resolved.

As mention in Section 3, the pulsed optical vortices illumination forms spiral metal structures, in which the melted metal is collected in the dark core after the vortex pulse has gone. Only an optical absorption force $\mathbf{F}^{l,s}$ will then act as the mass transport driving force through the laser ablation process. The resulting general relationship given by

$$F^{\ell-1,1} = F^{\ell,0} \quad \ell \geq 2 \tag{10}$$

is established.

Two optical vortices with the same total angular momentum J should provide the same azimuthal optical radiation force and they are degenerate. This theoretical analysis can also support well the experimental results obtained previously in the metal with pulsed optical vortices illumination.

5. Conclusion

The optical vortex carries an annular intensity profile and an orbital angular momentum owing to a helical wave front. We discovered, for the first time, that the optical vortex with orbital angular momentum enables us to twist a melted metal, silicon and a photo-isomerized polymer to establish monocrystalline achiral or chiral structures (i.e., microneedles, microspheres, chiral nanostructures and chiral surface relief). The chirality of the structured materials is determined selectively by the handedness of the optical vortex. A spin angular momentum arising from to a helical electric field of the circular polarization then accelerates (or decelerates) the chiral materials formation. Two-dimensional chiral nanostructures array can be easily fabricated merely by translating the target and then irradiating the optical vortex onto the target.

http://dx.doi.org/10.5772/67073

We also found that chiral surface relief formation in the azo-polymers requires the irradiation with continuous-wave optical vortices with a positive product of orbital and spin angular momenta (constructive spin-orbital angular momentum coupling); therefore, the degeneracy among the optical vortices with the same total angular momentum is resolved.

In contrast, spiral metal structures formation by the pulsed optical vortices illumination, the melted metal receives only an optical absorption force as the mass transport driving force through the laser ablation process. Thus, optical vortices with the same total angular momentum J are degenerate and they create the same chiral structures.

Such chiral nanostructures will open potentially the door to develop various advanced material sciences and technologies, such as selective identification of chiral chemical composites, enhancement of chiral chemical reactions, plasmonic metasurfaces, nanoimaging systems, energy-saving displays and biomedical nanoelectromechanical systems.

Acknowledgements

The authors acknowledge a JST-CREST Grant and a JSPS KAKENHI Grant Number JP 16H06507 in Scientific Research on Innovative Areas "Nano-Material Optical-Manipulation".

Author details

Takashige Omatsu[1]*, Katsuhiko Miyamoto[2] and Ryuji Morita[3]

*Address all correspondence to: omatsu@faculty.chiba-u.jp

1 Graduate School of Advanced Integration Science, Chiba University, Chiba, Japan

2 Molecular Chirality Research Center, Chiba University, Chiba, Japan

3 Graduate School of Engineering, Hokkaido University, Sapporo, Japan

References

[1] L. Allen, M. W. Beijersbergen, R. J. C. Spreeuw and J. P. Woerdman, "Orbital angular momentum of light and the transformation of Laguerre–Gaussian laser modes," Phys. Rev. A, **45**(11), 8185–8189 (1992).

[2] M. Padgett, J. Courtial and L. Allen, "Light's orbital angular momentum," Phys. Today, **57**(5), 35–40 (2004).

[3] M. J. Padgett, F. M. Miatto, M. P. J. Lavery, A. Zeilinger and R. W. Boyd, "Divergence of an orbital-angularmomentum-carrying beam upon propagation," New J. Phys., **17**(2), 023011 (2015).

[4] T. Kuga, Y. Torii, N. Shiokawa, T. Hirano, Y. Shimizu and H. Sasada, "Novel optical trap of atoms with a doughnut beam," Phys. Rev. Lett., **78**(25), 4713–4716 (1997).

[5] M. Padgett and R. Bowman, "Tweezers with a twist," Nat. Photonics, **5**, 343–348 (2011).

[6] D. G. Grier, "A revolution in optical manipulation," Nature **424**, 810–816 (2003).

[7] J. Wang, J. Yuan Yang, I. M. Fazal, N. Ahmed, Y. Yan, H. Huang, Y. X. Ren, Y. Yue, S. Dolinar, M. Tur and A. E. Willner, "Terabit free-space data transmission employing orbital angular momentum multiplexing," Nat. Photonics, **6**, 488–496 (2012).

[8] N. Bozinovic, Y. Yue, Y. Ren, M. Tur, P. Kristensen, H. Huang, A. E. Willner, S. Ramachandran, "Terabit-scale orbital angular momentum mode division multiplexing in fibers," Science, **340**(6140), 1545–1548 (2013).

[9] G. Molina-Terriza, J. P. Torres and L. Torner, "Twisted photons," Nat. Phys., **3**(5), 305–310 (2007).

[10] I. Heller, G. Sitters, O. D. Broekmans, G. Farge, C. Menges, W. Wende, S. W. Hell, E. J. G. Peterman and G. J. L. Wuite, "STED nanoscopy combined with optical tweezers reveals protein dynamics on densely covered DNA," Nat. Methods, **10**(9), 910–916 (2013).

[11] T. Watanabe, Y. Iketaki, T. Omatsu, K. Yamamoto, M. Sakai and M. Fujii, "Two-point-separation in super-resolution fluorescence microscope based on up-conversion fluorescence depletion technique," Opt. Express, **11**(24), 3271 (2003).

[12] Y. Iketaki, T. Watanabe, S. Ishiuchi, M. Sakai, T. Omatsu, K. Yamamoto, M. Fujii, T. Watanabe, "Investigation of the fluorescence depletion process in the condensed phase; application to a tryptophan aqueous solution," Chem. Phys. Lett. **372**(5), 773–778 (2003).

[13] B. A. Parviz, D. Ryan and G. M. Whitesides, "Using self-assembly for the fabrication of nano-scale electronic and photonic devices," IEEE Trans. Adv. Packag., **26**(3), 233–241 (2003).

[14] S. Khumpuang, M. Horade, K. Fujioka and S. Sugiyama, "Microneedle fabrication using the plane pattern to cross-section transfer method," Smart Mater. Struct., **15**(2), 600–606 (2006).

[15] Y. Tang, A. E. Cohen, "Enhanced enantioselectivity in excitation of chiral molecules by superchiral light," Science, **332**(6027), 333–336 (2011).

[16] V. K. Valev, J. J. Baumberg, C. Sibilia, T. Verbiest, "Chirality and chiroptical effects in plasmonic nanostructures: fundamentals, recent progress and outlook," Adv. Mater., **25**, 2517–2534 (2013).

[17] T. Omatsu, K. Chujo, K. Miyamoto, M. Okida, K. Nakamura, N. Aoki, R. Morita, "Metal microneedle fabrication using twisted light with spin," Opt. Express, **18**(17), 17967–17973 (2010).

[18] P. Vettiger, M. Despont, U. Drechsler, U. Durig, W. Haberle, M. I. Lutwyche, H. E. Rothuizen, R. Stutz, R. Widmer and G. K. Binnig, "The "Millipede"—more than one thousand tips for future AFM data storage," IBM J. Res. Develop., **44**(3), 323–340 (2000).

[19] Q. H. Wang, A. A. Setlur, J. M. Lauerhaas, J. Y. Dai, E. W. Seelig and R. P. H. Chang, "A nanotube-based fieldemission flat panel display," Appl. Phys. Lett., **72**(22), 2912 (1998).

[20] L. Lin, A. P. Pisano, "Silicon-processed microneedles," J. Microelectromech. Syst., **8**(1), 78–84 (1999).

[21] D. V. McAllister, P. M. Wang, S. P. Davis, J-H Park, P. J. Canatella, M. G. Allen and M. R. Prausnitz, "Microfabricated needles for transdermal delivery of macromolecules and nanoparticles: fabrication methods and transport studies," PNAS, **100**(24), 13755–13760 (2003).

[22] E. Larrañeta, R. E. M. Lutton, A. D. Woolfson, R. F. Donnelly, "Microneedle arrays as transdermal and intradermal drug delivery systems: materials science, manufacture and commercial development," Mater. Sci. Eng. R: Rep., **104**, 1–32 (2016).

[23] S. Nolte, C. Momma, H. Jacobs, A. Tünnermann, B. N. Chichkov, B. Wellegehausen, H. Welling, "Ablation of metals by ultrashort laser pulses," J. Opt. Soc. Am. B, **14**(10), 2716–2722 (1997).

[24] X. Liu, D. Du, G. Mourou, "Laser ablation and micromachining with ultrashort laser pulses," IEEE J. Quantum Electron., **33**(10), 1716–1716 (1997).

[25] M. C. Gower, "Industrial applications of laser micromachining," Opt. Express, **7**(2), 56–67 (2000).

[26] J. Hamazaki, R. Morita, K. Chujo, Y. Kobayashi, S. Tanda, T. Omatsu, "Optical-vortex laser ablation", Opt. Express, **18**(3), 2144–2151 (2010).

[27] S.M. Barnett, "Optical angular momentum flux", J. Optics B Quantum Semiclassical Optics, **4**, S7–S16 (2002).

[28] A. T. O'Neil, I. MacVicar, L. Allen, M. J. Padgett, "Intrinsic and extrinsic nature of the orbital angular momentum of a light beam," Phys. Rev. Lett., **88**, 053601 (2002).

[29] M. W. Beijersbergen, R. P. C. Coerwinkel, M. Kristensen, J. P. Woerdman, "Helical-wavefront laser beams produced with a spiral phase plate," Opt. Commun., **112**(5–6), 321–327 (1994).

[30] F. Takahashi, K. Miyamoto, H. Hidai, K. Yamane, R. Morita, T. Omatsu, "Picosecond optical vortex pulse illumination forms a monocrystalline silicon needle," Sci. Rep., **6**, 21738, (2016).

[31] A. Blanco, E. Chomski, S. Grabtchak, M. Ibisate, S. John, S. W. Leonard, C. Lopez, F. Meseguer, H. Miguez, J. P. Mondia, G. A. Ozin, O. Toader and H. M. Driel1 "Large-scale synthesis of a silicon photonic crystal with a complete three-dimensional bandgap near 1.5micrometres," Nature **405**, 437–440 (2000).

[32] Y. A. Vlasov, X. Z. Bo, J. C. Sturm and D. J. Norris, "On-chip natural assembly of silicon photonic bandgap crystals," Nature **414**, 289–293 (2001).

[33] L. Hu and G. Chen, "Analysis of optical absorption in silicon nanowire arrays for photovoltaic applications," Nano Lett., **7**, 3249–3252 (2007).

[34] J. Bonse, S. Baudach, J. Krüger, W. Kautek, M. Lenzner, "Femtosecond laser ablation of silicon–modification thresholds and morphology," Appl. Phys. A, **74**, 19–25 (2002).

[35] J. Bonse, K. W. Brzezinka and A. J. Meixner, "Modifying single-crystalline silicon by femtosecond laser pulses: an analysis by micro Raman spectroscopy, scanning laser microscopy and atomic force microscopy," Appl. Surf. Sci., **221**, 215–230 (2004).

[36] A. B. Brailovsky, S. V. Gaponov and V. I. Luchin, "Mechanisums of melt droplets and solid-particle ejection from a target surface by pulsed laser action," Appl. Phys. A, **61**, 81–86 (1995).

[37] R. Evans, S. Camacho-López, F. G. Pérez-Gutiérrez, G. Aguilar, "Pump-probe imaging of nanosecond laser-induced bubbles in agar gel," Opt. Express, **16**(10), 7481–7492 (2008).

[38] K. Toyoda, K. Miyamoto, N. Aoki, R. Morita and T. Omatsu, "Using optical vortex to control the chirality of twisted metal nanostructures," Nano Lett., **12**, 3645–3649 (2012).

[39] K. Toyoda, F. Takahashi, S. Takizawa, Y. Tokizane, K. Miyamoto, R. Morita and T. Omatsu, "Transfer of light helicity to nanostructures," Phys. Rev. Lett., **110**, 143603 (2013).

[40] F. Takahashi, S. Takizawa, H. Hidai, K. Miyamoto, R. Morita and T. Omatsu, "Optical vortex pulse illumination to create chiral monocrystalline silicon nanostructures," Physica Status Solidi (a), **213**(4), 1063–1068, (2016).

[41] M. Park, M. Kim, J. Joo, K. Kim, J. Kim, S. Ahn, Y. Cui and J. Cho, "Silicon Nanotube Battery Anodes," Nano Lett., **9**(11), 3844 (2009).

[42] D. Zschech, D. Kim, A. P. Milenin, R. Scholz, R. Hillebrand, C. J. Hawker, T. P. Russell, M. Steinhart and U. Gösele, "Ordered arrays of <100>-oriented silicon nanorods by CMOS-compatible block copolymer lithography," Nano Lett., **7**(6), 1516 (2007).

[43] M. Watabe, G. Juman, K. Miyamoto and T. Omatsu, "Light induced conch-shaped relief in an azo-polymer film," Sci. Rep., **4**, 4281 (2014).

[44] A. Natansohn, P. Rochon, "Photoinduced motions in azo-containing polymers," Chem. Rev. **102**, 4139–4175 (2002).

[45] H. Ishitobi, M. Tanabe, Z. Sekkat, S. Kawata, "The anisotropic nanomovement of azo-polymers," Opt. Express, **15**, 652–659 (2007).

[46] N. K. Viswanathan, D. Y. Kim, S. Bian, J. Williams, W. Liu, L. Li, L. Samuelson, J. Kumar and S. K. Tripathy, "Surface relief structures on azo polymer films. J. Mater. Chem., **9**, 1941–1955 (1999).

[47] C. J. Barrett, A. L. Natansohn and P. L. Rochon, "Mechanism of optically inscribed high-efficiency diffraction gratings in azo polymer films," J. Phys. Chem., **100**, 8836–8842 (1996).

[48] A. Ambrosio, L. Marrucci, F. Borbone, A. Roviello, P. Maddalena, "Light-induced spiral mass transport in azo-polymer films under vortex-beam illumination," Nat. Commun., **3**, 989 (2012).

[49] K. Harada, H. Inoue, M. A. El-Morsy, M. Itoh, S. Umegaki, T. Yatagai, "Holographic recording and control of diffraction efficiency using photoinduced surface deformation on azo-polymer films," Jpn. J. Appl. Phys., **41**, 1851–1854 (2002).

[50] G. J. Fang, J. E. Maclennan, Y. Yi, M. A. Glaser, M. Farrow, E. Korblova, D. M. Walba, T. E. Furtak, N. A. Clark, "Athermal photofluidization of glasses," Nat. Commun., **4**, 1521 (2013).

[51] D. Barada, G. Juman, I. Yoshida, K. Miyamoto, S. Kawata, S. Ohno, T. Omatsu, "Constructive spin-orbital angular momentum coupling can twist materials to create spiral structures in optical vortex illumination," Appl. Phys. Lett., **108**, 051108 (2016).

[52] A. Ashkin, J. M. Dziedzic, J. E. Bjorkholm, S. Chu, "Observation of a single-beam gradient force optical trap for dielectric particles," Opt. Lett., **11**(5), 288–290 (1986).

[53] Y. Harada, T. Asakura, "Radiation forces on a dielectric sphere in the Rayleigh scattering regime," Opt. Commun., **124**, 529–541 (1996).

[54] A. Ambrosio, P. Maddalena, L. Marrucci, "Molecular model for light driven spiral mass transport in azopolymer films," Phys. Rev. Lett., **110**, 146102 (2013).

[55] J. Bin, W. S. Oates, "A unified material description for light induced deformation in azobenzene polymers," Sci. Rep., **5**, 14654 (2015).

[56] L. Marrucci, "Soft matter and optical vortices: A good match for new science and technology," Mol. Cryst. Liq. Cryst. **595**(1), 9–20 (2014).

Chapter 5

Dynamical Particle Motions in Vortex Flows

Steven Wang and Naoto Ohmura

Additional information is available at the end of the chapter

http://dx.doi.org/10.5772/66315

Abstract

Circular vortex flows generate interesting self-organizing phenomena of particle motions, that is, particle clustering and classification phenomena. These phenomena result from interaction between vortex dynamics and relaxation of particle velocity due to drag. This chapter introduces particle clustering in stirred vessels and particle classification in Taylor vortex flow based on our previous research works. The first part of this chapter demonstrates and explains a third category of solid-liquid separation physics whereby particles spontaneously localize or cluster into small regions of fluids by taking the clustering phenomena in stirred vessels as an example. The second part of this chapter discusses particle classification phenomena due to shear-induced migration. Finally, this chapter discusses about process intensification utilizing these self-organizing phenomena of particle motions in vortex flows.

Keywords: solid-liquid flow, particle classification, particle clustering, chaotic-mixing field, process intensification

1. Introduction

Vortex motions play an important role for the enhancement of transport phenomena in chemical processes. Utilizing appropriate vortex motions can significantly enhance the energy efficiency of unit operations and make chemical processes more compact and safer. Some process intensification methods are, therefore, closely related to vortex dynamics. A continuous oscillatory-baffled reactor (OBR) proposed by Mackely and Ni [1], which first appeared in 1991, is one of successful novel tubular reactors that offers the prospect of a compact plug-flow reactor with uniform, controllable mixing. The continuous OBR comprises tubes fitted with equally spaced, low-constriction orifice plate baffles having an oscillatory motion superimposed upon the net flow of the process fluid [2]. The combination of the baffles and the oscillatory motion creates vortical flow patterns conductive to efficient heat and mass transfer, whilst maintaining plug flow. On the upstroke of the piston, toroidal vortices are formed above the baffle. The vortices dissipate on the following downstroke of the piston, as similar vortices are formed below the baffle. This constant creation and destruction of vortices results in intensifying local mixing. On the other hand, the baffles are obstacles to axial dispersion, which give a plug-flow residence time distribution. Harvey and Stonestreet [3] conducted a case study of an OBR for a batch saponification process. According to their estimation, the OBR can drastically reduce reactor volume from 75 m^3 of the current batch reactor to 0.5 m^3, and reaction time required can be reduced from 2 h to 12 min by intensifying local mixing.

A Taylor-Couette flow reactor (TCFR) is another typical example of plug-flow-type reactors. Appearance of pairs of counter-rotating Toroidal vortices called Taylor vortices due to hydrodynamic instability provides a minimal axial dispersion combined with a local mixing intensity in a TCFR. In this sense, a TCFR has the potential for an ideal plug-flow reactor with low shear stress [4], which is suitable for processes dealing with shear-sensitive materials such as crystallizations [5], emulsion polymerization [6], bioprocesses [7] and so on.

Circular vortex flows generate another interesting self-organizing phenomena of particle motions, that is, particle clustering and classification phenomena. These phenomena result from interaction between vortex dynamics and relaxation of particle velocity due to drag. As mentioned above, it has been considered that vortex motions contribute to intensification of mixing. When we do away with this fixed idea, however, we might have a chance to develop a novel process intensification method. This chapter introduces particle clustering in stirred vessels and particle classification in Taylor vortex flow based on our previous research works.

The first part of this chapter demonstrates and explains a third category of solid-liquid separation physics whereby particles spontaneously localize or cluster into small regions of fluids by taking the clustering phenomena in stirred vessels as an example. In the second part of this chapter, particle classification phenomena due to shear-induced migration are discussed. Finally, this chapter discussed about process intensification utilizing these self-organizing phenomena of particle motions in vortex flows in 'Conclusions' section.

2. Particle trapping in stirred flow

In this section, we aim to propose an unexpected category of solid-liquid separation physics where weakly buoyant particles can migrate into two specific regions of non-turbulent fluid in a mixing vessel. We apply a novel three-dimensional (3D) particle visualization method to expose the real-time trajectories of inertial particles, and the results suggest that particles could deviate from underlying flows, and subsequently form dense clusters. The new finding provides a sound basis for solid-liquid separation technologies.

2.1. Experimental set-up and measurements

Figure 1 shows the schematic of the stirred system, consisting of the overhead motor (Heidolph RZ2102), a Rushton turbine and a cylindrical Perspex vessel. A digital camera (Canon Legria HFR28) was used to take sequential images. In the flow visualization experiments, two 1000W

http://dx.doi.org/10.5772/66315

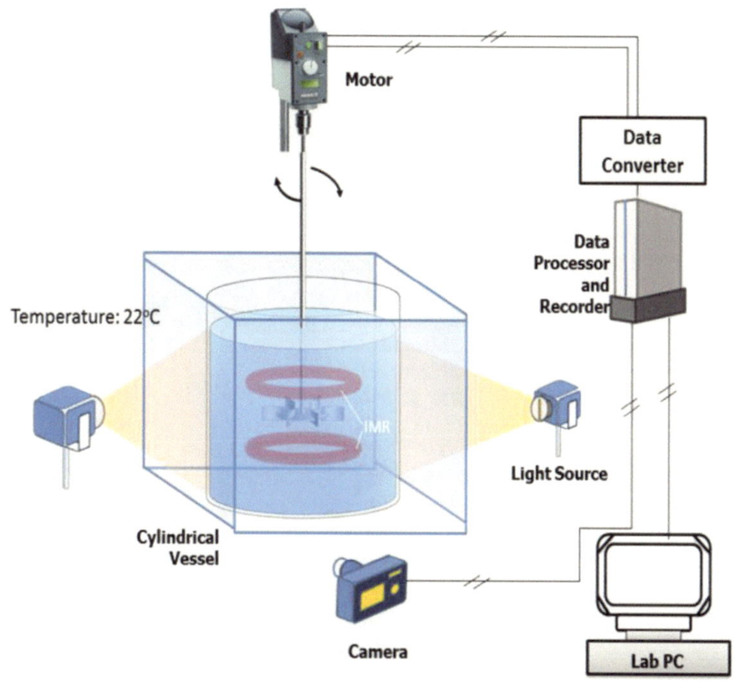

Figure 1. Experimental set-up used in this study [8].

Arri IP23 lamps were used [8]. The Rushton impeller is 7 cm in diameter and placed centrically in the vessel along a single shaft. The details of all other impellers are shown in **Figure 2**.

All experiments are completed at modest values of the Reynolds number *Re* (here $Re = \rho N D^2/\mu < 200$ where ρ is the fluid density, N is the rotating speed or in other words the rotating frequency having the dimension of [1/s], D is the impeller diameter and μ is the kinematic viscosity of the fluid.

We use a wide range of particles for the trapping experiments, and $0.35 \le \rho_p/\rho_f \le 1.19$, and $0.65 \le a \le 7$mm, where ρ_p is the particle density, ρ_f is the fluid density and a is the particle diameter. Particles used in the study include polystyrene, ion-exchange resin, polymethyl methacrylate (PMMA), urea resin, rubber particles and wooden particles. Glycerin is used as the working fluids, and is provided by APS Healthcare (Nat Oleo, 99.7% grade). A Bohlin CVO 50-controlled stress rheometer is used to measure the viscosity of the working liquid.

2.2. Passive particle motion

Multiphase phase mixing is an important process that takes place throughout the chemical, mineral, food and water industries. Homogenization is the ultimate goal of the mixing devices. In particular, mechanical agitation tanks have been the first mixing-device option for a number of processes essentially due to its versatility, comparative simplicity and effectiveness in providing fast mixing. While multiphase mixing is of highly industrial interests, we

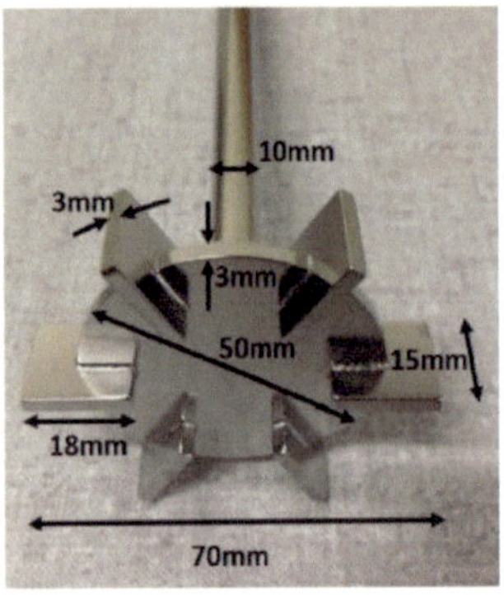

(a) Rushton impeller (0.07m diameter)

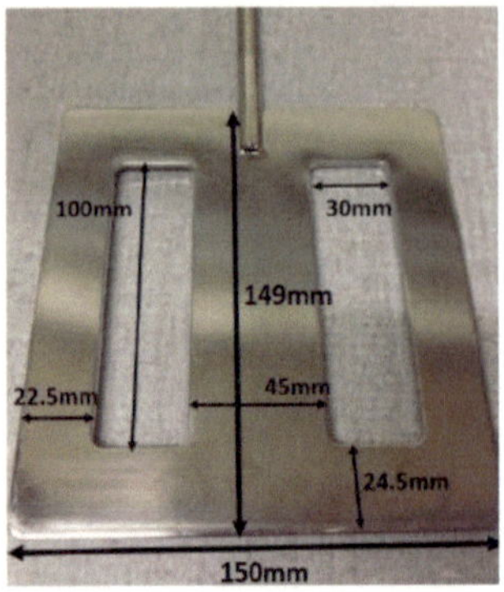

(b) Large impeller (0.15m diameter)

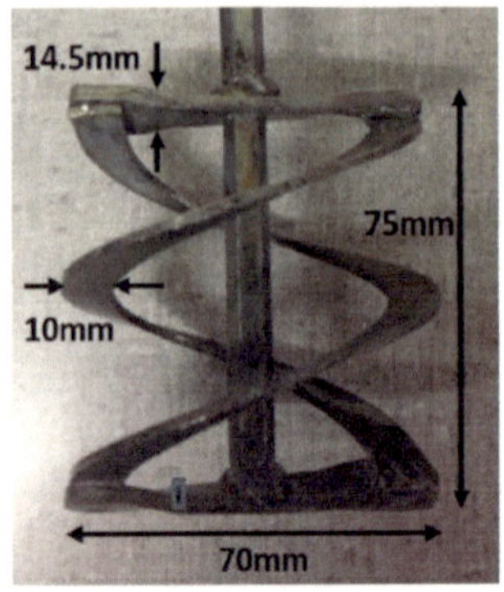

(c) Helical impeller (0.07m diameter)

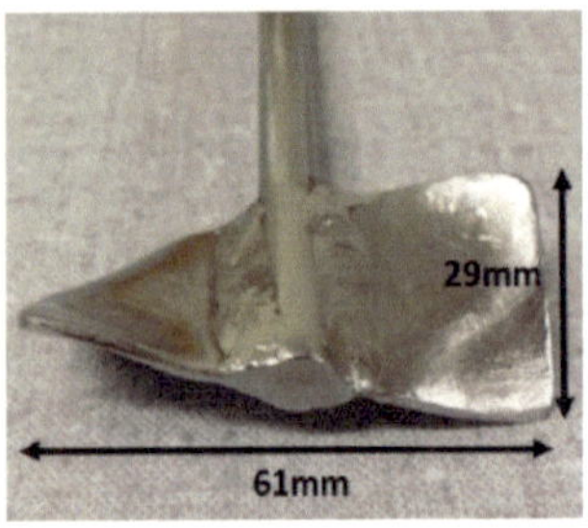

(d) 2 bladed impeller (0.061m diameter)

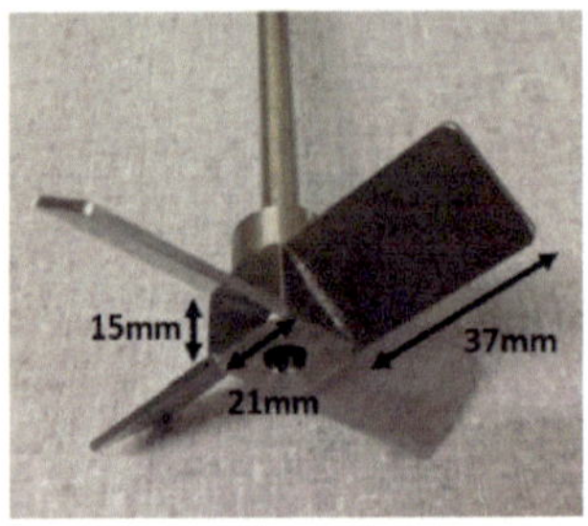

(e) 4 bladed impeller (0.095m diameter)

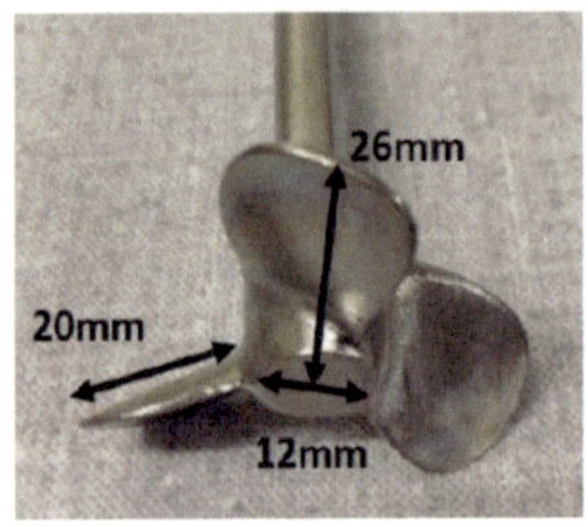

(f) Marine style impeller (0.052m diameter)

Figure 2. Impeller photographs and their dimensions. (a) Rushton impeller; (b) Large impeller; (c) Helical impeller; (d) Two bladed impeller; (e) Four bladed impeller; (f) Marine style impeller.

realize that the prior scientific contributions deal almost exclusively with single-phase fluid-fluid mixing. In this section, we disclose the typical flow mechanism of a `standard` stirred tank that is operated in a low *Re* regime.

Laminar stirred tank can produce the complex path lines via chaotic advection. The passing of blades causes the onset of chaos sea by bringing in small perturbations while producing two isolated mixing regions, which are called 'Kolmogorov-Arnold-Moser (KAM)' tubes in dynamical system. KAM tubes are the regions of confined mixed zones, segregated by well-defined boundary layer. At low Reynolds numbers, these KAM tubes are present in the form of toroidal vortices. From **Figure 3a**, we can clearly see the chaos sea (stretching and folding areas) as well as the vortex tube, by means of dye-visualization experiment. It is also clearly seen from this picture that the vortex structure is highly complex, consisting of island and filaments. It should be noted that there is no material/fluid exchange between the KAM tubs and the surrounding areas.

In **Figure 3a**, we can consider the dye particles as passive particles. In laminar flows, the Lagrangian motion of a passive particle at location $\mathbf{X}_p = (X, Y, Z)$ moving according to fluid velocity field $\mathbf{u} = (u, v, w)$ is mathematically captured by the following kinematic equations:

http://dx.doi.org/10.5772/66315

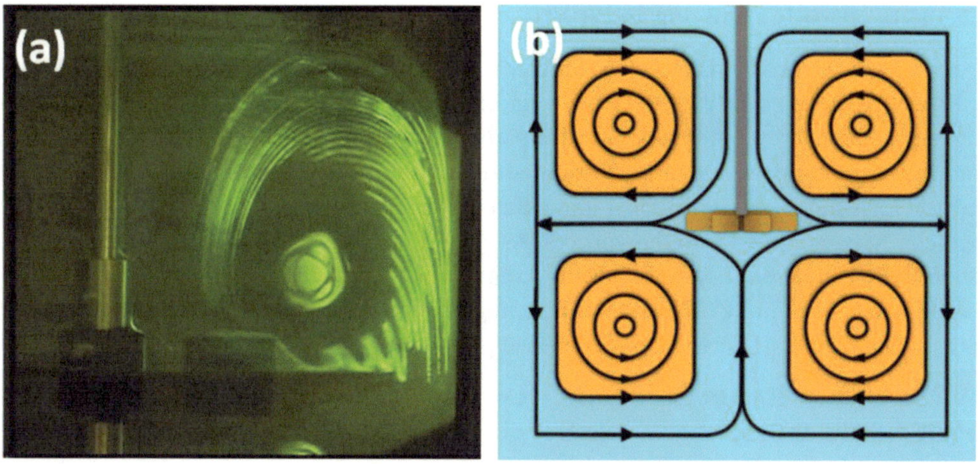

Figure 3. (a) Chaotic sea and a KAM tube in the laminar flow system. (b) Fluid particle system for (laminar) stirred tank at low *Re*; two tori present above and below the impeller. Yellow area: KAM tubes (low strain area); the rest of the system: chaotic flow area (high strain area) [8].

$$\begin{aligned} \frac{dX}{dt} &= u\left(\mathbf{X}_p, t\right), \\ \frac{dY}{dt} &= v\left(\mathbf{X}_p, t\right), \\ \frac{dZ}{dt} &= w\left(\mathbf{X}_p, t\right). \end{aligned} \tag{1}$$

The basic assumption in obtaining Eq. (1) is that the small particles do not deviate from the fluid velocities and do not alter the fluid velocities. For passive particles in this low-Reynolds-number flow condition, they always stick to the streamlines in the chaos sea, and do not move into the KAM tubes. The dye-visualization experiment can confirm that, when particles are tiny and spherical, particles move passively along with the respective streamlines. **Figure 3b** shows the conclusive fluid particle system that consists of two tori and surrounding chaotic flow regions.

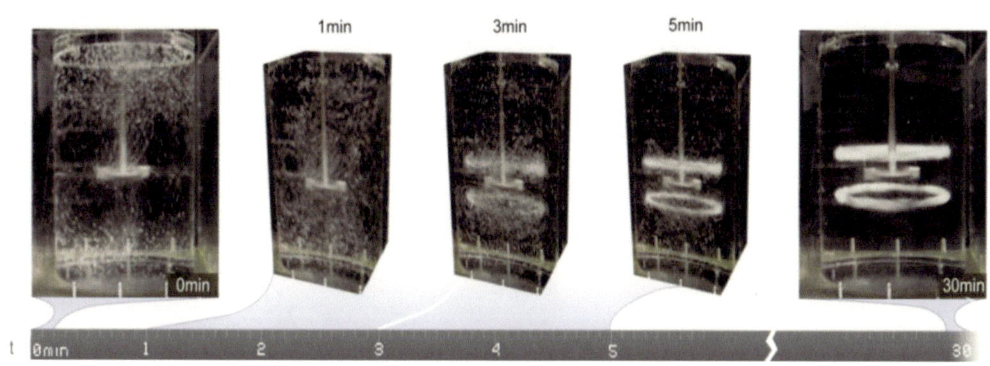

Figure 4. Particle trapping in a stirred vessel. $\rho_p/\rho_f = 0.80$, $a = 1.4$ mm [8].

2.3. Inertial particle clustering and criteria

Inertial particles behave completely different from passive ones. As illustrated in **Figure 4**, spherical particles (with $\rho_p/\rho_f = 0.80$, a = 1.4 mm) can deviate their respective streams, and move into the nearby KAM tubes. In this particular case, at Re = 133, particles were poured into the stirred tank and swiftly distributed throughout the system except the KAM tubes. Particle focusing takes pace almost immediately, and after a few minutes certain amount of particles have migrated into the vortices. As shown in **Figure 4**, two particle bands occupy 10–15% of the total fluid system, causing a highly inhomogeneous condition. This behaviour has been discussed in Ref. [8], and this trapping phenomenon was found for a number of different systems with different types of inertial particles.

We also conduct 3D particle trajectory velocimetry for a single finite-sized particle, and **Figure 5** shows the Lagrangian trajectory of a single one. We start the measurement at the moment when we introduce the particles into the system. The large dot is the particle position at $t = 0$. It is obvious that the particle follows a helical pattern through the impeller until it reaches the KAM tubes (as shown in **Figure 5a, c**). The particles move along the vortex tubes after they are 'sucked' into these specific regions (**Figure 5b**). We also calculate the residence time at the core of the vortex tube, and it is about 0.32 s [9].

Our theory is that when a particle is tiny, spherical and neutrally buoyant, it strictly follows the fluid streamline; by contrast, a large inertial particle can deviate the fluid flow essentially due to added inertia. The significance of size and density of the particle is captured by an inertia number

$$\sigma = \frac{St}{R} \text{ with } St = \frac{2}{9}\left(\frac{a}{L}\right)^2 Re, \quad R = \frac{1}{\frac{1}{2} + \frac{\rho_p}{\rho_L}} \tag{2}$$

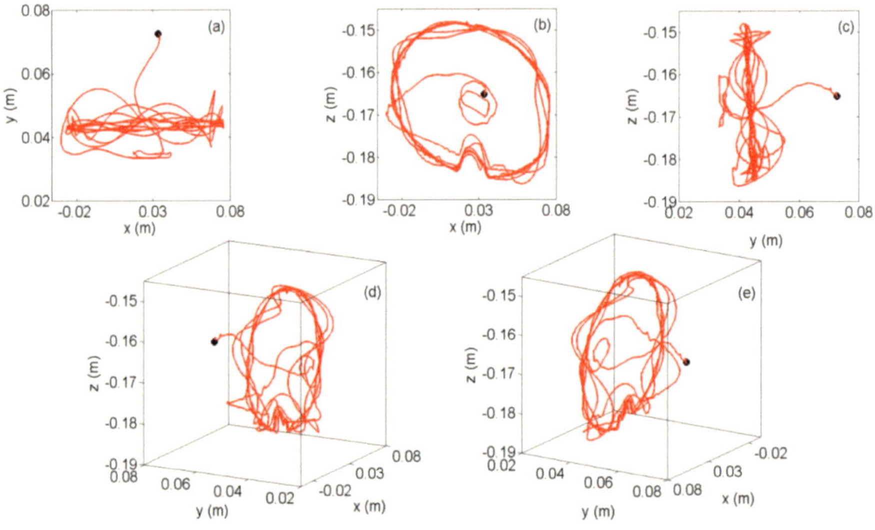

Figure 5. Tracking of Lagrangian positions of a single particle for a time interval of 7s from different views [9]. (a) x-y plane; (b) x-z plane; (c) y-z plane; (d)-(e) 3D trajectory

where a is the particle radius, L is the flow-length scale, St is the Stokes number, Re is the Reynolds number and R is the particle/fluid density ratio. The inertial factor σ is the ratio of the particle relaxation time and the typical time scale of the flow. σ is high, indicating that particle moves inertially and away from the underlying fluid streamline, while σ is smaller, suggesting that a particle moves passively. Sapsis and Haller derived a criterion for scattering regions (where particles do not follow their respective streamlines) in the flow in terms of the eigenvalue of a tensor. With further calculation, we have the scattering criteria:

$$\sigma\gamma > 1 \tag{3}$$

where $\dot{\gamma}$ is the local non-dimensional strain rate and $\sigma\dot{\gamma}$ is interpreted as inertial stress. We note that the strain along streamline is proportional to the radius of curvature of the streamline. A particle can either follow a fluid streamline if it curves less, while scatters from the streamline when the streamline curves too much. As Eq. (3) is a local criterion, we further obtain a global instability criterion by using the strain-Re relationship in the mixing tank. In laminar mixing tank, we have $\dot{\gamma} = kRe$, where $k = 5$ for the impeller we use in the study. The global flow instability criterion is

$$k\sigma Re > 1 \tag{4}$$

We then substitute $Re = \rho_f N L^2/\mu$ into Eq. (4), and then we can calculate the minimum speed $N_{\min}$ required to trigger the scattering and trapping phenomenon

$$N_{\min} = \frac{\mu}{aL\rho_f}\sqrt{\frac{9}{2}\frac{R}{k}} \times 60 \tag{5}$$

Figure 6a gives the minimum speeds required to trigger the focusing effect for particles with $\rho_p/\rho_f = 0.5$, 1 and 2, respectively, in the system. We can observe from **Figure 6a** that denser and smaller particles need higher speeds to trigger the instability.

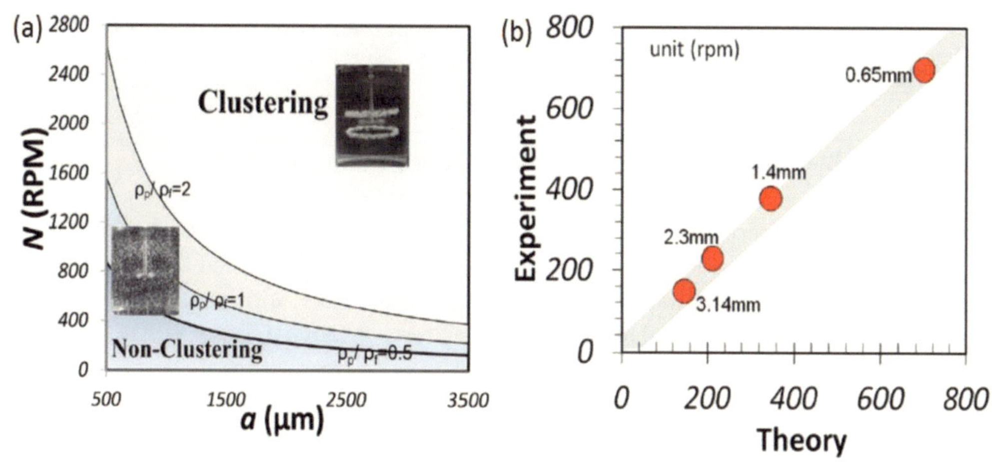

Figure 6. (a) Theory curves for critical speeds; (b) experiments versus theory.

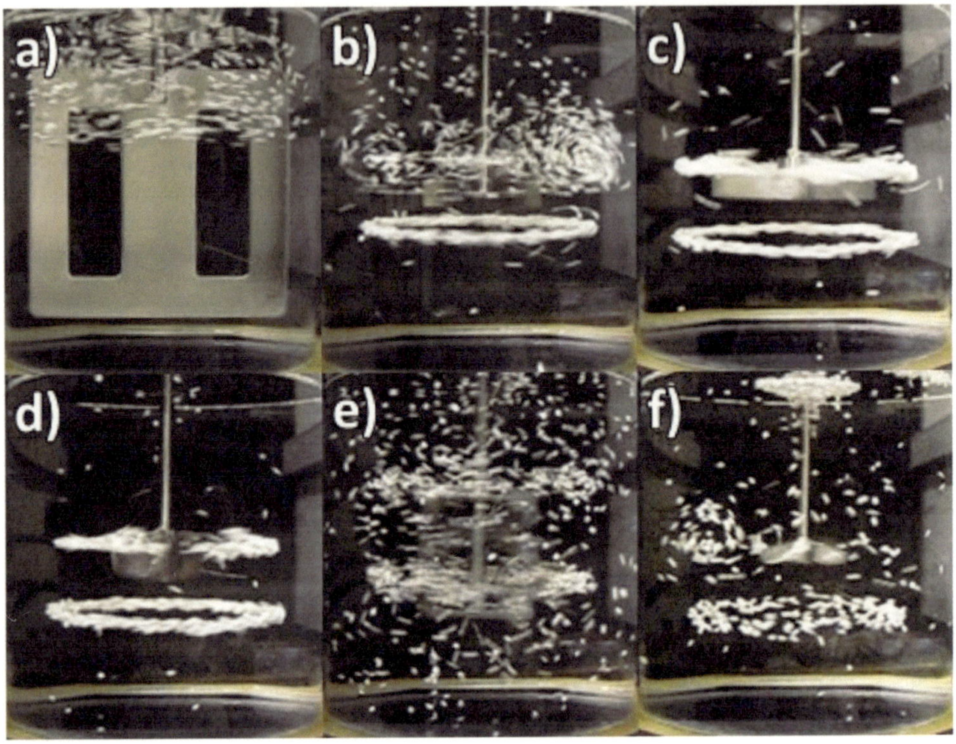

Figure 7. Photographs of the stirred vessel with particles after 15 min at highest speed for respective impeller. (a) Large impeller; (b) Rushton turbine; (c) Four-bladed impeller; (d) Two-bladed impeller; (e) Helical impeller; (f) Martin turbine.

The minimum clustering speed can be theoretically predicted from Eq. (5), and we compare the theory and experimental results in **Figure 6b**. It is remarkable that focusing of particles can occur when rotation speed reaches a predictable speed that relies on the particle and fluid properties. One more striking finding is that, we find that as long as the vortex tubes are present in the agitation system, clustering can take place in the mixing tank, irrespective of the impeller used for agitation. This is evident in **Figure 7**.

3. Particle classification phenomena in Taylor-Couette flow

Chemical and biochemical processes often have to separate, classify or disperse particles having various size and density. Wereley and Lueptow [10] numerically revealed that the motion of dilute, rigid and spherical particles with density greater than the fluid was determined by the interplay between the centrifugal force and the drag force in laminar Taylor vortex flow. They also found that particles approached a limit cycle orbit in *r-z* cross-sectional plane and the limit cycle orbit of particle depended on the density and size of particle. Furthermore, even neutrally buoyant particles did not follow fluid streamlines exactly. Leighton and Acrivos [11] demonstrated in their experiments using a Couette device that neutrally buoyant particles might migrate across streamlines and against concentration gradients when undergoing bounded shear flow. This chapter focuses on the particle classification phenomena in Taylor-Couette flow system that are discussed in the two cases of sedimenting and floating particles.

3.1. Experimental and numerical methods

Detailed information of experimental and numerical methods appears in our previous literature [12, 13]. **Figure 8** shows the experimental set-up with a measuring system. The concentric cylinder system consists of an inner cylinder of stainless steel and an outer cylinder of plexiglass in order to observe particle motions visually [12]. The outer diameter of the inner cylinder has 50 mm and the outer cylinder has 70 mm of the inner diameter. The length of cylinder system is 300 mm and the effective volume of the annular space is 550 cm^3. The critical Reynolds number for the present radius ratio (η = 0.714) can be estimated to be Re_C = 80.6 by the linear stability theory [14]. The working fluid is an aqueous solution of glycerin having the density, ρ_f = 1050–1210 kg/m^3. Each experimental observation was started after having passed enough time to form a stable vortex structure.

Fluorescent green dye was used as a passive tracer to visualize vortex structure. After adding the fluorescent green dye, the cross section of vortices was illuminated by a plane sheet of semiconductor laser light and the sequential visual data were taken by a digital video camera.

Polymethyl methacrylate (PMMA) particles suspended in the same aqueous solution of glycerol as the working fluid were fed into the top of the cylinder system at a very low flow rate whose axial Reynolds number (Re_{ax}) was 0.42. The particle size was initially distributed from 10 to 80 μm. Particle density, ρ_p, was distributed from 1200 to 1250 kg/m^3. Hence, the density ratio, $\beta = \rho_p/\rho_f$, was 1.04–1.05, which means the density of the particle was slightly heavier than the fluid. The inlet concentration of particles was 0.26 wt% and the suspended solution was withdrawn using a hypodermic needle at regular intervals at 30 mm above the bottom of the apparatus. The sample solution was taken at both near the outer wall and in the vortex core.

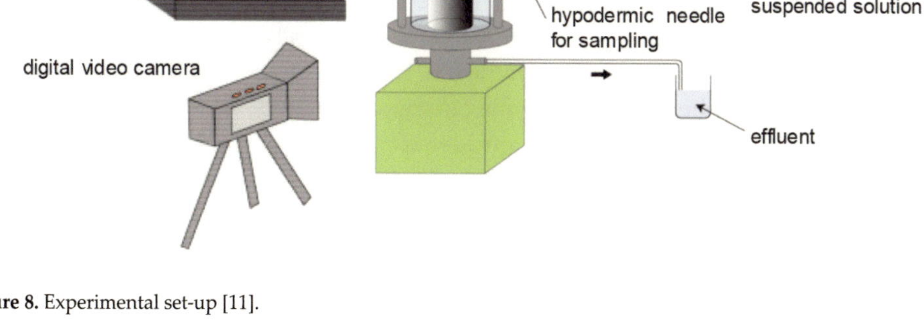

Figure 8. Experimental set-up [11].

The particle size distribution (PSD) of the sampled solution was obtained by measuring the size of particles with a microscope (KEYENCE, VHX-100N) on a number basis of percentage. About 200 particles were counted in the PSD measurements.

On the other hand, in the case of particle classification of floating particles, ion-exchanger acrylic resin particles (Amberlite, Organo Co.) were used [13]. The axial flow was not added in order to spatially fix axial classification patterns. They were sieved and classified into two groups with mean diameters of 710 and 974 μm, respectively. Their densities ρ_p were measured to be 1190 kg/m^3 from their buoyancy in an aqueous solution of glycerine (the density ratio, $\rho_p/\rho_f = 0.98$). In order to distinguish between these two groups, the smaller and larger particles were dyed with red and blue, respectively. Two grams of each group of particles was initially added to the fluid from the top of the cylinder system. After that, the inner cylinder was rotated at a specific rotational speed [13]. After a steady-state fluid flow was established, the behaviour of the particles was observed using a digital video camera (NP-F960, Sony Co.) and the RGB intensity was measured at a certain time by using image-editing software (ImagePro Plus, Nippon Roper Co.) to characterize particle dispersion [13].

In the numerical simulation, a commercial computer fluid dynamics (CFD) code RFLOW (Rflow Co., Ltd.) was used for the simulation of the motion of particles. This numerical code is based on a finite-volume method and is enabled to simulate particle motion by a particle-tracking method and a (DEM) [13]. Numerical simulations were conducted under the same conditions as those used in the experiment, except the particle size and volumetric fraction of particles. Our works were conducted by using two different numerical simulation methods, namely a particle-tracking method and a DEM. For more detailed information about numerical simulation methods, refer to our previous works [12, 13].

3.2. Results and discussion

Figure 9 shows cross-sectional views presenting axial diffusion process of tracer with measuring time from the injection of green dye. The Reynolds number and the axial Reynolds numbers were 200 and 0.42, respectively. As shown in **Figure 9**, the flow-visualization experiment using a laser-induced fluorescence method clearly revealed that there exist two distinct mixing regions in laminar Taylor vortex flow. The tracer near the vortex cell boundary was rapidly transported downward in the axial direction owing to the bypass flow effect. On the other hand, the fluid element was confined to the vortex core region without being exchanged with the outer flow region. In the case of PMMA particles whose density is slightly heavier than the working fluid, the centrifugal force or the shear-induced particle migration due to the rotation of inner cylinder displaces particles towards the wall of the outer cylinder and the larger particles tend to be located more on the outside in the vortex cell. Owing to the effect of the bypass flow, the larger particles in the outermost region of vortex cells more quickly disperse in the axial direction. Hence, we considered that particles can be classified by taking advantage of the bypass flow effect. **Figure 10** shows diagrams of PSD in the sampled solution and photographs of the particles corresponding to the PSDs [12]. Initially, the particle size distribution is relatively homogeneous in the range from 10 to 80 μm. The population of particles smaller than 40 μm is 52%, while that of particles larger than 40 μm is 48%. Particles smaller than 50 μm are dominant in the vortex core region at 42 min after feeding the suspended

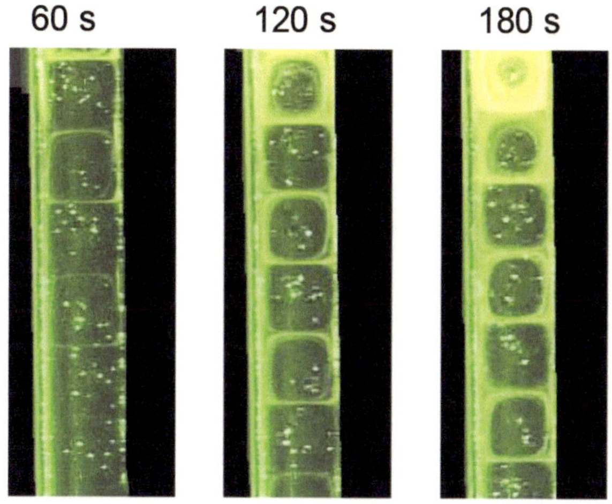

Figure 9. Cross-sectional views of axial diffusion of tracer green dye ($Re = 200$ and $Re_{ax} = 0.42$) [12].

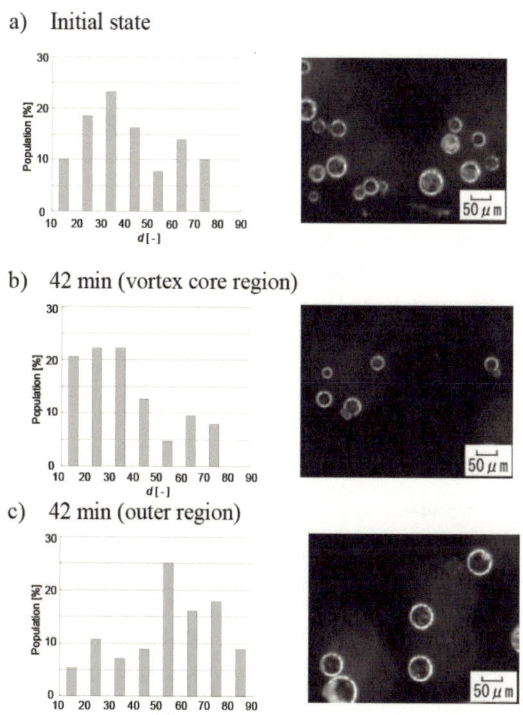

Figure 10. Particle size distribution and photographs [12]. a) Initial state; b) 42 min (vortex core); c) 42 min (outer region).

solution. The population of particles smaller than 50 μm is more than 75%. On the other hand, particles larger than 50 μm are dominant in the outer region of vortex and the population of particles larger than 50 μm almost reaches 70%. It can be, therefore, considered that the particles are classified by the size of 50 μm under this experimental condition. **Figure 11** shows

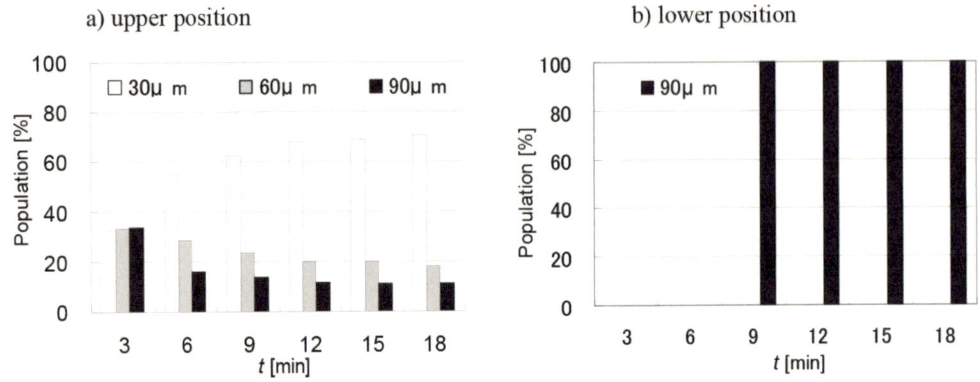

Figure 11. Time variation of PSD obtained by numerical simulation [12]. a) Upper position; b) Lower position.

the time variation of PSD numerically obtained in the upper portion (0–100 mm from the top) and the lower portion (200–300 mm from the top), respectively [12]. In the upper portion, the population of the smallest particles (30 μm) increases with time and reaches about 65% after 18 min. On the other hand, in the time variation of PSD obtained in the lower portion, only the largest particles can be seen even after 18 min. These numerical results suggest that particles can be classified and collected by installing several outlet ports in the axial direction even when the density difference is very small [12].

Figures 12 and **13** show spatially fixed axial particle dispersion patterns and their corresponding RGB intensity variation in the axial direction in the case of floating particles [13]. **Figure 12** indicates that the particles penetrate deeper into the Taylor vortex flow region with increasing Reynolds number. In **Figure 12(a)**, the coloured region is classified into two layers and the first layer from the top dominantly contains blue particles (larger particles), and the second

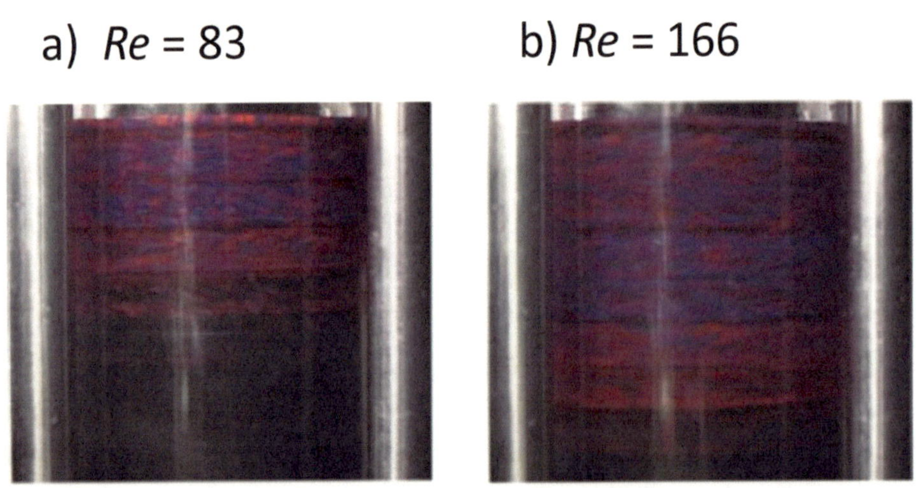

Figure 12. Front views of particle dispersion [13]. a) *Re* = 83; b) *Re* = 166.

http://dx.doi.org/10.5772/66315

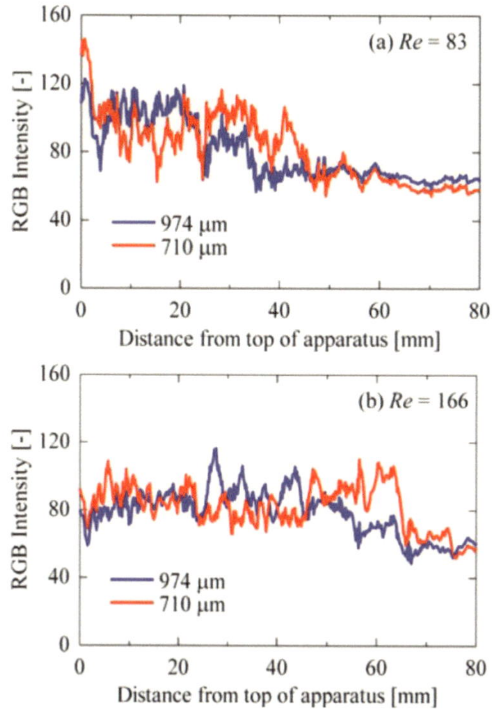

Figure 13. RGB intensity against axial distance from the top of apparatus [13]. (a) *Re* = 83; (b) *Re* = 166.

layer dominantly contains red ones (smaller particles). On the other hand, in **Figure 12(b)**, the coloured region is classified into three layers. The first layer seems to have blue and red particles evenly. The second layer contains more blue particles and the third layer contains more red ones.

These images show that the red and blue particles were segregated. To clarify this behaviour, the results of a qualitative evaluation by variation of RGB intensity are shown in **Figure 13** [13]. For *Re* = 83, the high-intensity region for blue lies between 10 and 30 mm, while that for red lies between 30 and 50 mm. *Re* increases to 166, the high-intensity region for blue shifts downward to about 30–50 mm, while that for red shifts to about 50–70 mm [13]. However, the width of the high-intensity region for each colour remains constant at about 20 mm, irrespective of the rotational speed [13]. As the gap between the inner and outer cylinders is 10 mm, the vertical vortex cell length, which corresponds to half of the axial wavelength, is about 10 mm. Hence, 20 mm of the axial length corresponds to the length of a pair of two counter-rotating vortices. Flow visualization also supported that these regions correspond to pairs of two counter-rotating vortices surrounded by inflow cell boundaries. These results indicate that particles of the same size tend to aggregate within a certain pair of vortices and smaller particles can penetrate deeper into the Taylor vortex flow region than larger particles.

Figure 14 shows the *r-z* cross-sectional plane of particle orbits which was numerically obtained for each size [13]. Contrary to the previous case of sedimenting particles, the smaller

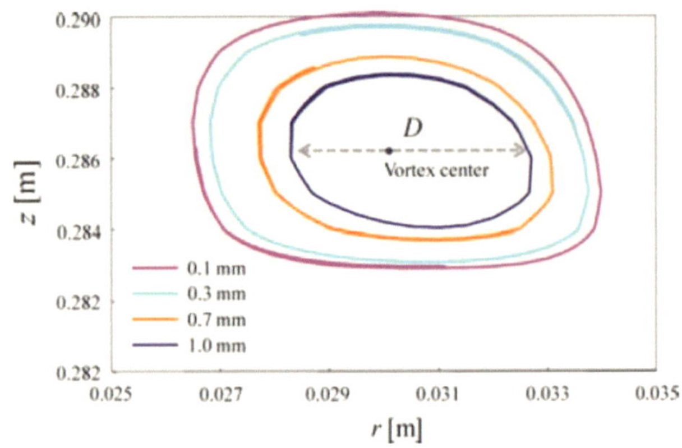

Figure 14. *r*-*z* cross sections of particle orbits in the particle-tracking method at *Re* = 207.5 for four particles of different sizes [13].

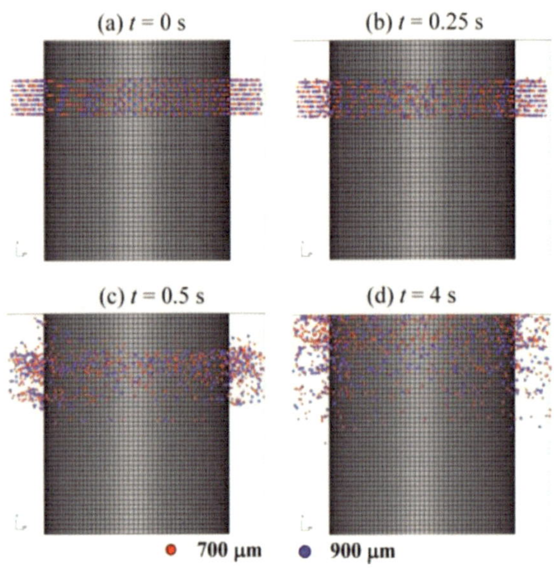

Figure 15. Particle diffusion process obtained by the DEM (*Re* = 166) (axial distance range: 0.24–0.30 m) [13]. (a) $t = 0$ s; (b) $t = 0.25$ s; (c) $t = 0.5$ s; (d) $t = 4$ s

particles move along the outer orbit on a torus. This result indicates that small particles may be exchanged between vortices owing to a bypass flow that exists in the outer region of each vortex as previously mentioned. However, the effect of bypass flow in the case without axial flow is smaller than in the case with axial flow. The DEM revealed two possible mechanisms for axial transport of particles in the axial direction. One is that the axial particle dispersion can result from the unsteadiness due to vortex development from the rest as shown in **Figure 15** [13]. The other is that particle collision can induce axial particle transportation. **Figures 16** and **17** show drag force acting on a particle and correlation between interparticle collision

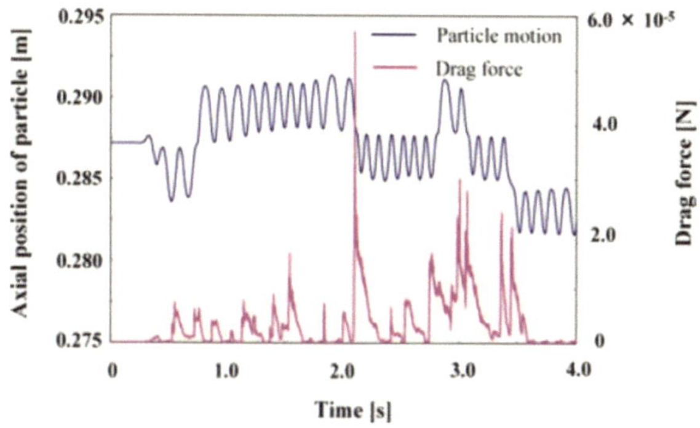

Figure 16. Effect of drag force acting on a particle at different axial position [13].

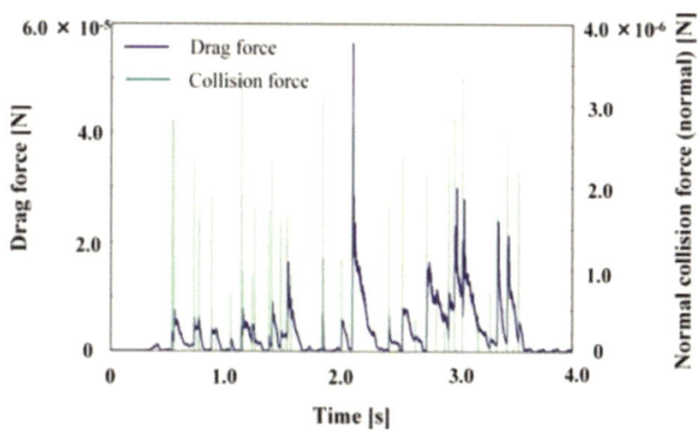

Figure 17. Correlation between interparticle collision and drag force [13].

and drag force [13]. In these figures, a particle is arbitrarily chosen by the DEM simulation. **Figure 16** shows the relation between the axial position of a 700-μm particle and the drag force acting on the particle. It has been found that a large drag force is generated during the motion of the particle over a vortex cell boundary, especially during motion towards a lower vortex. The relation between the drag force acting on the particle and interparticle collision is shown in **Figure 17**. The peaks corresponding to normal collision force indicate that the particle collides with another particle. As seen in **Figure 17**, the drag force acts on the particle just after collision in almost all the cases. Therefore, particle transfer across a vortex cell boundary is inferred to occur at the moment of a particle collision.

4. Conclusions

This chapter introduced particle clustering in stirred vessels and particle classification in Taylor vortex flow based on our previous research works [8, 12, 13]. For particleclustering phenomena, to the authors' best knowledge, no universal agreement has been made on the tendency of inertial particle motion in laminar flow system. In this study, we found that inertial particles can move away from their streamlines in the chaotic flow regions and subsequently migrate into the vortex tubes. The chaos-vortex flow structure can be easily triggered in many stirred systems; we thus expect that this trapping phenomenon can be encountered in a number of systems. For particle classification in Taylor vortex flow, on the other hand, particles having various sizes can be classified by taking account of the vortex motion and the bypass flow near the outer edge of vortices. In this case, the centrifugal force and the shear-induced migration due to the rotation of inner cylinder play the important role for moving larger particles more outside in the vortex cell.

From the viewpoints of process intensification, the following applications may be considered. Firstly, clustering particles can easily eliminate isolated mixing regions, which are an obstacle for global mixing in a stirred vessel [15, 16]. The advantage of this method is simpler and less reliant on energy input, because shorter mixing time can be obtained at very low Reynolds number even without inserting baffles. Secondly, Wang et al. [8] proposed a novel solid-liquid separation technique using a conventional mixing vessel, as shown in **Figure 18**. The conceptual design can find utility in a number of applications including the production of celluloid, fat and oils, cellulose ethers, gelatin and starch. To achieve more effective separations, stirred separators should be connected either in series or in parallel to maximize the throughput [8].

As for particle classification phenomena, an integrated process with granulation and classification can be considered. In fact, Kim et al. [17] proposed a novel crystallizer which simultaneously classifies the crystals utilizing vortex moving. Larger particles are moving

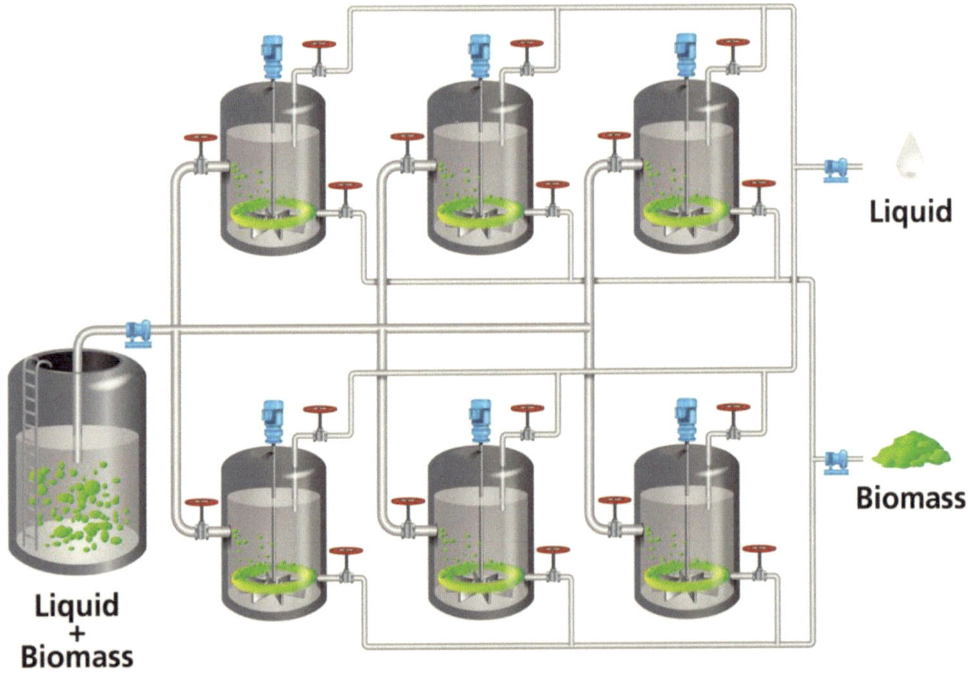

Figure 18. A conceptual design for biomass-liquid separation applied in biorefinery [8].

downward and smaller particles are moving upward. Small and large particles are obtained separately from the top outlet and bottom outlet, respectively. The flow characteristics of Taylor vortex flow are quite similar to the Dean vortex flow. Ookawara et al. [18] proposed a micro-separator/classifier having a curved microchannel with a rectangular cross section. This micro-separator/classifier takes advantages of the effects of centrifugal force and Dean vortices generated at the curve channel. This clearly indicates that the particle classification phenomena can be applied to micro-flow processes.

Author details

Steven Wang[1] and Naoto Ohmura[2*]

*Address all correspondence to: ohmura@kobe-u.ac.jp

1 Department of Chemical Engineering and Advanced Materials, Newcastle University, Newcastle upon Tyne, United Kingdom

2 Department of Chemical Science and Engineering, Kobe University, Kobe, Japan

References

[1] Mackely M. R., Ni. X. Mixing and dispersion in a baffled tube for steady laminar and pulsatile flow. Chemical Engineering Science. 1991;**46**(12):3139–3151. DOI: 10.1016/0009-2509(91)85017-R

[2] Reay D., Ramshaw C., Harvey A. Process Intensification. 1st ed. Oxford, UK: Butterworth-Heinemann; 2008. 444 p.

[3] Harvey A. P., Stonestreet P. A mixing-based design methodology for continuous oscillatory baffled reactors. Chemical Engineering Research and Design. 2002;**80**(1):31–44. DOI: 10.1205/026387602753393204

[4] Kataoka K., Doi H., Hongo T., Futagawa M. Ideal plug-flow properties of Taylor vortex flow. Journal of Chemical Engineering of Japan. 1975;**8**(6):472–476. DOI: 10.1252/jcej.8.472

[5] Jung W. M., Kang S. H., Kim W-S., Choi C. K. Particle morphology of calcium carbonate precipitated by gas–liquid reaction in a Couette–Taylor reactor. Chemical Engineering Science. 2000;**55**(4):733–747. DOI: 10.1016/S0009-2509(99)00395-4

[6] Kataoka K., Ohmura N., Kouzu M., Simamura Y., Okubo M. Emulsion polymerization of styrene in a continuous Taylor vortex flow reactor. Chemical Engineering Science. 1995;**50**(9):1409–1416. DOI: 10.1016/0009-2509(94)00515-S

[7] Giordano R. L. C., Giordano R. C., Cooney C. L.. Performance of a continuous Taylor–Couette–Poiseuille vortex flow enzymic reactor with suspended particles. Process Biochemistry. 2000;**35**(10):1093–1101. DOI: 10.1016/S0032-9592(00)00143-6

[8] Wang S., Metcalfe G., Stewart R. L., Wu J., Ohmura N.Feng X. et al.. Solid-liquid separation by particle-flow-instability. Energy & Environmental Science. 2014;**7**(12):3982–3988. DOI: 10.1039/C4EE02841D

[9] Gulan U., Holzner M., Jiang M., Windhab E., Wang S. Visualization of particle focusing in chaotic flow. Journal of Visualized Experiments, in press, 2016

[10] Wereley S. T., Lueptow R. M. Velocity field for Taylor–Couette flow with an axial flow. Physics of Fluids. 1999;**11**(12):3637–3649. DOI: 10.1063/1.870228

[11] Leighton D., Acrivos A. The shear-induced migration of particles in concentrated suspensions. Journal of Fluid Mechanics. 1987;**181**:415–439. DOI: 10.1017/S0022112087002155

[12] Ohmura N., Suemasu T., Asamura Y. Particle classification in Taylor vortex flow with an axial flow. Journal of Physics: Conference Series. 2005;**14**:64–71. DOI: 10.1088/1742-6596/14/1/009, Suemasu

[13] Saomoto K., Horie T., Kumagai N., TakigawaT., Noui-Mehidi M. N., Ohmura N.. Dispersion of floating particles in a Taylor vortex flow reactor. Journal of Chemical Engineering of Japan. 2010;**43**(4):319–325. DOI: 10.1252/jcej.09We07

[14] Di Prima R. C., Swinney H. L. Instabilities and transition in flow between concentric rotating cylinders. In: Swinney H. L., Gollub J. P., editors. Hydrodynamic Instabilities and the Transition to Turbulence. 1st ed. Berlin Heidelberg, Germany: Springer-Verlag; 1981. pp. 139–180.

[15] Alatengtuya, Nishioka N., Horie T., Noui-Mehidi M. N., Ohmura N. Effect of particle motion in isolated mixing regions on mixing in stirred vessel. Journal of Chemical Engineering of Japan. 2009;**42**(10):459–463. DOI: 10.1252/jcej.09we087

[16] Takahashi K., Motoba M. Chaotic mixing created by object inserted in a vessel agitated by an impeller. Chemical Engineering research and Design. 2009;**87**(4):386–390. DOI: 10.1016/j.cherd.2009.01.003

[17] Kim S. J., Kim D. H., Gu b., Kim, D. Y., Yang D. R.. Simulation of Taylor–Couette reactor for particle classification using CFD. Journal of Crystal Growth. 2013;**373**:106–110. DOI: 10.1016/j.jcrysgro.2012.12.006

[18] Ookawara S., Higashi R., Dtreet D., Ogawa K. Feasibility study on concentration of slurry and classification of contained particles by microchannel. Chemical Engineering Journal. 2004;**101**(1–3):171–178. DOI: 10.1016/j.cej.2003.11.008

Chapter 6

Numerical Simulation of Vortex-Dominated Flows Using the Penalized VIC Method

Seung-Jae Lee

Additional information is available at the end of the chapter

http://dx.doi.org/10.5772/65371

Abstract

Vorticity plays a key role in determining fluid flow dynamics, especially in vortex-dominated flows. Vortex methods, which are based on the vorticity-based formulation of the Navier-Stokes equations, have provided deeper insight into physical reality in a variety of flows using vorticity as a primary variable. The penalized vortex-in-cell (VIC) method is a state-of-the-art variant of vortex methods. In the penalized VIC method, Lagrangian fluid particles are traced by continuously updating their position and strength from solutions at an Eulerian grid. This hybrid method retains beneficial features of pure Lagrangian and Eulerian methods. It offers an efficient and effective way to simulate unsteady viscous flows, thereby enabling application to a wider range of problems in flows. This article presents the fundamentals of the penalized VIC method and its implementations.

Keywords: vorticity, unsteady viscous flow, pressure, vortex-in-cell method, penalization method, vorticity transport equation

1. Introduction

Vorticity is an important derived variable playing both mathematical and physical roles in the solution and understanding of problems [1], irrespective of whether the flow is laminar or turbulent. Vortex methods (VMs) utilize such a vorticity as a main variable of fluid dynamics. VMs are essentially a grid-free approach in which Lagrangian fluid particles are used as basic computational elements. The computational domain is discretized into a set of N fluid particles, which carry time-evolving vorticity.

VMs are based on the vorticity-based formulation of the Navier-Stokes equations; namely, the vorticity transport equation (VTE). There are several advantages in using the vorticity-based formulation. Firstly, it allows a purely kinematical problem to be decoupled from the pressure term. Thereby, the need for solving the pressure is eliminated. It means that pressure computation is not part of the solution procedure. If the pressure field is desired, it can be obtained in an explicit manner using the resolved vorticity and velocity fields. Secondly, the vorticity-based formulation does not contain any frame-dependent terms (for detailed information, refer to [2]). In rotating frame, for example, VTE is only altered by the addition of a constant forcing function in the form of 2 Ω_ω where Ω_ω is a time-dependent angular velocity. In the primitive-variable formulation, on the other hand, the mathematical character of the equations changes depending on whether or not the reference frame is inertial. Gresho [3] offers more comprehensive information on the fundamental formulations for incompressible flow simulations. Finally, the vorticity-based formulation is inherently able to better resolve finer scales because vorticity is one order higher than velocity. Also, computations are performed where vorticity-carrying particles exist, thus permitting a substantial reduction in the computational domain.

VMs have a history as old as finite differences. Many inherent benefits were discussed in comprehensive reviews of [4–8]. Admittedly, the primitive-variable and vorticity-based formulations, each have their own advantages and disadvantages. Their numerical characteristics cannot be a measure of whether one formulation is better than another. The choice of the numerical methodology is a matter of intended application. However, it is obvious that VMs have been an attractive alternative to conventional numerical methods based on the primitive-variable formulation, especially in terms of vortex-dominated flow simulations.

VMs track numerically the evolution of vortices in unsteady flow. Their most serious difficulty comes from a computational time of particle-particle interactions. Particles' velocity computed by the interaction can be quantified according to the familiar Biot-Savart law. This requires $O(N^2)$ operations, which is expensive for large N. A reduction in the computational complexity has been achieved using the fast multipole method (FMM) with hierarchical data structures, which have had a giant impact in computations using a Lagrangian framework. FMM, which was first introduced by Greengard and Rokhlin [9], is one of the ten algorithms with the greatest influence of the development and practice of science and engineering in the twentieth century [10]. FMM and its variants significantly reduce the cost of computing the interaction with N particles; namely, $O(N^2)$ to $O(NlogN)$ or $O(N)$ operations (refer to [11] for a short review).

Another approach for speed up in computation of particle velocities is the vortex-in-cell (VIC) method, which was originally developed by Christiansen [12]. This numerical method evaluates a velocity field, instead of the velocity of individual particles. It involves interpolation of physical quantities (vorticity and velocity) between Lagrangian particles and an Eulerian grid. Hence, the VIC method is often referred to as a hybrid method. In the VIC method, velocity field is computed by solving a Poisson equation, $\nabla^2\psi = -\omega$ or $\nabla^2 u = -\nabla \times \omega$. A fast Poisson solver accelerates the computation sufficiently. For example, a fast Fourier transform (FFT)-based solver reduces an operation count to $O(MlogM)$, where M is the number of grid points. Although the amount of computation is problem dependent, the VIC method is known to be faster than FMM to simulate unsteady viscous flows (refer to [13]).

For treating vorticity diffusion problem in VMs, several different techniques were developed [14]; the core spreading method [15], the random-walk method [16], the particle strength exchange method [17], the deterministic method [18], and the vortex redistribution method [19]. Interestingly, Graham [20] successfully solved the diffusion term of VTE using an Eulerian grid instead of complicated Lagrangian operators (also refer to [21–23]). This means that in the VIC method both convection and diffusion parts can be evaluated on a regular grid. Another important point of concern about VMs is boundary conditions for wall-bounded flows. In recent years, the VIC method has been successfully combined with the penalization method [24], which is useful to enforce boundary conditions at a solid wall (for example, refer to [25–33]). The combination, hereafter called the penalized vortex-in-cell (pVIC) method, greatly simplifies an entire numerical procedure required to construct a solution algorithm, thus leading to a well-performing parallel program. As a consequence, the evolution of both velocity and vorticity fields are evaluated on an Eulerian grid, and then the values for Lagrangian fluid particles are determined from the grid solution using an interpolation scheme. This offers an efficient and effective way to simulate unsteady viscous flows.

The pVIC method to simulate unsteady viscous flows will be discussed further in the following sections. In Section 2, the basic formulations and numerical methods are described. Section 3 describes implementations of the numerical methods and post-procedures for obtaining pressure and force from numerical solutions. Section 4 introduces improved implementations to further efficiently simulate a flow. Section 5 presents numerical results and discussions to validate the presented numerical method. A summary and conclusion is included in Section 6.

2. Basic formulation and numerical methods

For an incompressible unsteady flow of a viscous fluid, conventional VTE is expressed as follows

$$\frac{\partial \omega}{\partial t} + u \cdot \nabla \omega = \left(\omega \cdot \nabla\right) u + \nu \nabla^2 \omega \tag{1}$$

where ν is the kinematic viscosity. The evolution of a flow can be evaluated in a fractional step manner. The algorithm of viscous splitting consists of sub-steps in which the convective and diffusive effects are considered. It is thus expressed in a Lagrangian frame as

$$\frac{Dx}{Dt} = u, \frac{D\omega}{Dt} = 0 \tag{2}$$

- Convection part
- Diffusion part

$$\frac{Dx}{Dt} = 0, \frac{D\omega}{Dt} = (\omega \cdot \nabla)u + \nu\nabla^2\omega \tag{3}$$

where $D/Dt = \partial/\partial t + u \cdot \nabla$ is the material derivative. N discrete fluid particles are linearly superposed to approximate the vorticity field as

$$\omega(x,t) = \sum_p^N \Gamma_p \zeta(x - x_p) \tag{4}$$

where ζ is a mollification of the Dirac-delta function. Each particle is characterized by its position x_p and strength Γ_p, *i.e.* a circulation $\Gamma_p = \int \omega dV \cong \omega_p V_p$ where V_p is a volume (an area in 2D) occupied by a fluid particle.

In the pVIC method, the velocity of the particles and vorticity evolution are evaluated on a uniform grid. For doing that, the particles' own vorticity is first transferred onto the grid by

$$\omega(x_g,t) = \frac{1}{h^3}\sum_p^N \Gamma_p M_4'\left(\frac{x_g - x_p}{h}\right) \tag{5}$$

where h is the grid spacing. The function M_4' is the third order interpolation kernel [34] defined as

$$M_4'(x) = \begin{cases} 0 \quad for|x| > 2 \\ \frac{1}{2}(2-|x|)^2(1-|x|) \quad for 1 < |x| \le 2 \\ 1 - \frac{5}{2}|x|^2 + \frac{3}{2}|x|^3 \quad for|x| < 1 \end{cases} \tag{6}$$

in each coordinate direction. This kernel conserves the 0th, 1st, and 2nd order moments. Vorticity of a single fluid particle is transferred to the nearest 16 grid nodes in 2D and 64 grid nodes in 3D. The resultant vorticity at grid nodes is obtained by summing the contribution of all the particles.

2.1. Convection via vortex-in-cell (VIC) method

In the pVIC method, a velocity vector can be expressed as

$$u = U_\infty + u_\omega = U_\infty + \nabla \times \psi \tag{7}$$

where U_∞ is a free stream velocity and u_ω represents a rotational velocity. ψ is called as a vector potential in 3D and a stream function in 2D. Taking the curl of Eq. (7) yields the Poisson equation $\nabla^2\psi = -\omega$ linking vorticity to the vector potential and in turn to the velocity. ψ is computed on a uniform grid using a FFT-based solver. The grid for solving the Poisson equation can define a compact computational domain with non-homogeneous Dirichlet boundary conditions. The boundary condition of a domain Ω can be approximated using a Green's function approach [28], which is given by

$$\psi_b = \frac{1}{4\pi}\int_\Omega \frac{\omega_p}{\left|x_p - x_b\right|} dV = \frac{1}{4\pi}\sum_p^N \frac{\Gamma_p}{\left|x_p - x_b\right|} \tag{8}$$

where $x_p \in \Omega$, $x_b \in \partial\Omega$, and $x_p \neq x_b$. In 2D, it follows that

$$\psi_b = \frac{-1}{2\pi}\sum_p^N \Gamma_p ln\left(\left|x_p - x_b\right|\right) \tag{9}$$

Finally, rotational velocities on the nodes of the grid are computed from the definition $u_\omega = \nabla \times \psi$ using a finite difference scheme.

In 2D, two unknowns (two components of a velocity vector) are reduced to a single unknown ψ_z (the scalar stream function) without any loss of generality. There are no problems with the continuity equation; the velocity field is automatically divergence-free. In 3D, however, a total of six unknowns are required to be solved instead of the usual four of the primitive-variable formulation. Unlike the stream function in 2D, the vector potential in 3D is far from being uniquely defined. The 3D formulation requires that the vorticity, velocity, and vector potential are all divergence-free; $\nabla \cdot \omega = 0$, $\nabla \cdot u = 0$, and $\nabla \cdot \psi = 0$. This is a new difficulty for 3D flow simulation. Among these three conditions, the second is satisfied by letting $u = \nabla \times \psi$ and the third is a consequence of the first. To enforce the first, the well-known projection scheme can be applied. The divergence-free of a vorticity field is accomplished by $\omega - \nabla F$, where F is a scalar field which is obtained by solving $\nabla^2 F = \nabla \cdot \omega$.

2.2. Diffusion via penalization method

The penalization method has been used to take into account a solid body immersed in a flow. Among fictitious domain methods, the penalization method is very easy to implement, robust and efficient. This consists only in adding a penalty term in the momentum equations to represent a solid body. For example, the L^2 penalization consists in adding a damping term on the velocity in the Navier-Stokes equations [24]. The penalty velocity satisfies Darcy's law in terms of a Neumann boundary condition on the pressure. Angot et al. [24] rigorously showed that the penalized equations can be used with confidence since penalty error is always negligible in the face of the error of approximation. In the penalization method, a solid body is regarded as a porous medium where the permeability is infinite in the fluid part and tends to zero in the solid part. Both the solid and fluid domains are solved directly without addition of any boundary condition at a solid wall. The penalization method can replace complicated algorithms to implement boundary conditions at a solid wall.

In the pVIC method, a penalty term is added into VTE as follows

$$\frac{D\omega}{Dt} = (\omega \cdot \nabla)u + \nu\nabla^2\omega + \nabla \times \left[\lambda\chi(u_s - u)\right] \tag{10}$$

where u_s is the velocity of the solid body and χ denotes a mask function that yields 0 in the fluid and 1 in the solid. The penalty parameter λ is equivalent to an inverse permeability. In the case of explicit Euler time discretization, λ must satisfy $\lambda\Delta t < O(1)$ to ensure stability. Alternatively, an implicit penalty term can expressed as

$$\frac{\Delta\omega}{\Delta t} = \nabla \times \left[\lambda\chi\left(u_s^n - \tilde{u}^{n+1}\right)\right] \tag{11}$$

where

$$\tilde{u}^{n+1} = \frac{u^n + \lambda\Delta t\chi u_s^n}{1 + \lambda\Delta t\chi} \tag{12}$$

The implicit scheme is unconditionally stable [27]. This allows to use large λ regardless of Δt for accurate solutions near a solid body. Based on implicit penalization scheme, the VTE is rewritten as

$$\frac{D\omega}{Dt} = (\omega \cdot \nabla)u + \nu\nabla^2\omega + \nabla \times \left[\frac{\lambda\chi}{1 + \lambda\Delta t\chi}(u_s - u)\right]. \tag{13}$$

Each term can be solved using a finite difference scheme. For example, the penalty term can be discretized by a first-order centred difference scheme. To reduce an error, one can use a smoothed mask function [35]. The stretching term, taking its alternative form $(\omega \cdot \nabla^T)u$ (the so-called transpose scheme), can be computed with the fourth-order central difference scheme. The transpose scheme is advantageous due to its conservative and accurate properties [36].

The diffusion term, $\nu\nabla^2\omega$, can be discretized by a 27-point isotropic Laplacian scheme which is less sensitive to the grid orientation. In 2D, the diffusion term can be discretized by a classical 9-point finite difference scheme.

2.3. Numerical procedure

For high performance computing, the message passing interface (MPI) can be used on a distributed-memory parallel system. For parallel computing, a computational domain Ω is decomposed by splitting it into several subdomains, with one subdomain per processor. Fluid particles are distributed among the subdomains depending on their positions.

In the pVIC method, convection, penalization, diffusion, and stretching (only in 3D) parts are sequentially evaluated on each subdomain. A numerical procedure follows that

a. Each processor interpolates the vorticity of its own particles into grid nodes through the M'_4 interpolation kernel.

b. Each processor computes ψ_b with its own particles' vorticity using the Green's function approach. Then, ψ_b for solving the Poisson equation is determined by a summation of the values obtained by each processor.

c. The Poisson equation $\nabla^2\psi = -\omega$ is solved using the FFT-based solver. Velocities on the grid nodes are computed from the resultant ψ, using a finite difference scheme.

d. Each processor evaluates the penalization term on the nodes using a finite difference scheme, and then the diffusion and stretching terms are computed. Divergence-free condition of vorticity field is enforced using the projection method.

e. Each processor interpolates the velocity and vorticity on the grid back to its own particle positions through the M'_4 interpolation kernel. Finally, the particles' position and strength are updated using an appropriate time integrator.

Some particles that have left a subdomain are assigned to an adjacent processor. If needed, the computational domain is increased depending on particle distribution. The whole process is repeated for the next time-step.

Particle redistribution is conducted every few time steps to ensure particle overlap. Concurrently, with this step, particles with negligibly small vorticity are discarded to prevent unnecessary increase in particle population. The resultant particles are filtered with a relative criterion; $\Gamma_i < C_\omega |\Gamma|_{max}$ where C_ω is a cut-off criterion. The discarded strengths were shared to the particles remaining in order to guarantee the conservation of circulation in the flow. It is found that the cut-off criterion has a significant influence on global force evaluation, rather than on vorticity and velocity fields. This will be further discussed in Section 3.4.

2.4. Numerical stability

In the pVIC method, fluid particles are traced by continuously updating their position and strength (vorticity or circulation) from the grid solution. The method retains the key feature of pure Lagrangian vortex methods; namely, the convection part disappears from VTE. As noted in [13], the particle convection is linearly unconditionally stable. The convection part does not impose any stability constraint but relies on the Lagrangian particle convection. The nonlinear stability condition that particles do not collide during one time-step (Δt) based on the rate of strain of a flow reads $\Delta t \leq C_1 |\nabla u|^{-1} \approx C_1 |\omega|^{-1}$. Also, the maximum allowable Δt is constrained by the diffusion Courant-Friedrichs-Lewy (CFL) condition, $\Delta t \leq C_2 h^2/\nu$, because the vorticity field is computed on a grid using a finite-difference scheme. This is less restrictive than the convective CFL condition for pure finite-difference methods, $\Delta t \leq Ch|u|^{-1}$. In practice, the nonlinear stability condition for the convection is much less demanding since the redistribution of particles to regular positions is periodically carried out. Δt for the diffusion is regarded as the physical time-step for an entire flow simulation. This permits the use of a relative large time-step compared to pure finite-difference methods.

3. Implementations

3.1. Domain decomposition for parallel computing

To decompose a computational domain in parallel, there are two typical strategies. One is that a computational domain is equidistantly decomposed. The other is that a computational domain is decomposed so as to assign the same number of particles to each processor; consequently, the subdomains have different sizes (for detailed information, refer to [37]). In our 2D flow simulations, the latter was typically faster than the former, whereas in 3D the former outperformed the latter. This is related to both balancing of computation load for Dirichlet boundary conditions and communication load of grid assembly to solve a Poisson equation.

In our pVIC code, each processor has all the boundary nodes to minimize communications among processors and boundary conditions at are determined by summing the contribution of all the particles. The same number of particles among processors ensures near-perfect balanced computation and communication loads, even if it causes unbalanced communication for grid because of different domain sizes. Since local computation of boundary conditions in 2D cases was typically more expensive than inter-processor communication, we found the second approach to be much faster. In 3D cases, however, balancing of communication for grid assembly becomes more important than that of computation of boundary conditions. This is due to a significant reduction in computing time by the spline approximation method (which will be introduced in Section 4.2). As a result, the equidistant domain decomposition can be a better choice for 3D flow simulations.

3.2. Time integrator

To move fluid particles with their own velocity, predictor-corrector methods are usually used. **Table 1** shows a summary of time integration schemes used in the literature, but the detailed scheme was not explained in most of papers.

Author(s)	Convection part	Diffusion part	Remarks
Ould-Salihi et al. [38]	RK2	EE	2D and 3D VIC
Coquerelle and Cottet [39]			2D and 3D pVIC
Rasmussen et al. [33]			2D pVIC
Lee at al. [32]			2D pVIC
Mimeau et al. [30]			2D pVIC
Morency et al. [40]	RK4	EE	2D pVIC
Cottet and Poncet [13]			3D VIC
Kosior and Kudela [41]			3D VIC
Cocle et al. [23]	RK2 and LF2	RK2 and AB2	3D VIC
Lonfils and Winckelmans [42]			3D VIC

The second-order Runge-Kutta method (RK2), the fourth-order Runge-Kutta method (RK4), the second-order Leapfrog method (LF2), the second-order Adams-Bashforth method (AB2), the first-order explicit Euler method (EE)].

Table 1. Comparison of numerical time-integration schemes reported in the available literature.

A popular Runge-Kutta method is known as a single stepping scheme with multiple stages per step. For example, the method can be expressed as

$$\boldsymbol{x}_p^* = \boldsymbol{x}_p^n + \boldsymbol{u}_p^n\left(\boldsymbol{x}_p^n, \boldsymbol{\omega}_p^n\right)\Delta t \tag{14}$$

$$\omega_p^{n+1} = \omega_p^n + \frac{d\omega_p^n}{dt}\Delta t \tag{15}$$

$$\boldsymbol{x}_p^{n+1} = \boldsymbol{x}_p^n + 0.5\left[\boldsymbol{u}_p^n\left(\boldsymbol{x}_p^n, \boldsymbol{\omega}_p^n\right) + \boldsymbol{u}_p^*\left(\boldsymbol{x}_p^*, \boldsymbol{\omega}_p^{n+1}\right)\right]\Delta t \tag{16}$$

where the superscript * indicates an intermediate stage [32, 43]. Note that the velocity field is evaluated twice in finding the new position of x_p^{n+1}. If a higher time integration scheme is used, more involved computations are needed. It is obvious that a higher-order integrator helps minimize error in particle tracking, compared to one evaluation. However, more evaluations of velocity field cause an increase in computational time.

Interestingly, Rossinelli et al. [44] evaluated particles' velocity using a modified explicit scheme as follows

$$\begin{aligned} \boldsymbol{x}_p^* &= \boldsymbol{x}_p^n + 0.5\Delta t \boldsymbol{u}_p^n\left(\boldsymbol{x}_p^n, \boldsymbol{\omega}_p^n\right) \\ \boldsymbol{x}_p^{n+1} &= \boldsymbol{x}_p^n + \Delta t \boldsymbol{u}_p^n\left(\boldsymbol{x}_p^*, \boldsymbol{\omega}_p^n\right) \end{aligned} \tag{17}$$

and a notable feature is that velocity field is evaluated only once although this scheme can be still considered as a kind of the Runge-Kutta method. Particles' velocity at the intermediate stage is computed using interpolation. This provides a considerable saving in computational time and memory consumption. Our research group investigated the feasibility of different integrators such as the Euler scheme with several stages, the midpoint rule, and the Simpson rule, as follows

$$\text{–Euler method} \quad x_p^{n+1} = \sum_{i=0}^{n-1}\left[x_p\left(t+i\delta t\right)+\delta t x_p u_p^n\left(t+i\delta t\right)\right] \tag{18}$$

where $\delta t = \Delta t/n'$

$$\text{–Midpoint rule} \quad x_p^{n+1} = x_p^n + \frac{\Delta t}{2}[u_p^n(x(t)) + u_p^n(x(t+\Delta t))] \tag{19}$$

$$\text{–Simpson rule} \quad x_p^{n+1} = x_p^n + \frac{\Delta t}{6}[u_p^n(x(t)) + u_p^n(x(t+\Delta t/2)) + u_p^n(x(t+\Delta t))] \tag{20}$$

Akin to Rossinelli et al. [44], particles' velocity at intermediate position was computed using interpolation from the values at neighbouring grid nodes. A benchmark problem was a flow around a sphere at Reynolds numbers of 50, 100, and 200. Overall computational time was about two times faster in the midpoint and Simpson rules, compared with the typical RK2 method in which velocity field is computed twice. The midpoint and Simpson rules provided comparable accuracy, and the difference in drag force was typically <0.3%. This approach seems to be useful. However, an essential prerequisite is that velocity field is little or not changing during Δt. This can require a smaller Δt, especially in high Reynolds number flows. Since the accuracy of a time integrator for tracing particles is dependent upon the change of a flow field in both time and space, a time integration scheme should be determined carefully.

3.3. Fast Poisson solvers

To solve a Poisson equation, the fast Fourier transform (FFT) and multigrid (MG) are utilized most widely. For a uniform grid, the behaviour of both solvers is well-known. In serial computation, a FFT solver requires $O(M\log M)$ for M grid points and a MG solver requires $O(M)$ operations. In an ideal parallel computation with zero cost communication, the FFT and MG solvers require $O(\log M)$ and $O(\log^2 M)$, respectively (for detailed information, refer to [45]). Here, the performance of two solvers is compared under exactly the same condition; a rectangular computational domain with equidistant nodes. Both solvers were developed in our research group. The FFT solver is based on an open-source library called Fastest Fourier Transform in the West (FFTW) [46], and the MG solver is based on full multigrid algorithms provided in Section 19.6 of the Numerical Recipes [47] (also refer to [41]). Here, the algorithm is briefly introduced. The full multigrid algorithm starts with the coarsest possible grid and then proceeds to finer grids. It performs one or more V-cycles at each level before proceeding down to the next finer grid. This means that the full multigrid algorithm produces a solution for each level. We used the red-black Gauss-Seidel scheme as a smoothing operator, bilinear interpolation for a prolongation operator, and half-weighting for a restriction operator. The full multigrid cycle can be seen in **Figure 1**.

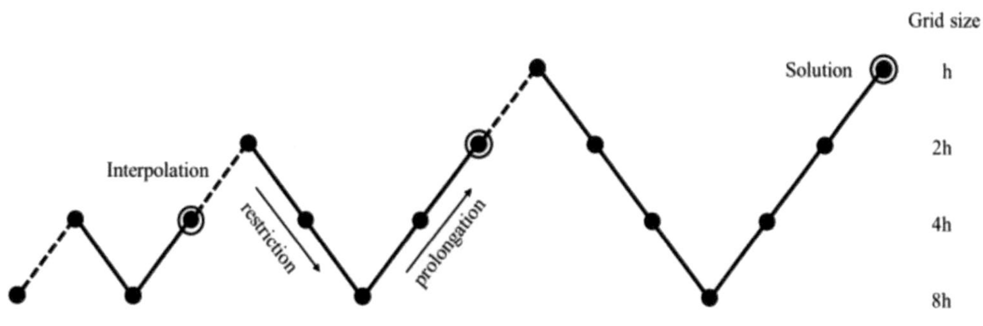

Figure 1. A computational cycle used in the MG solver.

To compare the two solvers, a 2D flow around a circular cylinder at a Reynolds number of 550 was considered as a benchmark problem. The MG solver with four levels was tuned by choosing one pre-smoothing sweep and two post-smoothing sweeps. **Figure 2a** shows a comparison of numerical results between the FFT and MG solvers. The MG solver performed comparably to the FFT solver in terms of accuracy. As expected, however, the MG solver was slower than the FFT solver as shown in **Figure 2b**. This is due that the MG solver is a kind of iterative method and thus it is an order magnitude more expensive compared to the FFT solver as a non-iterative method. Since, so far, most of VIC codes solve the Poisson equation in a computational domain of simple shape such as a square or cube, the FFT solver as a non-iterative method is thought to be the best choice. However, it is worth noting that the MG solver can be applied to locally refined grids in a straightforward way. Adaptive mesh refinement is useful both to increase accuracy and to decrease computational cost in comparison with a uniform grid.

3.4. Force calculations

From the fact that the integral of vorticity moments becomes the change of the momentum, fluid force exerted on the solid body (F_s) is expressed as

$$F_S = -\rho \frac{d}{dt}\left(\frac{1}{N-1} \int_V x \times \omega dV \right) \tag{21}$$

where N is the dimension of the space. The volume integral can be replaced by the summation of the moments for all the particles. Also, the adoption of the penalization method introduces another approach for the force evaluation. One can obtain the force with the velocity field inside the body as follows:

$$F_s = -\rho \int_V \lambda \chi (u_s - u) dV \tag{22}$$

where the penalization term itself is included in the integral as the time-change of the momentum. For the implicit penalization scheme, it can be rewritten as

$$F_s = -\rho \int_V \frac{\lambda \chi}{1 + \lambda \chi \Delta t} (u_s - u) dV \tag{23}$$

The integral over the whole domain is actually confined to the interior of the body due to the definition of χ.

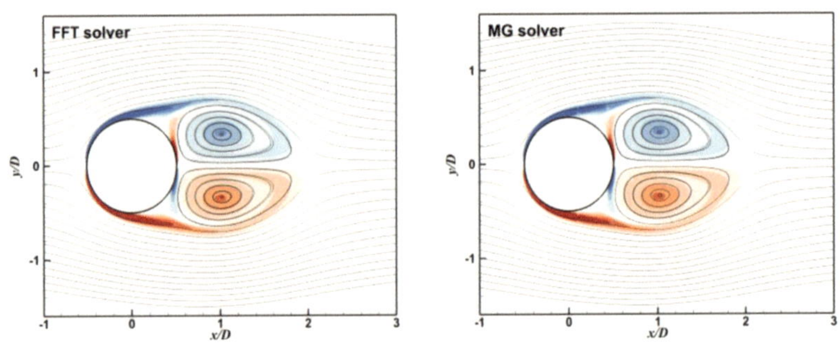

(a) vorticity contours and streamlines

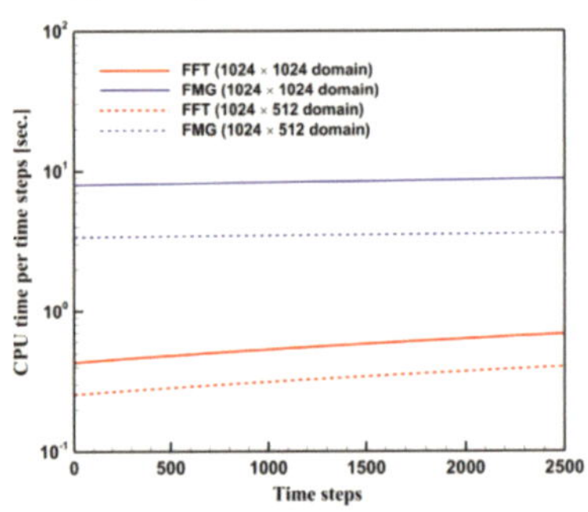

(b) CPU times per time steps

Figure 2. Comparison of flow fields and CPU times between the FFT and MG solvers. (a) Vorticity contours and streamlines and (b) CPU times per time steps.

In 2D flow simulations, the drag coefficients have been reported in the literatures [30, 37, 44, 48]. They have evaluated the drag coefficient using one or both of the approaches. Rasmussen et al. [33] and Gazzola et al. [31] addressed that the discrepancy in the drag coefficient between the two approaches were not significant. In 3D, however, it was found that the method based on the vorticity moment is unstable occasionally; that is, a large C_ω can make it incorrect even if there is little difference both in vorticity and velocity fields. **Figure 3** shows the drag coefficient (C_D) of a sphere at a Reynolds number of 100 computed by the two different approaches employing three different cut-off criteria; $C_\omega = 10^{-4}$, 10^{-5}, and 10^{-6}. C_D is defined by $F_x/(0.5\rho U_\infty A)$ with a sectional area $A = \pi R^2$. The drag coefficient computed with vorticity moments were converged at $C_\omega = 10^{-6}$, whereas the drags computed with the penalized velocity were nearly the same in all the tested cases. This describes that force evaluation using the vorticity moment should be taken with special care. As a result, the force evaluation with the penalized velocity can be a better option, especially in 3D simulations. This enables one to avoid an excessive increase in the number of particles, thus reducing the computational time and memory. In the cases with $C_\omega = 10^{-4}$, 10^{-5}, and 10^{-6}, there were 1.0, 1.9, and 3.5 million particles, respectively.

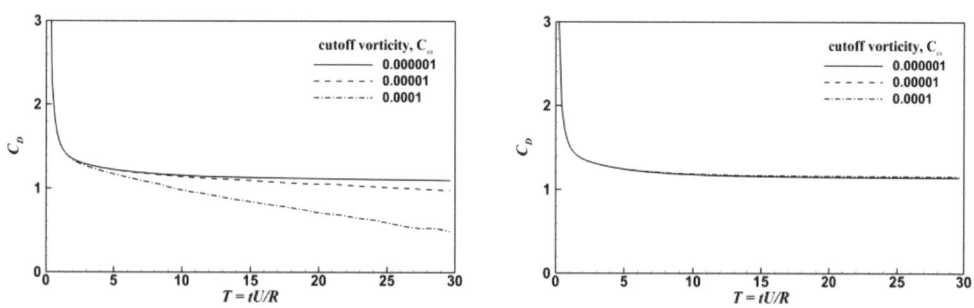

(a) with the vorticity moment

(b) with the penalized velocity

Figure 3. Drag coefficients calculated with two different approaches: (a) vorticity moment and (b) penalized velocity.

3.5. Pressure field

As discussed previously, pressure field is independently computed from the entire solution procedure and explicitly obtained by solving a pressure Poisson equation, $\nabla^2 H = \nabla \cdot (u \times \omega)$. Lee [49] introduced an integral approach to obtain the pressure field at any fixed time as follows:

$$H = \int_S \left(\frac{\partial H}{\partial n} G - H \frac{\partial G}{\partial n} \right) dS + \int_V \nabla \cdot (u \times \omega) G dV \tag{24}$$

where G is the Green function solution. H is the Bernoulli function defined as

$$H = \frac{p - P_\infty}{\rho} + \frac{1}{2}\left(u^2 - U_\infty^2\right) \tag{25}$$

where p is the static pressure above the reference pressure P_∞ and ρ is fluid density. The value of H on the body surface is computed using a boundary integral approximation. A mathematical identity for a vector or scalar field is used to define field values of a quantity of interest, which involves an integral of singularities distributed over the surface and throughout the field. The boundary integral approach has been successfully established. However, it has disadvantages such as the presence of singular Green's kernels. Special attention is needed to accurately compute boundary integrals around a singular point. The higher mathematical complexity is needed to get a usable computational formulation. The matrices that result from the integral method are asymmetrical, and they are not easy to solve. Furthermore, this approach takes long time of computation.

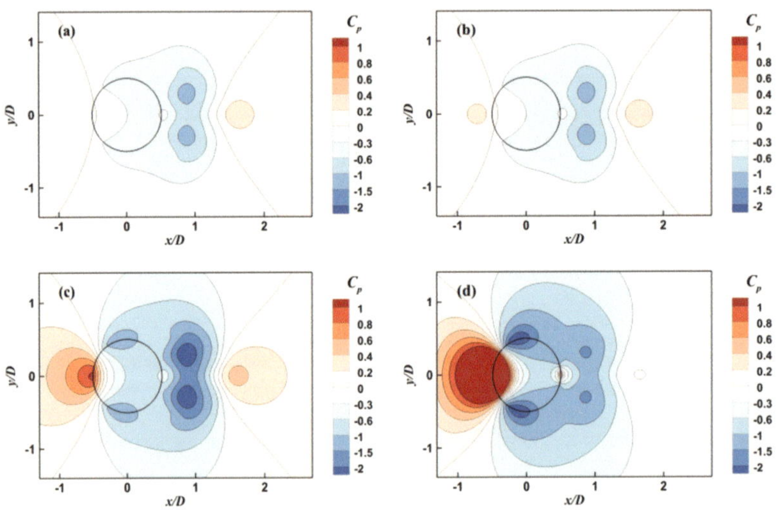

Figure 4. Pressure distributions computed with different penalty parameters for pressure; (a) $\lambda' = 0$, (b) $\lambda' = 0.1/\Delta t$, (c) $\lambda' = 1/\Delta t$, and (d) $\lambda' = 10/\Delta t$. Note that the figures are reprinted with permission from [28].

Alternatively, pressure field can be evaluated based on the penalized Navier-Stokes equation [28]. Satisfying the continuity, the Poisson equation for pressure can be expressed as

$$\nabla^2 H = \nabla \cdot (u \times \omega) + \lambda' (u_s - u) \cdot \nabla \chi \tag{26}$$

where λ' is distinguished from the λ by its order of magnitude. This equation is solved on the grid using a fast Poisson solver. The boundary condition for H can be given as homogeneous Dirichlet type. **Figure 4** shows pressure distributions around a circular cylinder for a Reynolds number of 550, which are computed using different orders of magnitude for λ'; $\lambda' = 0, 0.1/\Delta t,$

http://dx.doi.org/10.5772/65371

$1/\Delta t$, and $10/\Delta t$. The assumption of $\lambda' \cong 1/\Delta t$ is thought to be reasonable. With the assumption, Lee et al. [28] successfully computed pressure field around a 2D cylinder as shown in **Figure 5**. It is still valid in 3D flow simulations as shown in **Figure 6**. This approach is quite efficient, but an accurate approximation of H at domain boundaries still remains a problem.

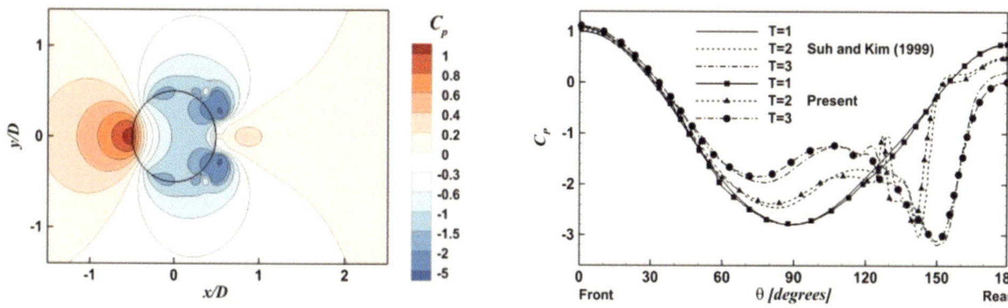

(a) pressure contour at $T = t\,U_\infty/R = 3$ (b) pressure coefficient at a solid surface

Figure 5. Pressure distribution around a circular cylinder at a Reynolds number of 9500. Note that the figures are reprinted with permission from [28]. (a) Pressure contour at $T = tU_\infty/R = 3$ and (b) pressure coefficient at a solid surface.

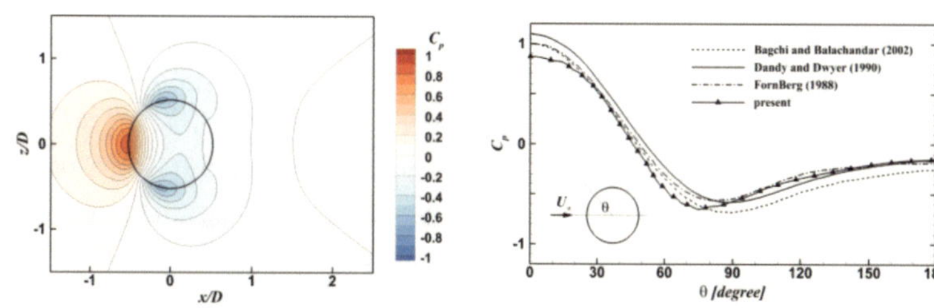

(a) pressure contour (b) pressure coefficient at a solid surface

Figure 6. Pressure distribution around a sphere for a Reynolds number of 100 at $T = tU_\infty/R = 30$. (a) Pressure contour and (b) pressure coefficient at a solid surface.

4. More efficient implementations

4.1. Multiple domains

In the pVIC method, the size of a computational domain relies on the distribution of fluid particles. A large domain leads to an increase in computational memory required for a grid. A multiple domain approach can be considered to handle much more particles with a limited computational memory. The entire domain Ω can be defined as the union of physical subdomains, covering all the fluid particles; that is, $x_p \in \Omega$ where $\Omega = \Omega_1 \cup \cdots \cup \Omega_{N_D}$. The number of small domains, N_D, is not constant and depends on a spatially evolving flow. When the domain size exceeds a certain limit, a new domain is created. For example, the first domain Ω_1 includes a solid body and the others are sequentially located downstream from the body as shown in **Figure 7**. Note that this approach differs from the domain decomposition for parallel computations. That is, each small domain Ω_i is again decomposed into subdomains for parallel computations; $\Omega_i = \Omega_{i_1} \cup \cdots \cup \Omega_{i_{N_P}}$ where N_P is the number of processors

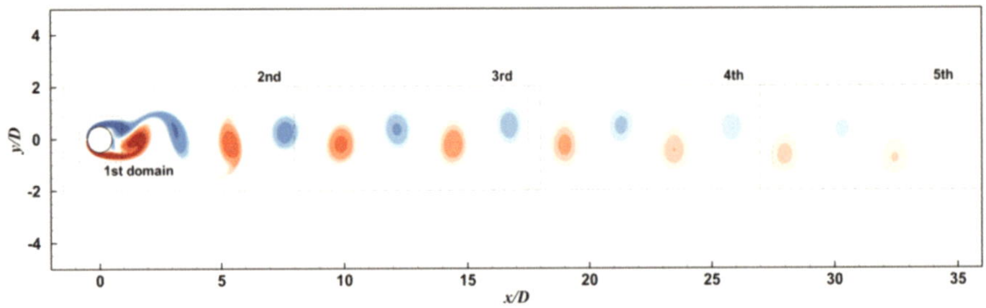

Figure 7. Vorticity contour around a circular cylinder for a Reynolds number of 185.

In Ω_i, vorticity and velocity fields are computed according to the numerical procedure given in Section 2.3. There is no dependency between the small domains since domain boundary conditions of Ω_i are computed by summing the contribution of all the particles. Hence, independent domains ensure a relatively small use of computational memory during a flow simulation, compared to an approach based on parent and child grids [23]. Also, the multiple domain approach enables to have different resolutions among small domains. For example, the grid spacing of Ω_i can be defined as either $h_i = \epsilon$ (single-resolution) or $h_i = 2^n\epsilon$ (multi-resolution) where ϵ denotes the particle size. The particle strength can be transferred into the nodes as follows:

$$\omega\left(x_g, t\right) = \frac{1}{h_i^3} \sum_p^N \Gamma_p M_4' \left(\frac{x_g - x_p}{h_i} \right) \tag{27}$$

Although this approach ensures the small consumption of computational memory, it causes an increase in computational time because of an increase in the number of domain boundaries. Fast computations of ψ at domain boundaries will be discussed in Section 4.2.

4.2. Approximation of far-field conditions

For simplicity, consider a single domain Ω. The resultant values at $\partial\Omega$ are determined by summing the values computed by all processors. Each processor requires $O(M_b N/P)$, where M_b and N/P denote the number of boundary nodes and its owned particles, respectively. Note that N/P can be reduced by an increase in the number of processors (P), whereas M_b is a constant depending on the size of Ω. M_b can be critical to determine the computational time. In practice, a substantial part of the overall computational time is spent on the calculation of ψ_b at domain boundaries for solving Poisson's equation. Lee et al. [37] attempted a fast computation of ψ_b using splines. They found that ψ_b varies very smoothly at boundaries of a domain. This feature permits the use of spline interpolation. In 2D, for example, one spline curve enables to represent one side of the computational domain. ψ_b at the part of the boundary nodes (M_{b_D}) is computed by pure direct summations, and then values at the other nodes (M_{b_S}) can be approximated using a spline curve. Here, $M_b = M_{b_D} + M_{b_S}$. Using spline, approximation leads to a reduction in M_{b_D}, which is directly linked to the reduction in the computational time.

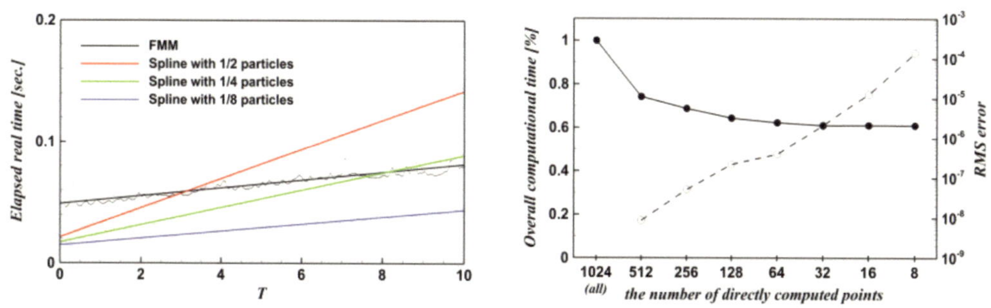

(a) computational time for boundary values

(b) overall computational time and error

Figure 8. CPU time in numerical simulation using the spline approximation. Note that the FMM is based on a tree level of 3 and an expansion degree of 10. The figures are reprinted from [50]. (a) Computational time for boundary values and (b) overall computational time and error.

Figure 8 shows CPU times elapsed in computations of boundary values using the spline approximation and the FMM [50]. The numerical simulation for a flow past a circular cylinder at Reynolds number of 550 was carried out using 16 CPUs with 4 GB of memory per processor. Once one-eighth of all the boundary points are equidistantly chosen as points for the direct summation, the spline approximation approach becomes faster than the FMM that is modified to compute only boundary values. In this case, the spline approximation approach has accuracy comparable to the FFM. A weak point of the FMM is that it has a quad-tree (oct-tree in 3D) data structure to hierarchically subdivide the computational domain. Each processor in a distributed memory parallel system must have a sufficient amount of memory for tree data structures. A required memory in the FFM depends on both the number of particles and grid nodes.

In 3D, the bi-cubic interpolation can be used instead of the bi-cubic spline (for detailed information on interpolation methods, refer to [47]). According to prior tests to assess this approach, the former was more accurate for our purpose. For example, for computing the stream functions at one side of the cube with 256 × 256 nodes from 3 million randomly distributed particles, the trial using the bi-cubic interpolation method took approximately 1 s when 16 × 16 nodes were used for direct summations, whereas the fully direct evaluations took 194 s in average. The error was typically <0.01%.

5. Numerical validations

5.1. Flows past an impulsive started circular cylinder

The impulsively started circular cylinder problem is a good prototype to validate a numerical method. The Reynolds number, $Re = U_\infty D/\nu$, based on the free stream velocity, U_∞ and the diameter of the cylinder, D, is selected to be 40, 550, 3000, or 9500. In the simulations, no symmetry constraint is imposed. The dimensionless time is based on the radius ($R = 0.5D$) of the cylinder; $T = tU_\infty/R$. The penalization parameter λ is fixed to 10^8.

5.1.1. Reynolds number of 40

The numerical parameters are $h = 0.04$ and $\Delta t = 0.016$, which are determined through the stability condition $\Delta t = h^2/\nu$. At a Reynolds number of 40, it is well-known that the flow reaches a steady state. A pair of stationary recirculating wakes develops behind the cylinder. The wake length, L/D, is 4.16 and separation angle, θ_s, is 51.7°, respectively. Fornberg [51] made a similar remark and gave $L/D = 4.48$ and $\theta_s = 51.5°$ experimentally. The drag coefficient is computed as $C_D = 1.483$. This is close to the experimentally measured value of 1.498 in [52]. **Figure 9** shows pressure coefficient (C_P) distribution around the cylinder body. The pressure is continuous through the cylinder body, and there is no Gibb's oscillation associated with the discontinuity at the body surface.

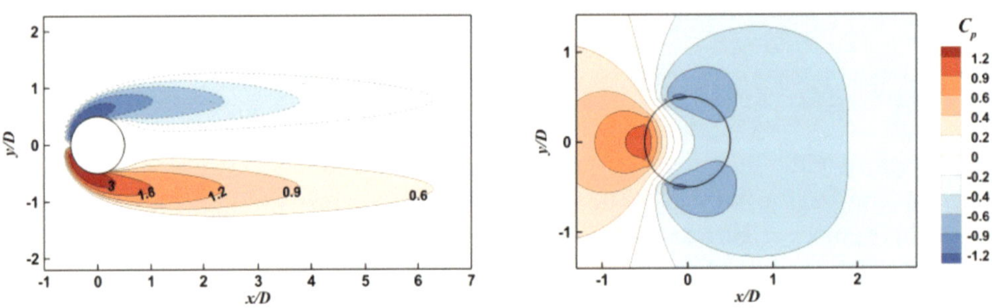

(a) vorticity contour (b) pressure contour

Figure 9. Vorticity and pressure contours around a circular cylinder at a Reynolds number of 40. (a) Vorticity contour and (b) pressure contour.

5.1.2. Reynolds number of 550

The simulation parameters are $h = 0.005$ and $\Delta t = 0.002$. Simulation was carried out until $T = 7$ to validate the present formulation in the early time stage after the impulsive start. The number of fluid particles ranged from approximately 30,000 to 150,000, and the total computation time is approximately 4 h on 8 CPUs (Intel Xeon64 3.3 GHz). Numerical results are presented for vorticity contours and streamlines in **Figure 10**. As time increases, a pair of secondary symmetric vortices appears and become stronger. The so-called bulge phenomena observed experimentally by Bouard and Coutanceau [53] is well captured by our numerical simulation. The vortex core position indicated by the coordinates a/D and b/D is investigated. In this flow simulation, the abscissa $a/D = 0.34$ and ordinate $b/D = 0.27$, and the wake length $L/D = 0.82$. In [53], $a/D = 0.36$, $b/D = 0.28$, and $L/D = 0.85$.

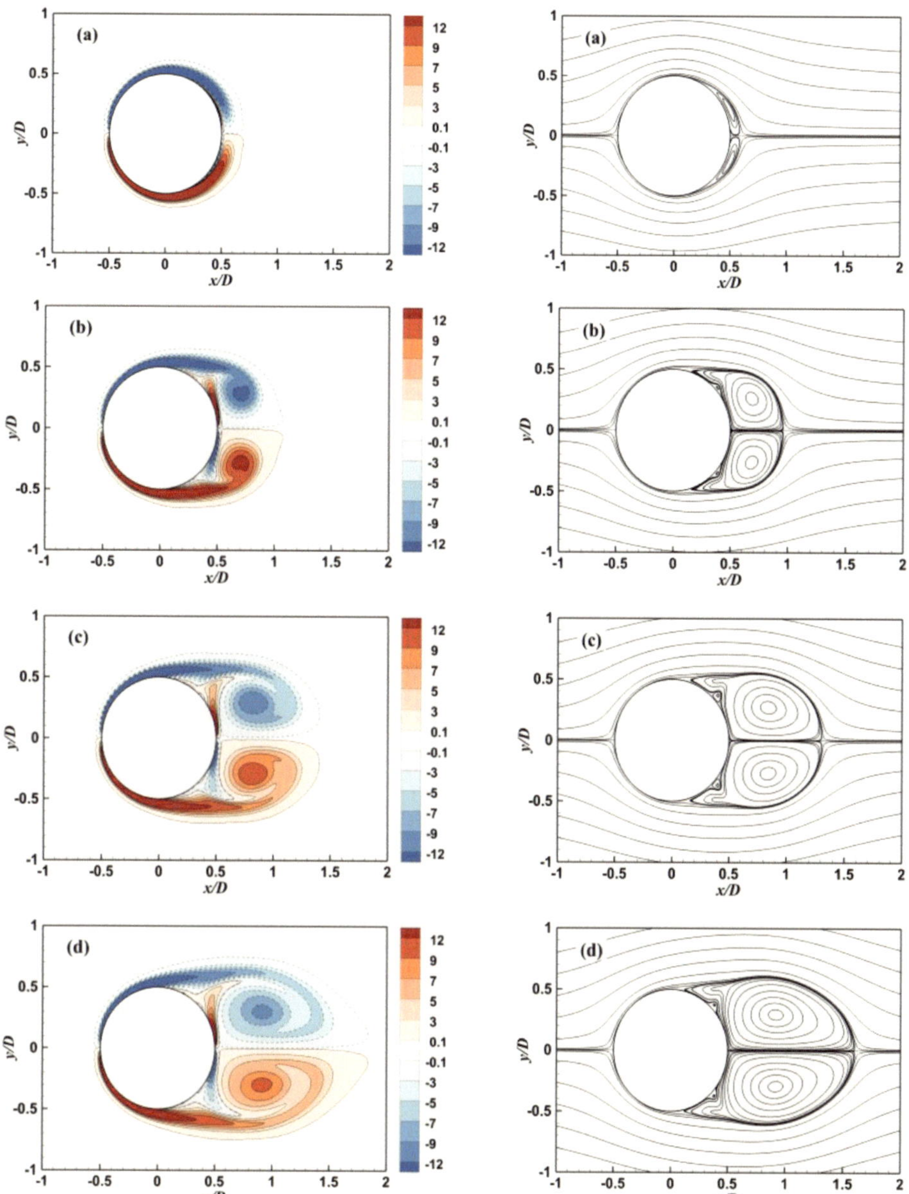

Figure 10. Vorticity contours (*left*) and streamlines (*right*) at a Reynolds number of 550; (a) $T = 1$, (b) $T = 3$, (c) $T = 5$, and (d) $T = 7$.

Figure 11 shows the time evolution of the drag coefficient for an impulsively started flow around a two-dimensional cylinder for Re = 550 calculated with the pVIC method, compared with results from the short time asymptotic solution of Bar-Lev and Yang [54], the vortex method result of Koumoutsakos and Leonard [55], and the VIC method result of Kudela and Kozlowski [56].

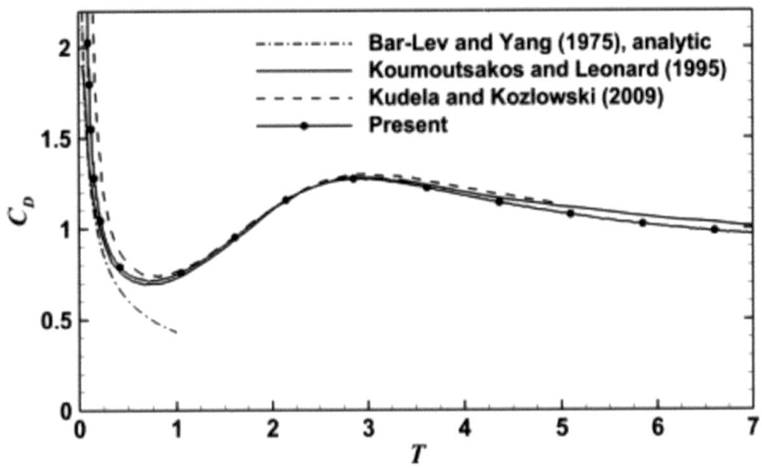

Figure 11. Time evolution of the drag coefficient for a cylinder at a Reynolds number of 550. Note that the figure is reprinted with permission from [28].

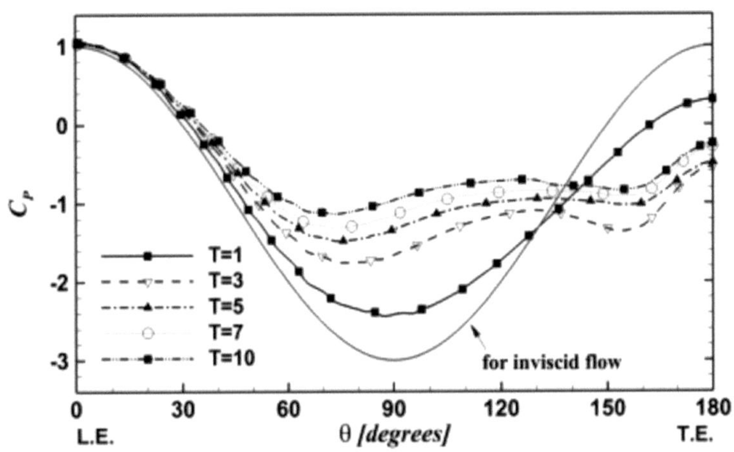

Figure 12. Surface pressure coefficients for a cylinder at a Reynolds number of 550. Note that the figure is reproduced with permission from [28].

The pressure distribution at the cylinder surface is shown in **Figure 12**. At very early stages in the numerical simulation, the surface pressure distribution is quite close to that for an ideal inviscid flow. As flow evolves, the minimum pressure coefficient gradually moves upstream.

The corresponding locations of $C_P = 0$ have the same upstream moving trends and is at the angular position $\theta = 35°$ as experimentally measured by Norberg [57].

5.1.3. Reynolds number of 3000

The simulation parameters are $h = 0.0025$ and $\Delta t = 0.002$. Simulation was carried out until T = 10 and the number of fluid particles ranged from approximately 130,000 to 420,000. The total run time is about 9 h on 16 CPUs (Intel Xeon64 3.3 GHz). **Figure 13** shows numerical results. Compared with a flow at Re = 550, the secondary vortices appear at an earlier time and grow larger. In the case of Re = 3000, the two secondary vortices formed are equivalent in size and in strength. It is the so-called α phenomena [53]. Also, the streamline computed for Re = 3000 compares with the flow visualization result of Loc and Bouard [58]. The present simulation correctly captures the expected physics of the flow at Re = 3000.

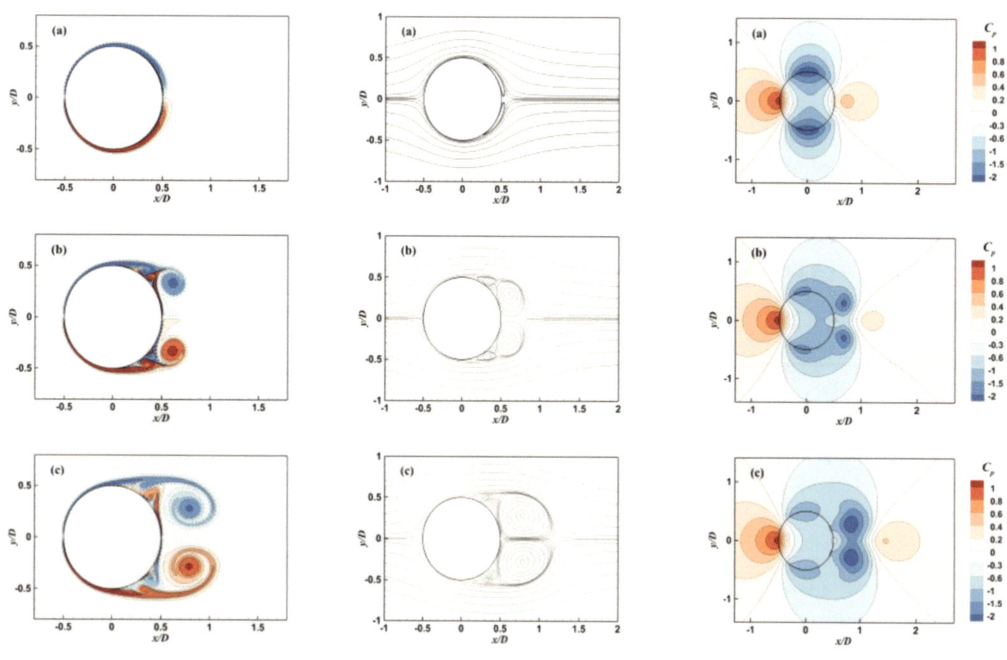

Figure 13. Numerical results for a circular cylinder at Re = 3000 at (a) $T = 1$, (b) $T = 3$, and (c) $T = 5$. Note that the figures are reprinted and reproduced with permission from [28]; vorticity contour (*left*), streamline (*middle*), pressure contour (*right*).

5.1.4. Reynolds number of 9500

The simulation parameters are $h = 0.001$ and $\Delta t = 0.002$. Flows were numerically simulated until T = 4, and the number of fluid particles ranged from approximately 800,000 to 1,200,000. The total run time is about 25 h on 24 CPUs (Intel Xeon64 3.3 GHz). **Figure 14** shows vorticity contours. They are in good agreement with the numerical results in [29].

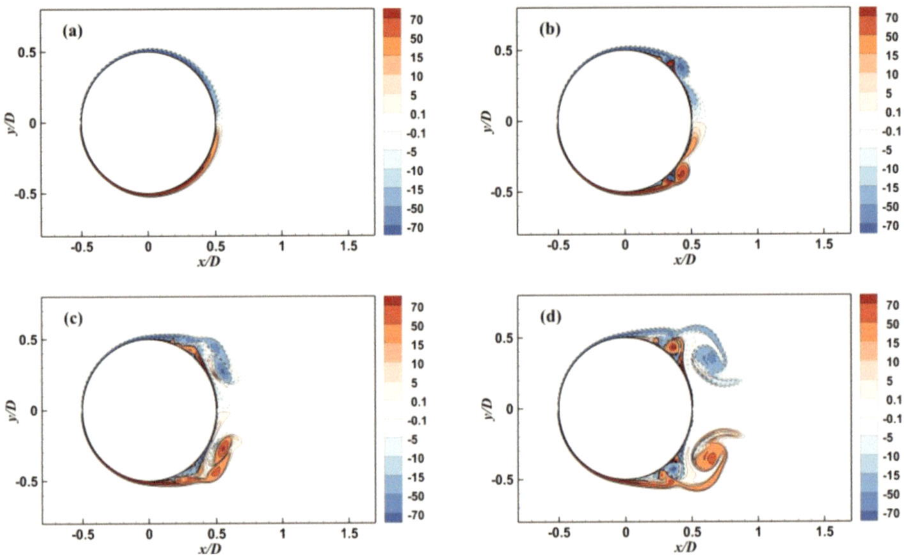

Figure 14. Vorticity contours at a Reynolds number of 9500; (a) $T = 1$, (b) $T = 2$, (c) $T = 3$, and (d) $T = 4$. Note that the figures are reprinted and reproduced with permission from [28].

As shown in **Figure 15**, the streamline patterns computed for Re = 9500 compare quite well with the flow visualization results in [58]. The α and β phenomena, which are observed experimentally by Bouard and Coutanceau [53], are well captured by the numerical simulation.

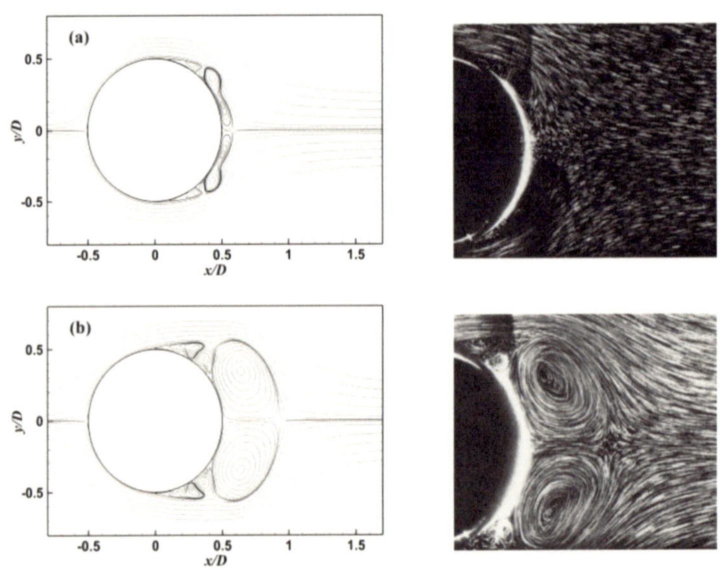

Figure 15. Streamlines at Reynolds number of 9500; (a) $T = 2$ and (b) $T = 4$ compared with flow visualization results of Loc and Bouard [58]. Note that the figures are reprinted with permission from [28].

5.2. Flows past an impulsive started sphere

When the pVIC method comes to a 3D implementation, a little attention should been paid to the quantitative validation as mentioned in Sections 2.1 and 3.4. An incompressible viscous flow past a sphere has extensively been studied by many researchers in theoretical, experimental, and numerical ways. It is well-known that the wake behind a sphere depends on the Reynolds number. Especially, the flows at 20 < *Re* < 210 are often investigated as a validation case by virtue of their steady and axisymmetric features in wake structure including separation. To demonstrate the feasibility of the pVIC method, we carried out numerical simulations of the impulsively started flow past a sphere at a Reynolds number of 100. **Figure 16** illustrates the configuration of the multiple domains. The sphere is immersed in a Cartesian grid that does not conform to its surface.

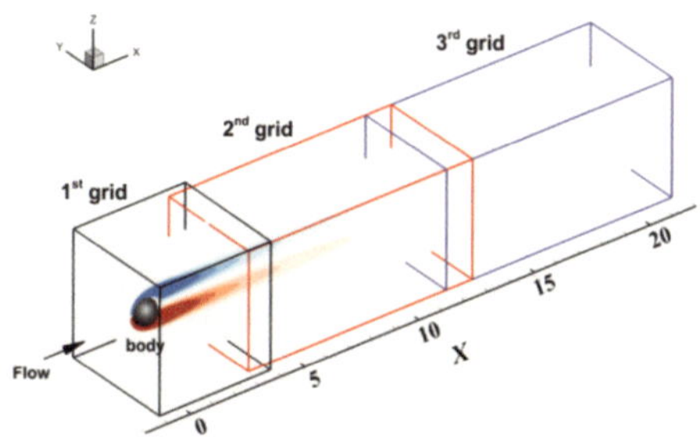

Figure 16. Configuration of computational domains.

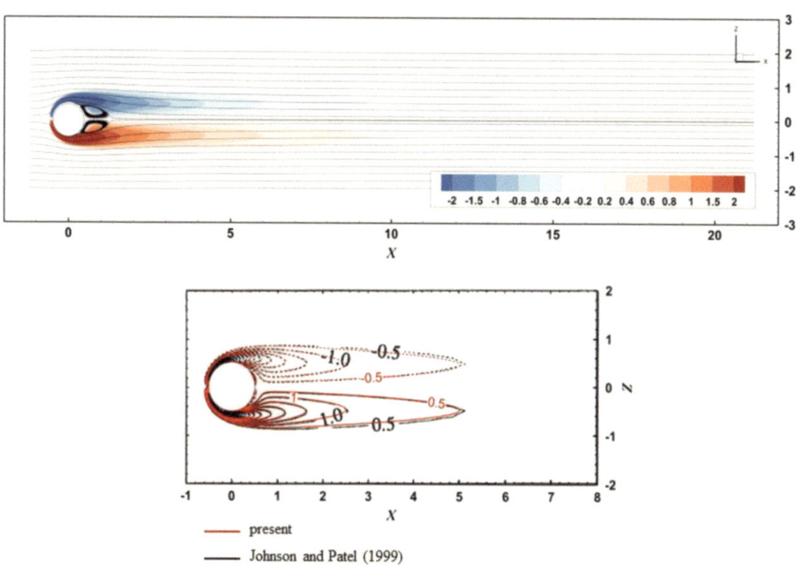

Figure 17. Contour of vorticity magnitude and streamline on *xz*-plane at a steady state.

The simulations were carried out until T = 30 to demonstrate the convergence of the present method and test its capability of long-time simulation. Numerical parameters were determined through the stability condition, $\nu\Delta t/h^2 < 1/6$, and the grid convergence was achieved with $h = 0.02$ and $\Delta t = 0.005$. It was concluded that giving $C_\omega = 10^{-5}$ is sufficient in terms of the force calculation. The number of fluid particles ranges from 2 to 5 million during the simulation, and the total computation time is roughly 67 h on 16 CPUs (Intel Xeon64 3.3 GHz). **Figure 17** shows the vorticity contour at the steady state. The vorticity contour is in excellent agreement with the numerical result in [59]. The wake shape and drag coefficient at the steady state are compared to the references in **Table 2**. The numerical results are good agreements with the reference results.

Author(s)	Wake centre (a/D, b/D)	Wake length (L/D)	Separation angle	Drag coefficient
Taneda [60], *exp.*	(0.75, 0.28)	0.89	52.3°	–
Johnson and Patel [59], *cal.*	(0.76, 0.29)	0.89	53.3°	–
Bagchi and Balachandar [61], *cal.*	–	0.87	53.2°	1.09
The presented pVIC method	(0.764, 0.282)	0.877	53.1°	1.09

Table 2. Comparisons of wake shape behind a sphere and drag coefficient with previous results.

5.3. Vortex shedding from a hydrofoil

We selected a National Advisory Committee for Aeronautics (NACA) 0009 cross section with a truncated trailing edge for the numerical simulations. This hydrofoil has the same as the experimental model used in [62–64]. Ten percent of the original chord c_o was removed from the trailing-edge region of the NACA 0009 hydrofoil. The hydrofoil geometry is further detailed in [63]. The maximum thickness, as normalized by the chord length c, is $t_{max}/c = 0.1$ and the thickness at the trailing edge is 0.0322. The numerical parameters $h = 0.0003$ and $\Delta t = 0.00015$ were chosen to simulate flow past the 2D hydrofoil at a Reynolds number of 2×10^6.

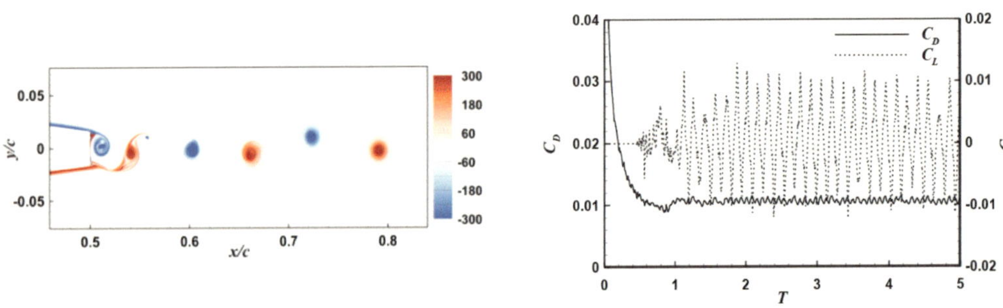

(a) vorticity contour (b) drag and lift coefficients

Figure 18. (a) Instantaneous vorticity contour and (b) a time history of drag and lift coefficients at a Reynolds number of 2×10^6.

The instantaneous contour of vorticity magnitude is plotted in **Figure 18a**. Vortices are regularly shed from the symmetric trailing-edge in the form of two trains of opposite-sign but equal-strength vortices. The Strouhal number, $St = fc/U_\infty$ = is about 7.0 from our numerical simulation, and Zobeiri et al. [64] made a similar remark experimentally gave 7.2 (also, refer to [63]). Regular vortex shedding causes periodic loading on the structure. **Figure 18b** shows the evolution of drag and lift coefficients. It has been well-known that the drag force oscillation during vortex shedding is much smaller than the lift force. Oscillations in drag force occur at twice the vortex shedding frequency owing to the fact that two vortices are shed from alternate sides during one full period of wake oscillation. Such features are well captured by our numerical simulation as shown in **Figure 18b**. The mean drag coefficient C_D is 0.0107 and the root-mean-square lift coefficient is 0.0056.

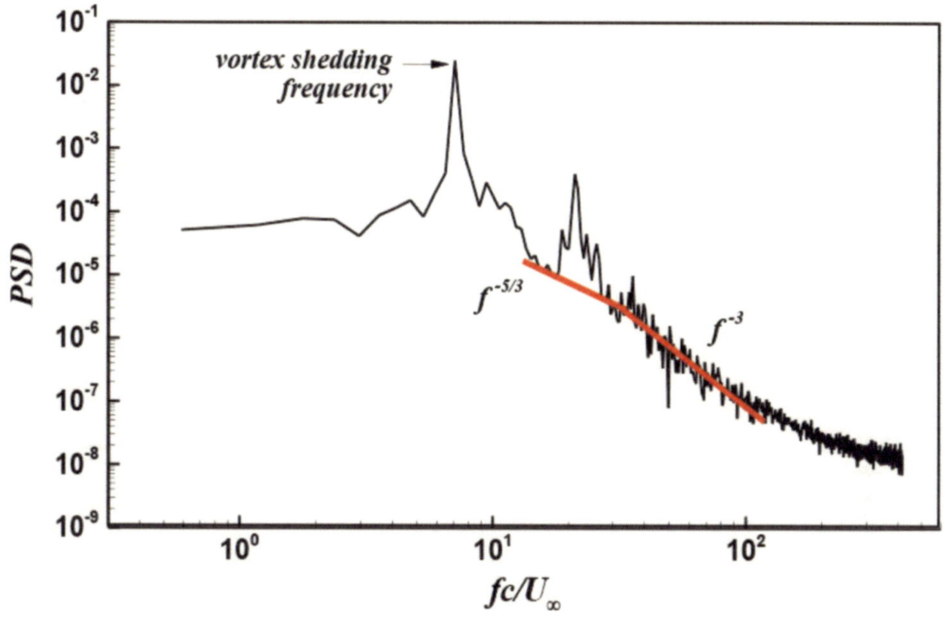

Figure 19. Temporal power spectral density (PSD) of vertical velocity fluctuations. Note that this figure is reprinted with permission from [32].

Figure 19 shows the temporal power spectrum of vertical velocity fluctuations measured in the near wake. The maximum peak is identical to the vortex shedding frequency, and the second peak area is found in the third harmonic of the shedding frequency. For $U_\infty < 5$, spectral levels are almost flat as f decreases. This trend is similar to experimental data of Bourgoyne et al. [65]. In their measurements, an important characteristic of the spectra is the presence of a clear region with a −5/3 slope; spectral power-law for high Reynolds number turbulent fluctuations. From our numerical simulation, however, the spectral density of the velocity fluctuations shows a decay of the form. A similar remark was made by Singh and Mittal [66]. This is due that the vortex stretching mechanism is absent in two-dimensional flows. More detail on fluid turbulence confined two spatial dimensions can be found in Boffetta and Ecke [67].

Figure 20 shows vortex shedding from a hydrofoil with different bevel angles in degrees. At $\beta = 0°$, the vortex shedding is regular and periodic. As the bevel angle β increases, vortex shedding becomes increasingly disorganized. The periodicity of both forces is almost lost at $\beta = 60°$.

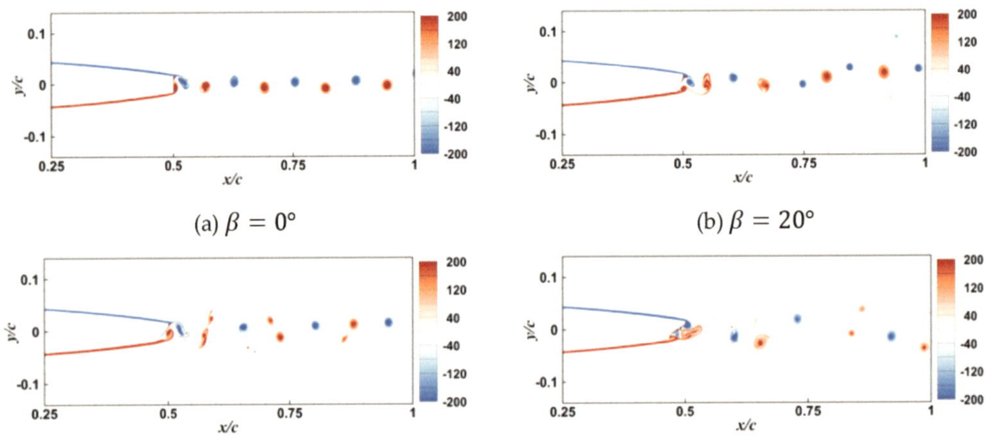

(a) $\beta = 0°$ (b) $\beta = 20°$

(c) $\beta = 40°$ (d) $\beta = 60°$

Figure 20. Vortex shedding from a hydrofoil with different bevel angles. Note that the figures are reproduced with permission from [37]. (a) $\beta = 0°$; (b) $\beta = 20°$; (c) $\beta = 40°$; (d) $\beta = 60°$.

Also, we conducted numerical simulations to investigate vortex shedding with respect to the sinusoidal motions of the free stream flow. The angle of attack varied periodically with time t; $\alpha = \alpha_o sin(2\pi f_\infty t)$ where f_∞ was the frequency of the free-stream flow oscillation and its oscillation amplitude was restricted at $\alpha_o = 2$ in degrees. The magnitude of the free-stream velocity was kept constant. The tested model is the NACA 0009 hydrofoil with $\beta = 60°$. Interestingly, at $f_\infty = 20$ vortices are regularly shed from the trailing edge, as shown in **Figure 21**. In the power spectral density (PSD) functions, drag oscillation induced by vortex shedding is clearly observed. This means that a particular oscillation frequency in the free stream velocity can cause regular and periodic vortex shedding.

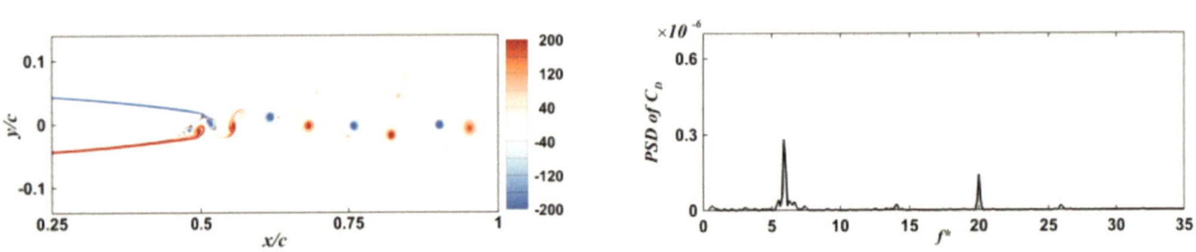

Figure 21. Vortex shedding at $f_\infty = 20$. Note that the figures reproduced and reprinted with permission from [37].

6. Concluding remarks

The pVIC method was applied to the simulation of an incompressible viscous flow past a solid body in 2D and 3D. The obtained results showed good agreement with experimental and numerical data from the published literature. This demonstrates the feasibility of the pVIC method. Obviously, the pVIC method offers a simple, efficient, and effective way to simulate unsteady flows. The fact that vortex-dominated flows are well-characterized by vortices makes the pVIC method more attractive. However, some hard problems still remain; for example, rigorous boundary conditions at solid walls and an efficient approximation of pressure boundary conditions at far field.

Author details

Seung-Jae Lee

Address all correspondence to: hul94@snu.ac.kr

Research Institute of Marine Systems Engineering, Seoul National University, South Korea

References

[1] B. R. Morton. The generation and decay of vorticity. Geophysical and Astrophysical Fluid Dynamics. 1984;28:277–308. doi:10.1080/03091928408230368

[2] C. G. Speziable. On the advantages of the vorticity-velocity formulation of the equations of fluid dynamics. Journal of Computational Physics. 1987;73(2):476–480. doi: 10.1016/0021-9991(87)90149-5

[3] P. M. Gresho. Incompressible fluid dynamics: some fundamental formulation issues. Annual Review of Fluid Mechanics. 1991;23:413–453. doi:10.1146/annurev.fl. 23.010191.002213

[4] A. Leonard. Vortex methods for flow simulation. Journal of Computational Physics. 1980;37(3):289–335. doi:10.1016/0021-9991(80)90040-6

[5] A. Leonard. Computing three-dimensional incompressible flows with vortex elements. Annual Review of Fluid Mechanics. 1985;17:523–559. doi:10.1146/annurev.fl.17.010185. 002515

[6] T. Sarpkaya. Computational methods with vortices-the 1988 freeman scholar lecture. Journal of Fluids Engineering. 1989;111(1):5–52. doi:10.1115/1.3243601

[7] K. Kamemoto. On attractive features of the vortex methods. In: M. Hafez and K. Oshima, editors. Computational Fluid Dynamics Review 1995. Wiley; 1995. p. 334–353.

[8] L. A. Barba, A. Leonard and C. B. Allen. Advances in viscous vortex methods—meshless spatial adaption based on radial basis function interpolation. International Journal for Numerical Methods in Fluids. 2005;47:387–421. doi:10.1002/fld.811

[9] L. Greengard and V. Rokhlin. A fast algorithm for particle simulations. Journal of Computational Physics. 1987;73(2):325–348. doi:10.1016/0021-9991(87)90140-9

[10] VJ. Dongarra and F. Sullivan. Guest editors introduction to the top 10 algorithms. Computing in Science and Engineering. 2000;2(1):22–23. doi:10.1109/MCISE. 2000.814652

[11] R. Beatson and L. Greengard. A short course on fast multipole methods. In: Wavelets, Multilevel Methods and Elliptic PDEs. Oxford University Press; 1997. p. 1–37. doi: 10.1.1.129.7826

[12] J. P. Christiansen. Numerical simulation of hydrodynamics by the method of point vortices. Journal of Computational Physics. 1973;13(3):363–379. doi:10.1016/0021-9991(73)90042-9

[13] G.-H. Cottet and P. Poncet. Advances in direct numerical simulations of 3D wall-bounded flows by Vortex-in-Cell methods. Journal of Computational Physics. 2004;193(1):136–158. doi:10.1016/j.jcp.2003.08.025

[14] T. K. Sheel. Development of a fast Vortex method for fluid flow simulation using special-purpose computers [thesis]. Japan: Keio University; 2008. 169

[15] HaK. Kuwahara and H. Takami. Numerical studies of two-dimensional vortex motion by a system of point vortices. Journal of the Physical Society of Japan. 1973;34(1):247–253. doi:10.1143/JPSJ.34.247

[16] A. J. Chorin. Numerical study of slightly viscous flow. Journal of Fluid Mechanics. 1973;57(4):785–796. doi:10.1017/S0022112073002016

[17] P. Degond and S. Mas-Gallic. The weighted particle method for convection-diffusion equations. Part 1: the case of an isotropic viscosity. Mathematics of Computation. 1989;53(188):485–507. doi:10.2307/2008716

[18] D. Fishelov. A new vortex scheme for viscous flows. Journal of Computational Physics. 1990;86(1):211–224. doi:10.1016/0021-9991(90)90098-L

[19] S. Shankar and L. van Dommelen. A new diffusion procedure for vortex methods. Journal of Computational Physics. 1996;127(1):88–109. doi:10.1006/jcph.1996.0160

[20] J. M. R. Graham. Computation of viscous separated flow using a particle method. In: K. W. Morton and M. J. Baines, editors. Numerical Methods for Fluid Dynamics III. Oxford: Oxford University Press; 1988. p. 310–317.

[21] J. T. Beale and A. Majda. Vortex methods. II: higher order accuracy in two and three dimensions. Mathematics of Computation. 1982;39(159):29–52. doi:10.2307/2007618

[22] C. H. Liu and D. J. Doorly. Vortex particle-in-cell method for three-dimensional viscous unbounded flow computations. International Journal for Numerical Methods in Fluids. 2000;32:29–50. doi:10.1002/(SICI)1097-0363(20000115)32:1%3C23::AID-FLD922%3E3.0.CO;2-O

[23] R. Cocle, G. Winckelmans, and G. Daeninck. Combining the vortex-in-cell and parallel fast multipole methods for efficient domain decomposition simulations. Journal of Computational Physics. 2008;227(21):9091–9120. doi:10.1016/j.jcp.2007.10.010

[24] P. Angot, C.-H. Bruneau, and P. Fabrie. A penalization method to take into account obstacles in viscous flows. Numerische Mathematik. 1999;81(4):497–520. doi:10.1007/s002110050401

[25] LiM. El Ossmani and P. Poncet. Efficiency of multiscale hybrid grid-particle vortex methods. Multiscale Modeling and Simulation. 2010;8(5):1671–1690. doi:10.1137/090765006

[26] R. Chatelin and P. Poncet. Hybrid grid–particle methods and penalization: a Sherman-Morrison-Woodbury approach to compute 3D viscous flows using FFT. Journal of Computational Physics. 2014;269(15):314–328. doi:10.1016/j.jcp.2014.03.023

[27] H. Beaugendre, F. Morency, F. Gallizio, and S. Laurens. Computation of ice shedding trajectories using Cartesian grids, penalization, and level sets. Modelling and Simulation in Engineering. 2011:274947. doi:10.1155/2011/274947

[28] S.-J. Lee, J.-H. Lee, and J.-C. Suh. Computation of pressure fields around a two-dimensional circular cylinder using the vortex-in-cell and penalization methods. Modelling and Simulation in Engineering. 2014:708372. doi:10.1155/2014/708372

[29] G.-H. Cottet, F. Gallizio, A. Magni, and I. Mortazavi. A vortex immersed boundary method for bluff body flows. In: Proceedings of 3rd Joint US-European ASME Fluids Engineering Summer Meeting; August 1-5; Montreal, Canada. 2010.

[30] C. Mimeau, F. Gallizio, G.-H. Cottet, and I. Mortazavi. Votex penalization method for bluff body flows. International Journal for Numerical Methods in Fluids. 2015;79:55–83. doi:10.1002/fld.4038

[31] M. Gazzola, C. Mimeau, A. A. Tchieu, and P. Koumoutsakos. Flow mediated interactions between two cylinders at finite Re numbers. Physics of Fluids. 2012;24:043103. doi:10.1063/1.4704195

[32] S.-J. Lee, J.-H. Lee, and J.-C. Suh. Further validation of the hybrid particle-mesh method for vortex shedding flow simulations. International Journal of Naval Architecture and Ocean Engineering. 2015;7:1034–1043. doi:10.1515/ijnaoe-2015-0072

[33] J. T. Rasmussen, G.-H. Cottet, and J. H. Walther. A multiresolution remeshed vortex-in-cell algorithm using patches. Journal of Computational Physics. 2011;230(19):6742–6755. doi:10.1016/j.jcp.2011.05.006

[34] J. J. Monaghan. Extrapolating B splines for interpolation. Journal of Computational Physics. 1985;60(2):253–262. doi:10.1016/0021-9991(85)90006-3

[35] W. Iwakami, Y. Yatagai, N. Hatakeyama, and Y. Hattori. A new approach for error reduction in the volume penalization method. Communications in Computational Physics. 2014;16(5):1181–1200. doi:10.4208/cicp.220513.070514a

[36] G. S. Winckelmans and A. Leonard. Contributions to vortex particle methods for the computation of three-dimensional incompressible unsteady flows. Journal of Computational Physics. 1993;109(2):247–273. doi:10.1006/jcph.1993.1216

[37] S.-J. Lee, J.-H. Lee, and J.-C. Suh. Numerical investigation on vortex shedding from a hydrofoil with a beveled trailing edge. Modelling and Simulation in Engineering. 2015:565417. doi:10.1155/2015/565417

[38] M. L. Ould-Salihi, G.-H. Cottet, and M. El Hamraoui. Bending finite-difference and vortex methods for incompressible flow computations. SIAM Journal on Scientific Computing. 2000;22(5):1655–1674. doi:10.1137/S1064827599350769

[39] M. Coquerelle and G.-H. Cottet. A vortex level set method for the two-way coupling of an incompressible fluid with colliding rigid bodies. Journal of Computational Physics. 2008;227(21):9121–9137. doi:10.1016/j.jcp.2008.03.041

[40] F. Morency, H. Beaugendre, and F. Gallizio. Aerodynamic for evaluation for ice shedding phenomenon using vortex in cell scheme, penalization and level set approaches. International Journal of Computational Fluid Dynamics. 2012;26(9–10):435–450. doi:10.1080/10618562.2012.739683

[41] A. Kosior and H. Kudela. Parallel computations on GPU in 3D using the vortex particle method. Computers and Fluids. 2013;80(10):423–428. doi:10.1016/j.compfluid.2012.01.014

[42] T. Lonfils and G. Winckelmans. Development of an immersed boundary method using boundary elements within a vortex-in-cell/parallel fast multipole method. In: J. C. F. Pereira and A. Sequeira, editors. European Conference on Computational Fluid Dynamics, ECCOMAS CFD; June 14–17; Lisbon, Portugal. 2010.

[43] P. Ploumhans and G. S. Winckelmans. Vortex methods for high-resolution simulations of viscous flow pat bluff bodies of general geometry. Journal of Computational Physics. 2000;165(2):354–406. doi:10.1006/jcph.2000.6614

[44] D. Rossinelli, M. Bergdorf, G.-H. Cottet, and P. Koumoutsakos. GPU accelerated simulations of bluff body flows using vortex particle methods. Journal of Computational Physics. 2010;229(9):3316–3333. doi:10.1016/j.jcp.2010.01.004

[45] U. Trottenberg, C. Oosterlee, and A. Schuller. Multigrid. Oxford, UK: Elsevier; 2001. P. 631.

[46] M. Frigo and S. G. Johnson. The design and implementation of FFTW3. Proceedings of the IEEE. 2005;93(2):216–231. doi:10.1109/JPROC.2004.840301

[47] W. H. Press, S. A. Teukolsky, W. T. Vetterling, and B. P. Flannery. Numerical Recipes in C++. Cambridge, UK: Cambridge University Press; 2002. p. 1002.

[48] M. M. Hejlesen, P. Koumoutsakos, A. Leonard, and J. H. Walther. Iterative Brinkman penalization for remeshed vortex methods. Journal of Computational Physics. 2015;280(1):547–562. doi:10.1016/j.jcp.2014.09.029

[49] S.-J. Lee. Numerical simulation of single-bubble dynamics with two-way coupling uing the Lagrangian vortex method [thesis]. Seoul, South Korea: Seoul National University; 2005. 193 p.

[50] S.-J. Lee and J.-C. Suh. Fast computation of domain boundary conditions using a cubic Spline for the penalized VIC method. International Journal of Computational Methods. Forthcoming.

[51] B. Fornberg. A numerical study of steady viscous flow past a circular cylinder. Journal of Fluid Mechanics. 1980;98(4):819–855. doi:10.1017/S0022112080000419

[52] A. S. Grove, F. H. Shair, and E. E. Petersen. An experimental investigation of the steady separated flow past a circular cylinder. Journal of Fluid Mechanics. 1964;19(1):60–80. doi:10.1017/S0022112064000544

[53] R. Bouard and M. Coutanceau. The early stage of development of the wake behind an impulsively started cylinder for $40 < Re < 10^4$. Journal of Fluid Mechanics. 1980;101(3): 583–607. doi:10.1017/S0022112080001814

[54] M. Bar-Lev and H. T. Yang. Initial flow field over an impulsively started circular cylinder. Journal of Fluid Mechanics. 1975;72(4):625–647. doi:10.1017/S0022112075003199

[55] P. Koumoutsakos and A. Leonard. High-resolution simulations of the flow around an impulsively started cylinder using vortex methods. Journal of Fluid Mechanics. 1995;296:1–38. doi:10.1017/S0022112095002059

[56] H. Kudela and T. Kozlowski. Vortex in cell method for exterior problems. Journal of Theoretical and Appiled Mechanics. 2009;47(4):779–796.

[57] C. Norberg. Pressure distribution around a circular cylinder in cross-flow. In: The Symposium on Bluff Body Wakes and Vortex-Induced Vibration (BBVIC); Queensland, Australia. 2002.

[58] T. P. Loc and R. Bouard. Numerical solution of the early stage of the unsteady viscous flow around a circular cylinder: a comparison with experimental visualization and

measurements. Journal of Fluid Dynamics. 1985;160:93–117. doi:10.1017/S0022112085003408

[59] T. A. Johnson and V. C. Patel. Flow past a sphere up to a Reynolds number of 300. Journal of Fluid Mechanics. 1999;378:19–70. doi:10.1017/S0022112098003206

[60] S. Taneda. Experimental investigation of the wake behind a sphere at low Reynolds numbers. Journal of the Physical Society of Japan. 1956;11:1104–1108. doi:10.1143/JPSJ.11.1104

[61] P. Bagchi and S. Balachandar. Shear versus vortex-induced lift force on a rigid sphere at moderate Re. Journal of Fluid Mechanics. 2002;473:379–388. doi:10.1017/S0022112002002628

[62] P. Ausoni, M. Farhat, X. Escaler, E. Egusquiza, and F. Avellan. Cavitation influence on Karman vortex shedding and induced hydrofoil vibrations. Journal of Fluid Engineering. 2007;129(8):966–973. doi:10.1115/1.2746907

[63] Ausoni. Turbulent vortex shedding from a blunt trailing edge hydrofoil [thesis]. École polytechnique fédérale de Lausanne, EPFL; 2009. 169 p. Available from: https://infoscience.epfl.ch/record/138935

[64] A. Zobeiri, P. Ausoni, F. Avellan, and M. Farhat. How oblique trailing edge of a hydrofoil reduces the vortex-induced vibration. Journal of Fluids and Structures. 2012;32:78–89. doi:10.1016/j.jfluidstructs.2011.12.003

[65] D. A. Bourgoyne, S. L. Ceccio, and D. R. Dowling. Vortex shedding from a hydrofoil at high Reynolds number. Journal of Fluid Mechanics. 2005;531:293–324. doi:10.1017/S0022112005004076

[66] S. P. Singh and S. Mittal. Flow past a cylinder: shear layer instability and drag crisis. International Journal for Numerical Methods in Fluids. 2005;47:75–98. doi:10.1002/fld.807

[67] G. Boffetta and R. E. Ecke. Two-dimensional turbulence. Annual Review of Fluid Mechanics. 2012;44:427–451. doi:10.1146/annurev-fluid-120710-101240

Chapter 7

Vortex Structures in Saturable Media

İlkay Bakırtaş

Additional information is available at the end of the chapter

http://dx.doi.org/10.5772/67074

Abstract

In this chapter, dipole and vortex solitons are computed using spectral renormalization method in the focusing two-dimensional saturable nonlinear Schrödinger (SNLS) equation with periodic and quasicrystal lattice potentials. The nonlinear stability of these multi-humped solitons is investigated using direct simulations of the SNLS equation. It is shown that multiple vortex structures on quasicrystal lattices can be nonlinearly stable as the saturation and the external lattice may prevent the collapse. These results may have applications to investigations of localized structures in nonlinear optics and Bose-Einstein condensates.

Keywords: saturable nonlinearity, vortex, dipole, Penrose quasicrystal, nonlinear stability

1. Introduction

Solitons with embedded vorticity, dipoles and other nonlinearly localized structures with a complex phase have attracted considerable interest in nonlinear optics research and in Bose-Einstein condensation (BEC). These types of multi-humped structures in the presence of an (optically or magnetically) induced potential have been reported in Bose-Einstein condensates (BECs) (cf. [1, 2]) and in optical Kerr media (cf. [3–8]).

Although these structures have not been truly observed in $\chi^{(3)}$ cubic media due to collapse, they appear as special numerical solutions of the focusing (2+1)D cubic nonlinear Schrödinger (NLS) equation with an external potential (lattice).

The stability of these solitons is important to applications. Rigorous stability theory of *fundamental* (positive) solitons is well established [9–12]. It is known that self-focusing media allow the stable propagation of one-dimensional (1+1)D spatial solitons, due to the balance between the linear diffraction and self focusing. However, two-dimensional (2+1)D optical solitons in

Kerr media with the instantaneous cubic nonlinearity are unstable, due to the self-focusing (collapse) at high powers. Nevertheless, saturation of the nonlinearity may prevent the collapse, securing stable soliton propagation.

In non-homogeneous media, it has been shown that solitons can undergo self-focusing and drift (tunneling) instabilities [13–16].

It has been previously shown that multi-humped soliton solutions of (2+1)D cubic NLS equation with a periodic external lattice can be stable in a certain parameter regime [17–19] and also in saturable (e.g. photorefractive) media [20–22]. However, the (computational) stability properties of multiple vortex solitons on background quasicrystal lattices have remained relatively unexplored. In Kerr media, in Ref. [23], it is computationally demonstrated that the two-dimensional cubic nonlinear Schrödinger (NLS) equation with a quasicrystal lattice potential admits multiple dipole and vortex solitons. In aforementioned work, it is numerically shown that certain multiple vortex structures on quasicrystal lattices tend to be nonlinearly stable if the vortex/dipole humps are located on lattice minima but they are nonlinearly unstable if the humps are located on lattice maxima.

The analytic stability theory of optical modes on lattices that possess a higher topological complexity, for example, lattices containing defects, dislocations and/or quasicrystal lattices, is not sufficiently developed. On the other hand, this issue is studied numerically in recent years [24] and it is reported that dipole and vortex solitons for the two-dimensional nonlinear Schrödinger (NLS) equation with external potentials that possess strong irregularities, that is, edge dislocations and a vacancy defects, exist and it is numerically observed that these multi-humped modes in the defect lattices can be stable or unstable depending on the lattice properties.

In this study, we compute soliton solutions of the focusing (2+1)D saturable nonlinear Schrödinger equation with periodic and quasicrystal (e.g. Penrose) potentials and study their nonlinear stability. The governing equation is

$$iu_z + \Delta u - \frac{e_0 u}{1 + V(x,y) + |u|^2} = 0, \tag{1}$$

where $\Delta u \equiv u_{xx} + u_{yy}$ is the diffraction term, z plays the role of time (or distance). In optics, e_0 corresponds to the applied DC bias field inducing the saturable nonlinearity, $u(x,y,z)$ is the complex-valued, slowly varying amplitude of the electric field in the xy-plane propagating in the z direction, $V(x,y)$ is an external optical potential (lattice) that can be formulated as the intensity of a sum of N phase-modulated plane waves [19], that is,

$$V(x,y) = \frac{V_0}{N^2}\left|\sum_{n=0}^{N-1} e^{i\vec{k}_n.\vec{x}}\right|^2. \tag{2}$$

where $V_0 > 0$ is constant and corresponds to the peak depth of the potential, that is, $V_0 = \max_{x,y} V(x,y)$, $\vec{x} = (x,y)$, $\vec{k}_n$ is a wave vector defined by $(k_x^n, k_y^n) = [K\cos(2\pi n/N), K\sin(2\pi n/N)]$. The external potential defined in Eq. (2) with $N = 2,3,4,6$ corresponds to periodic structures (crystals) and with $N = 5, 7$ correspond to quasicrystals, which have a local symmetry around the origin but, different than that of periodic crystals, these types of structures are not invariant under spatial translations.

In this work, as the external potential, we consider the periodic and quasicrystal lattices corresponding to $N = 4$ and $N = 5$ in Eq. (2), respectively. In particular, the quasicrystal with $N = 5$ is often called the Penrose tiling [25]. Contour images of these potentials are displayed all with $V_0 = 1$ and $K = 1$ in **Figure 1**. Freedman et al. observed solitons in Penrose and other quasicrystal lattices generated by the optical induction method [6].

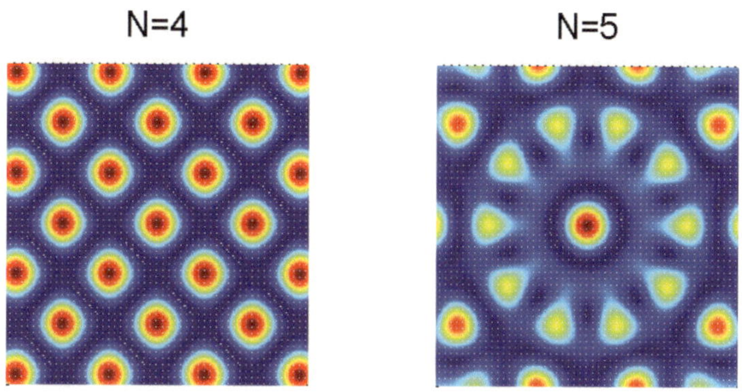

Figure 1. (Color online) Contour images of the lattices, all with $V_0 = 10$ and $K = 1$.

In order to compute localized solutions (i.e., soliton solutions) to nonlinear evolution equations arising in optics, various techniques have been used, for example, shooting and relaxation techniques and the self-consistency method. A different method is introduced by Petviashvili [26] to construct localized solutions in the two-dimensional Korteweg-de Vries equation (usually referred to as the Kadomtsev-Petviashvili equation).

Petviashvili's method aims to transform the governing equation to Fourier space and determines a convergence factor based upon the homogeneity of a single nonlinear term. This method has been used to find localized solutions in a wide range of nonlinear systems in many works. This method can be successfully applied to nonlinear systems only if the degree of the nonlinearity is fixed in the related evolution equation. It is a well-known fact that in nonlinear optics, many equations involve nonlinearities with different homogeneities, such as cubic-quintic, or even lack of homogeneity, as in saturable nonlinearity. A novel numerical scheme is proposed by Ablowitz and Musslimani [27] for computing solitons in nonlinear wave guides. This method is called spectral renormalization method and the idea behind the method is to transform the governing equation into Fourier space and find a nonlinear nonlocal integral equation coupled to an algebraic equation. The coupling prevents the numerical scheme from diverging. By this iteration scheme, the initial condition rapidly converges to an optical mode which is the numerical solution of the governing equation. This method can efficiently be applied to a large class of nonlinear wave problems including higher-order nonlinear terms with different homogeneities.

In this work, we use spectral renormalization method as explained below.

2. Spectral renormalization method

Spectral renormalization method is essentially a Fourier iteration method. The idea of this method was proposed by Petviashvili. Later, this method is improved by Ablowitz et al. and applied to (2+1)D NLS equation. In this subsection, the method is configured so that it can be applied to the saturable NLS equation.

Consider the saturable (2+1)D nonlinear Schrödinger equation with a potential in three-dimensional space:

$$iu_z(x,y,z) + u_{xx}(x,y,z) + u_{yy}(x,y,z) - \frac{e_0 u(x,y,z)}{1 + V(x,y) + |u(x,y,z)|^2} = 0 \tag{3}$$

Using the ansatz $u(x,y,z) = f(x,y)e^{-i\mu z}$ gives

$$\mu f e^{-i\mu z} + (f_{xx} + f_{yy})e^{-i\mu z} - \frac{e_0 f e^{-i\mu z}}{1 + V + |f|^2} = 0. \tag{4}$$

Multiplying both sides of this equation by $e^{i\mu z}$ results in

$$\mu f + f_{xx} + f_{yy} - \frac{e_0 f}{1 + V + |f|^2} = 0. \tag{5}$$

By applying Fourier transformation, one obtains

$$\mu \hat{f} - (k_x^2 + k_y^2)\hat{f} - \mathcal{F}\left(\frac{e_0 f}{1 + V + |f|^2}\right) = 0. \tag{6}$$

To prevent singularities in the denominator in future calculations, the term $r\hat{f}$ is added to and subtracted from Eq. (6):

$$(\mu + r)\hat{f} - (k_x^2 + k_y^2 + r)\hat{f} - \mathcal{F}\left(\frac{e_0 f}{1 + V + |f|^2}\right) = 0. \tag{7}$$

Solving for the $\hat{f}$ in the second term above yields

$$\hat{f} = \frac{(\mu + r)\hat{f} - \mathcal{F}\left(\frac{e_0 f}{1+V+|f|^2}\right)}{k_x^2 + k_y^2 + r}. \tag{8}$$

Making the substitution $f(x,y) = \lambda w(x,y)$ where λ is a non-zero constant gives

$$\lambda\hat{w} = \frac{(\mu + r)\lambda\hat{w} - \mathcal{F}\left(\frac{e_0\lambda w}{1+V+|\lambda|^2|w|^2}\right)}{k_x^2 + k_y^2 + r}. \tag{9}$$

Dividing by λ yields

$$\hat{w} = \frac{(\mu + r)\hat{w} - \mathcal{F}\left(\frac{e_0 w}{1+V+|\lambda|^2|w|^2}\right)}{k_x^2 + k_y^2 + r}. \tag{10}$$

When indexed, Eq. (10) can be utilized in an iterative method in order to find w. For this purpose, $\hat{w}$ can be calculated using the following iteration scheme:

$$\hat{w}_n = \frac{(\mu + r)\hat{w}_{n-1} - \mathcal{F}\left(\frac{e_0 w_{n-1}}{1+V+|\lambda|^2|w_{n-1}|^2}\right)}{k_x^2 + k_y^2 + r}, n \in Z^+ \tag{11}$$

with the initial condition taken as a Gaussian-type function (for the fundamental soliton)

$$w_0 = e^{-((x-x_0)^2+(y-y_0)^2)} \tag{12}$$

and the stopping criteria $|w_n - w_{n-1}| < 10^{-8}$. Here, the values of x_0 and y_0 define the location of the initial condition. In order to center the initial condition on the lattice maximum (the one that appears at the center of the lattice), one should, for example, take $x_0 = y_0 = 0$ and to center the initial condition on a lattice minimum (usually taken as one of the closest minima to the central maximum), one should take $x_0 = \pi, y_0 = 0$, for the periodic lattice $N = 4$.

However, λ is unknown and hence must be calculated for each iteration

Multiplying Eq. (**10**) by $k_x^2 + k_y^2 + r$ leads to

$$(k_x^2 + k_y^2 + r)\hat{w} = (\mu + r)\hat{w} - \mathcal{F}\left(\frac{e_0 w}{1 + V + |\lambda|^2|w|^2}\right). \tag{13}$$

After moving all terms to the left side, one has

$$(k_x^2 + k_y^2 - \mu)\hat{w} + \mathcal{F}\left(\frac{e_0 w}{1 + V + |\lambda|^2|w|^2}\right) = 0. \tag{14}$$

Multiplying by the conjugate of $\hat{w}$, that is, by $\hat{w}^*$ results in

$$(k_x^2 + k_y^2 - \mu)|\hat{w}|^2 + \mathcal{F}\left(\frac{e_0 w}{1 + V + |\lambda|^2|w|^2}\right)\hat{w}^* = 0. \tag{15}$$

Finally, by integrating this equation, one gets

$$\int_{-\infty}^{\infty}\int_{-\infty}^{\infty}(k_x^2+k_y^2-\mu)|\hat{w}|^2dk+\int_{-\infty}^{\infty}\int_{-\infty}^{\infty}\mathcal{F}\left(\frac{e_0 w}{1+V+|\lambda|^2|w|^2}\right)\hat{w}^*dk=0. \tag{16}$$

This is nothing but an equation of the form $F(\lambda)=0$ which can be solved for λ by employing the Newton-Raphson method, for instance. The Newton-Raphson method is a numerical method for finding roots of equations and is given by the following iteration scheme:

$$\lambda_n=\lambda_{n-1}-\frac{F(\lambda_{n-1})}{F'(\lambda_{n-1})},\quad n\in Z^+ \tag{17}$$

where $F'=\frac{dF}{d\lambda}$ and λ_0 are the initial guess for the root.

Here

$$F(\lambda)=\int_{-\infty}^{\infty}\int_{-\infty}^{\infty}(k_x^2+k_y^2-\mu)|\hat{w}|^2dk+\int_{-\infty}^{\infty}\int_{-\infty}^{\infty}\mathcal{F}\left(\frac{e_0 w}{1+V+\lambda^2|w|^2}\right)\hat{w}^*dk, \tag{18}$$

$$F'(\lambda)=\int_{-\infty}^{\infty}\int_{-\infty}^{\infty}(-2\lambda)\mathcal{F}\left(\frac{e_0 w|w|^2}{(1+V+\lambda^2|w|^2)^2}\right)\hat{w}^*dk, \tag{19}$$

$\lambda_0=1$ and the stopping criterion is $|\lambda_n-\lambda_{n-1}|<10^{-8}$.

Once $\hat{w}$ is obtained from Eq. (11) by means of Eq. (17) to calculate λ for each iteration, the desired soliton is $f(x,y)=\lambda w(x,y)=\lambda\mathcal{F}^{-1}(\hat{w})$.

In this work, we numerically find both multiple dipole and vortex solitons on periodic ($N=4$) and quasicrystal "Penrose" ($N=5$), background lattices. The nonlinear (in)stabilities are also examined for these localized structures by direct computations of Eq. (1). The initial conditions are taken to be a dipole or a vortex.

3. Numerical investigation of dipole and vortex solitons

In this section, we show the existence of both dipole and vortex solitons centered at the lattice maxima for both periodic and the Penrose potentials. Hereafter, the potential depth is set to $V_0=1$, the saturation parameter is set to $e_0=8$ and the propagation constant for all dipole and vortex structures is taken to be as $\mu=4$.

For the spectral renormalization, we used the following initial condition, centered at the lattice maxima for both periodic and the Penrose potentials

$$w_0(x,y,0)=A\sum_{n=0}^{M-1}e^{-[(x+x_n)^2+(y+y_n)^2]+i\theta_n} \tag{20}$$

where x_n, y_n represent the location of vortex solitons, θ_n is the phase difference, M corresponds to the number of humps and A is a positive integer.

Since the lattice solitons located on lattice maxima are found to be nonlinearly unstable for cubic nonlinearity case in Ref. [23], in this work, we deal with the dipoles and vortex solitons located on lattice maxima to demonstrate the stabilization effect of the saturation.

3.1. Dipole solitons

A dipole or a two-phase localized vortex is found for the periodic and Penrose lattices.

The dipole profile, its phase structure and the contour plot of the dipole humps superimposed on the underlying periodic potential are shown in **Figure 2**.

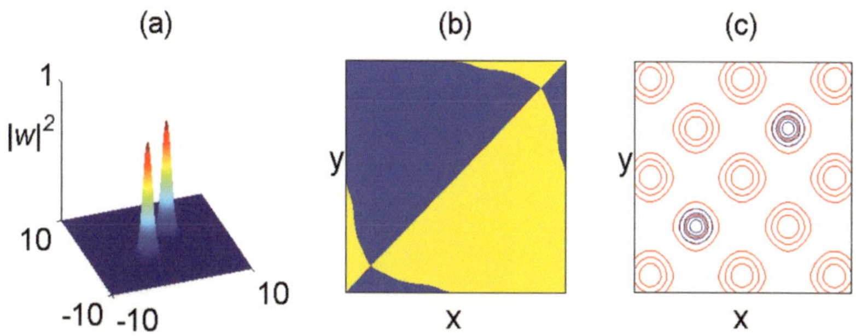

Figure 2. (a) A dipole profile centered at the lattice maxima of a periodic potential, (b) the phase structure of the dipole and (c) the contour plot of the dipole humps superimposed on the underlying periodic potential ($N = 4$).

Dipoles at the lattice maxima for quasicrystal (Penrose) potentials are also found numerically; here $A = 3$ and $r = 6.2504$. The dipole profile, its phase structure and the contour plot of the dipole humps superimposed on the underlying Penrose lattice are shown in **Figure 3**.

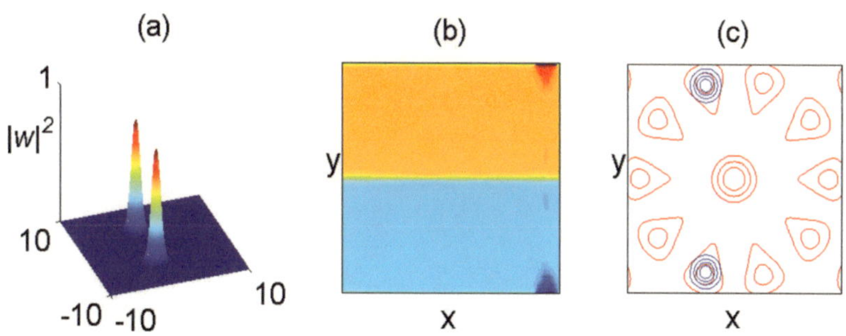

Figure 3. (a) A dipole profile centered at the lattice maxima, (b) the phase structure of dipole solitons and (c) the contour plot of dipole solitons superimposed on the underlying Penrose potential ($N = 5$).

3.2. Vortex solitons

Vortex solitons on lattice maxima including four-hump vortex solitons on a periodic lattice and both five- and 10-hump vortex solitons on a Penrose lattice were also investigated.

First, we consider the lattice which is periodic along the x and y directions which corresponds to $N = 4$ in Eq. (**2**). Four-humped vortex solitons were obtained by taking the initial maxima of the periodic lattice, x_n, y_n, θ_n to be

$$x_n = \frac{\sqrt{2\pi}}{2}\cos\theta_n, \quad y_n = \frac{\sqrt{2\pi}}{2}\sin\theta_n, \quad \theta_n = \frac{(2n-1)\pi}{N}, \quad n = 0,1,2,3. \tag{21}$$

Using the above initial condition, we numerically found a four-hump vortex centered at the periodic lattice maxima. The final vortex profile, the phase structure and the contour plot of the vortex solitons superimposed on the lattice maxima are shown in **Figure 4**.

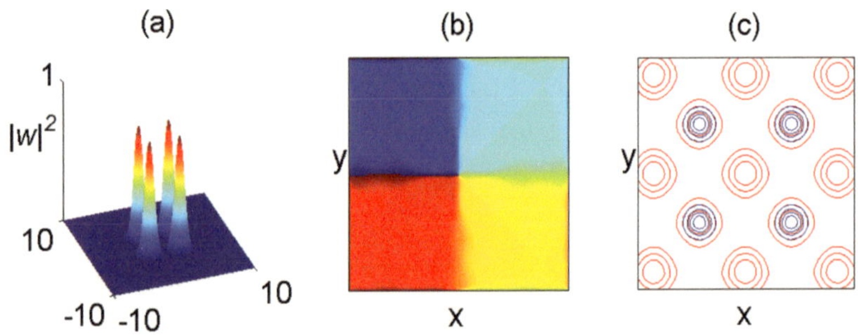

Figure 4. (a) Vortex profile centered at the periodic lattice maxima, (b) the phase structure of the vortex soliton and (c) the contour plot of the vortex soliton superimposed on the underlying periodic lattice ($N = 4$).

Similarly, a five-hump vortex centered at the maxima of the Penrose potential shown in **Figure 5** is obtained when $\theta_n = 2\pi n/5$. Both the five-hump and the 10-hump vortices are found.

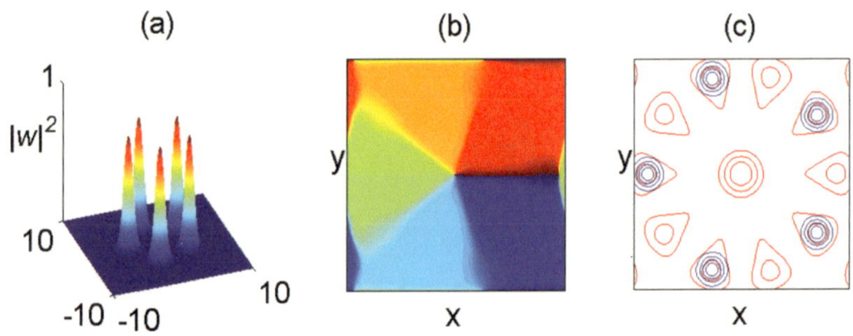

Figure 5. (a) Vortex profile centered at the Penrose lattice maxima, (b) the phase structure of the vortex soliton and (c) the contour plot of the vortex soliton superimposed on the underlying Penrose lattice ($N = 5$).

The 10-hump vortex located at the Penrose lattice maxima is shown in **Figure 6**; in this case $\theta_n = \pi n/5$.

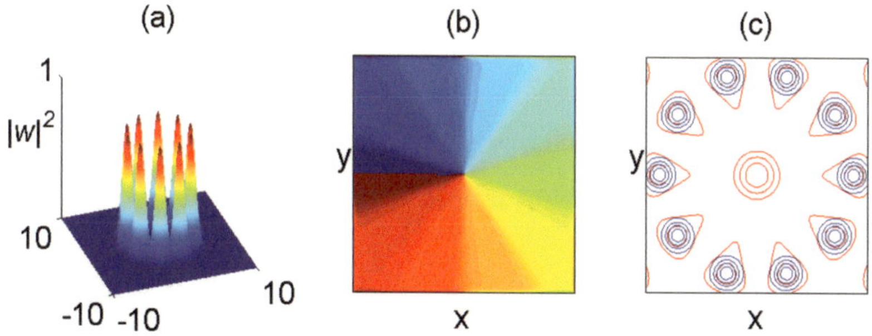

Figure 6. (a) Vortex profile centered at the Penrose lattice maxima, (b) the phase structure of the vortex soliton and (c) the contour plot of the vortex soliton superimposed on the underlying Penrose lattice ($N = 5$).

4. Nonlinear stability of vortex and dipole solitons

An important issue is the nonlinear stability of these vortex and dipole solitons. In order to examine the nonlinear stability of the vortex and dipole solitons found above, we directly compute Eq. (1), over a long distance ($z = 30$ is typically found to be sufficient) for both types of potentials. The initial conditions were taken to be a multi-phased optical mode, namely a dipole or a vortex.

The nonlinear stability of the dipoles and the vortex structures are investigated by monitoring the maximum amplitude versus the propagation distance, the change in the location of centers of mass and the phase of the dipole/vortex solitons.

A dipole/vortex is assumed to be stable if it preserves

1. Its peak amplitude, as opposed undergoing self-focusing and/or finite-distance collapse.
2. Its position on the lattice, that is, be drift-stable (drift-unstable solitons are typically characterized by "humps" that drift from lattice maxima toward nearby minima).
3. Its phase structure.

The center of mass is calculated as

$$(\langle x \rangle, \langle y \rangle) = \frac{1}{P}\int_{-\infty}^{\infty}\int_{-\infty}^{\infty} |u|^2 dxdy. \tag{22}$$

Here, $P \equiv P[u] := \int_{-\infty}^{\infty}\int_{-\infty}^{\infty} |u|^2 dxdy$ is the soliton power.

One should note that, for each case, the average displacement of the center of mass in the x-direction may change from negative to positive values depending on the initial locations of the dipole/vortex humps.

The nonlinear stability of vortex and dipole solitons is investigated in the following subsections separately.

4.1. Nonlinear stability of dipole solitons

In this subsection, the nonlinear stability properties of dipole solitons on periodic and Penrose lattice maxima are investigated.

The nonlinear stability properties of the dipole solitons on periodic lattice maxima are demonstrated in **Figure 7**. As the initial condition, we took the dipole solitons at periodic lattice maxima, as shown in **Figure 2**. The maximum amplitude, the location of the centers of mass of dipole solitons on periodic lattice maxima versus the propagation distance $z = 30$ are plotted in **Figure 7**. The dipole profile at the final point $z = 30$ is also depicted in the same figure. It is clearly seen that the dipole solitons on periodic lattice maxima are nonlinearly stable since the maximum amplitude oscillates with relatively small amplitude and the dipole solitons preserve their shape and location during the evolution.

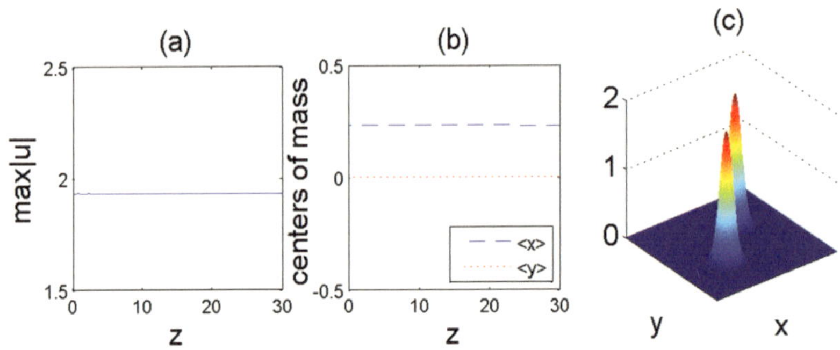

Figure 7. Nonlinear stability of a dipole on periodic lattice maxima. (a) Maximum amplitude as a function of propagation distance, (b) center of mass and (c) dipole profile at $z = 30$.

The contour plots and the complex phase structures of this dipole are also depicted along the propagation distance z in **Figure 8**. This figure reveals that the dipole located on periodic lattice maxima preserves its shape and complex phase along the evolution distance.

Next, we investigate the nonlinear stability of dipole solitons on the Penrose lattice maxima that is previously shown in **Figure 3**. Similar to the dipole on the periodic lattice maxima, these structures are also found to be nonlinearly stable due to the conservation of the location and the phase structure and at the same time, only showing small oscillations in the maximum amplitude during the evolution (see **Figures 9** and **10**).

We should note that, for cubic nonlinearity case, in Ref. [23], the nonlinear stability properties of the dipole solitons on both periodic and the Penrose lattice maxima are demonstrated and both dipole solitons are found to be nonlinearly unstable since they exhibit strong localization after a few diffraction lengths and breakup in their phase structures. The Penrose lattice dipole solitons also suffer from the drift instability since the dipole humps both move from the lattice maxima toward nearby lattice minima immediately.

http://dx.doi.org/10.5772/67074

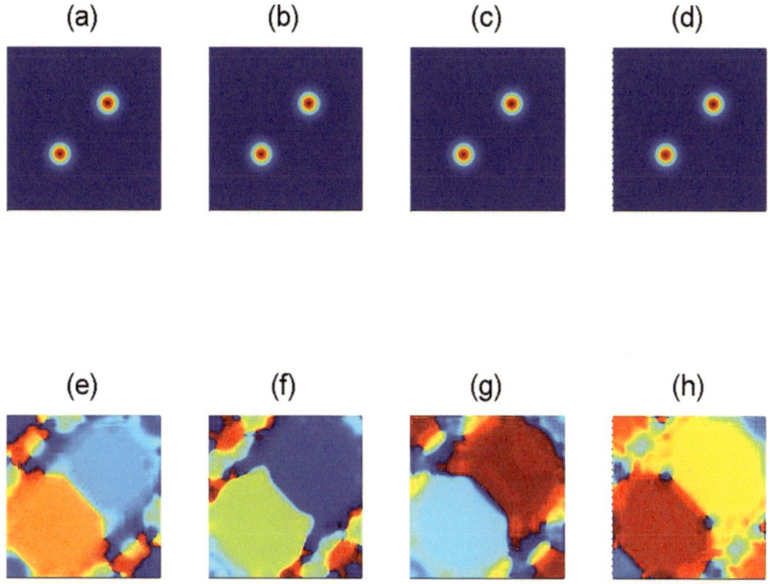

Figure 8. Nonlinear stability of a dipole on periodic lattice maxima. (a)–(d) Contour plots of the amplitude at $z = 10$, $z = 15$, $z = 20$, $z = 30$, respectively; (e)–(h) Contour plots of the complex phase at $z = 10$, $z = 15$, $z = 20$, $z = 30$, respectively.

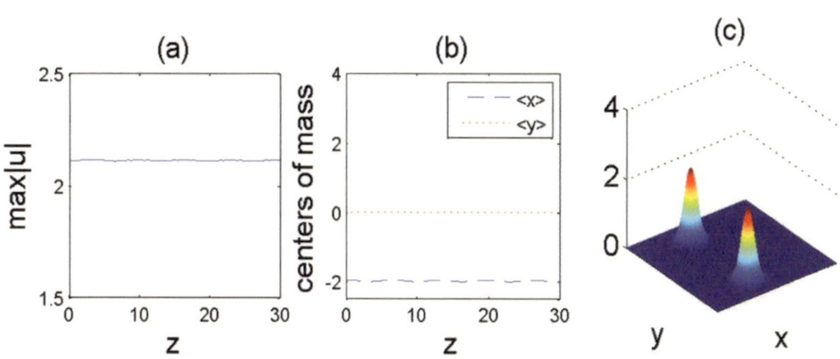

Figure 9. Nonlinear stability of a dipole on Penrose lattice maxima. (a) Maximum amplitude as a function of propagation distance, (b) center of mass and (c) dipole profile at $z = 30$.

4.2. Nonlinear stability of vortex solitons

The nonlinear stability properties of the vortex solitons on periodic and Penrose lattice maxima are also investigated in this study.

Following the same order with that of dipoles on lattice maxima case, we start with the four-hump vortex solitons on periodic lattice maxima. Taking the four-hump vortex on periodic lattice maxima (see **Figure 4**) as the initial condition, the nonlinear stability properties are examined. In **Figure 11**, we plot the maximum amplitude and the location of the centers of mass of the vortex humps versus the propagation distance $z = 30$ and the vortex profile at the final point $z = 30$. It is seen that the maximum amplitude of the vortex solitons oscillates with small amplitude. Furthermore, during the evolution, the vortex solitons do not move from their locations, meaning that there is no drift instability either.

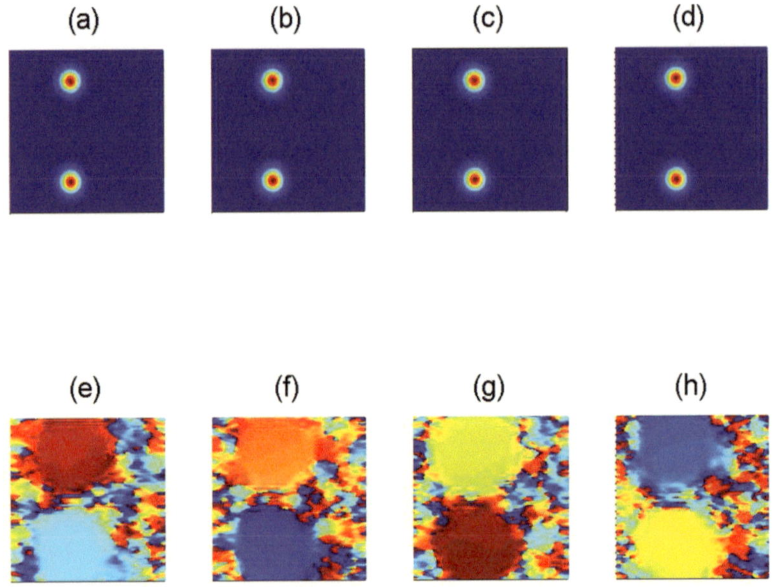

Figure 10. Nonlinear stability of a dipole on Penrose lattice maxima. (a)–(d) Contour plots of the amplitude at $z = 10$, $z = 15$, $z = 20$, $z = 30$, respectively; (e)–(h) Contour plots of the complex phase at $z = 10$, $z = 15$, $z = 20$, $z = 30$, respectively.

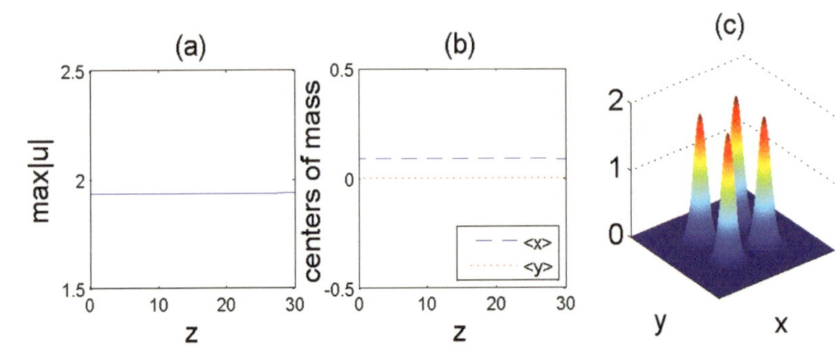

Figure 11. Nonlinear stability of four vortex solitons on periodic lattice maxima. (a) Maximum amplitude as a function of propagation distance, (b) center of mass and (c) vortex profile at $z = 30$.

The contour plots and the complex phase structures of this four-hump vortex are also depicted along the propagation distance z in **Figure 12**. This figure reveals that the four-hump vortex located on periodic lattice maxima perfectly preserves its shape and complex phase along the evolution distance. As a result of these facts, four-hump vortex solitons at periodic lattice maxima are found to be nonlinearly stable.

http://dx.doi.org/10.5772/67074

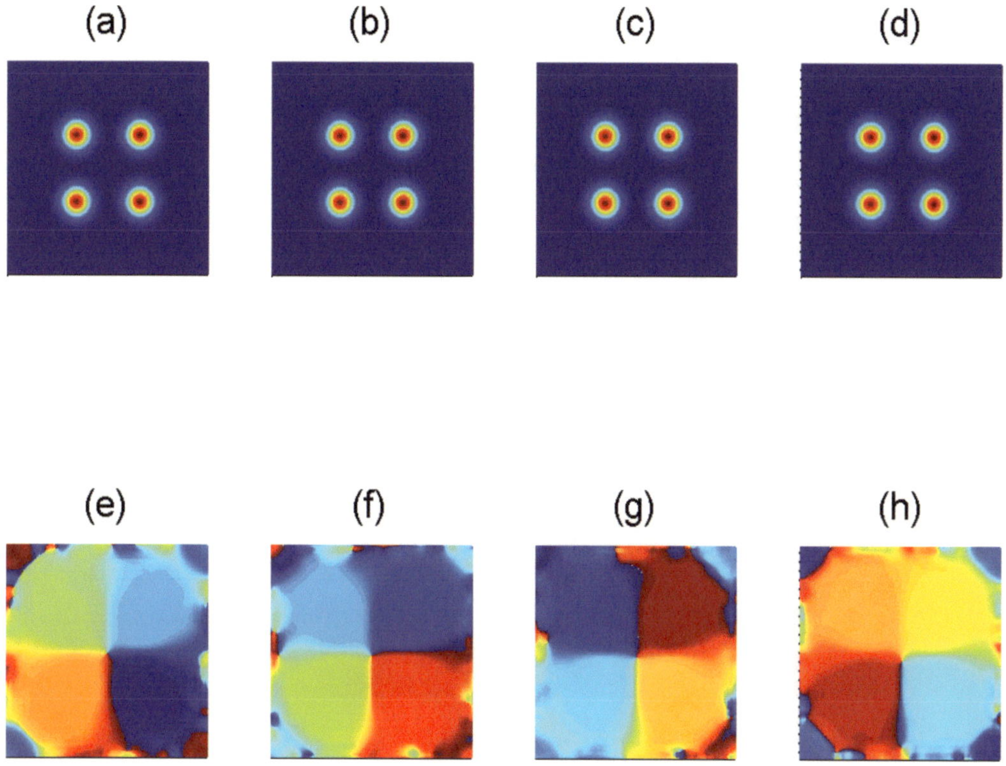

Figure 12. Nonlinear stability of a vortex on periodic lattice maxima. (a)–(d) Contour plots of the amplitude at $z = 10$, $z = 15$, $z = 20$, $z = 30$, respectively; (e)–(h) Contour plots of the complex phase at $z = 10$, $z = 15$, $z = 20$, $z = 30$, respectively.

Previously in Ref. [23], it is numerically demonstrated that, for the cubic nonlinearity case, the maximum amplitude of the four-hump vortex solitons centered at the periodic lattice maxima increases sharply, after $z = 0.3$ and due to this instant blow up in the maximum amplitude these vortex solitons are found to be nonlinearly unstable. It is shown that the phase structure also breaks up for the same vortex structure.

We investigated the nonlinear stability properties of the five- and 10-hump vortex solitons at the lattice maxima of Penrose potential by taking the five- (**Figure 5**) and 10-hump (**Figure 6**) vortices on Penrose lattice maxima as initial conditions, respectively.

As can be seen from **Figure 13**, the maximum amplitude of these five-hump vortex solitons oscillate with relatively small amplitude and vortex humps stay at the same place during the direct simulation (no drift instability). The contour plots and the complex phase structures of this five-hump vortex are also depicted along the propagation distance z in **Figure 14**. This figure reveals that the five-hump vortex located on Penrose lattice maxima preserves its shape and complex phase along the evolution distance. As a result of these facts, vortex solitons on the Penrose lattice maxima are found to be nonlinearly stable.

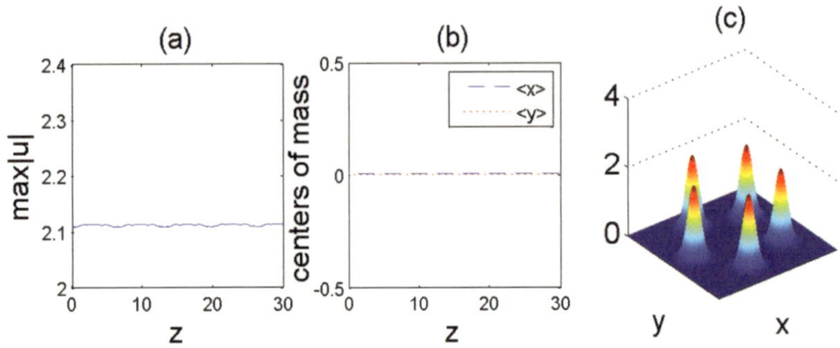

Figure 13. Nonlinear stability of five vortex solitons on Penrose lattice maxima. (a) Maximum amplitude as a function of propagation distance, (b) center of mass and (c) vortex profile at $z = 30$.

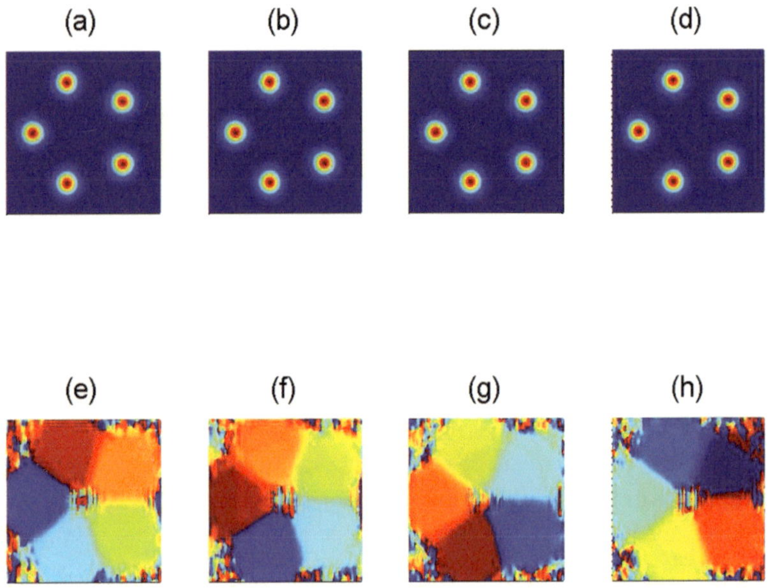

Figure 14. Nonlinear stability of a five vortex on Penrose lattice maxima. (a)–(d) Contour plots of the amplitude at $z = 10$, $z = 15$, $z = 20$, $z = 30$, respectively; (e)–(h) Contour plots of the complex phase at $z = 10$, $z = 15$, $z = 20$, $z = 30$, respectively.

Next, the nonlinear stability properties of vortex solitons on a Penrose lattice with 10 humps at lattice maxima are examined.

In **Figure 15**, we plot the maximum amplitude and the location of the centers of mass of a 10-hump vortex on Penrose lattice maxima versus the propagation distance $z = 15$. The contour plots and the complex phase structures of this 10-hump vortex are also depicted along the propagation distance z in **Figure 16**.

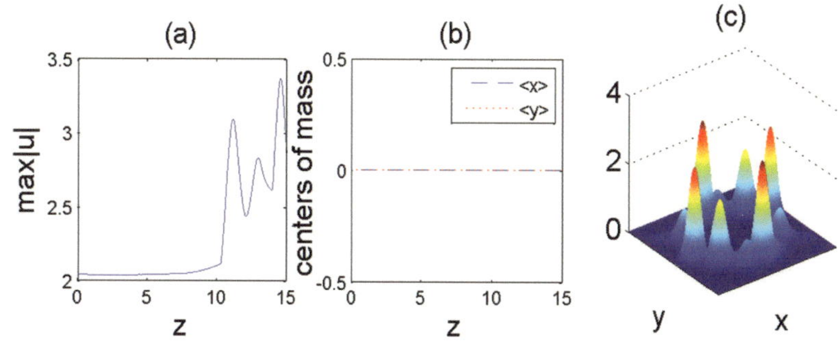

Figure 15. Nonlinear instability of 10 vortex solitons on Penrose lattice maxima. (a) Maximum amplitude as a function of propagation distance, (b) center of mass and (c) vortex profile at $z = 15$.

It is seen that the maximum amplitude of these vortex solitons increases somewhat until $z = 10$ and after this value, the maximum amplitude increases sharply. On the other hand, although the vortex humps preserve their locations, the phase structure starts to break up and amalgamation of four-vortex humps (two by two) is observed at around $z = 10$. At $z = 13$, the phase structure is entangled and amalgamation of vortex humps is very clear resulting in the disappearance of four-vortex humps. At $z = 15$, the phase structure is totally entangled and only four-vortex humps keep fairly acceptable amplitude and the others seem to disappear (see **Figure 16**). Therefore, as a result, we found the 10-hump vortex on Penrose lattice maxima as unstable.

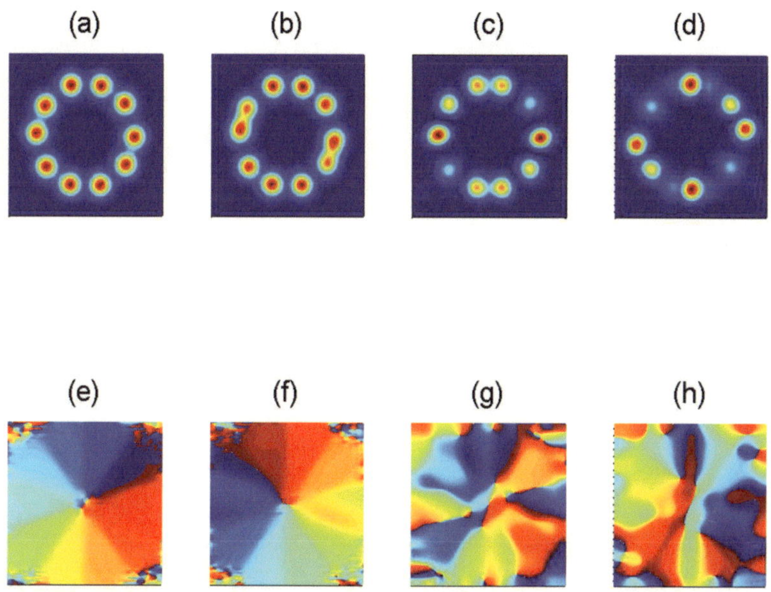

Figure 16. Nonlinear instability of 10 vortex solitons on Penrose lattice maxima. (a)–(d) Contour plots of the amplitude at $z = 8$, $z = 10$, $z = 13$, $z = 15$, respectively; (e)–(h) Contour plots of the complex phase at $z = 8$, $z = 10$, $z = 13$, $z = 15$, respectively.

In this case, it is clearly observed that the instability occurs due to amalgamation of vortex humps. In the Penrose lattice, as the number of the vortex humps increases, the humps get closer to each other which increases the possibility of amalgamation and this results in the instability. But still, the effect of saturation is very clear that the instability occurs later than that of cubic nonlinearity case. Again, one should note that, in previous cited work [23], for both cases, it is shown that the maximum amplitudes increase quite sharply after a short propagation distance (around $z = 0.8$ and $z = 0.75$ for five- and 10-hump cases, respectively) indicating nonlinear instability. The phase also breaks up these cases and moreover, during the evolution, both five-hump and 10-hump vortex solitons move from their initial locations (lattice maxima) to the lattice minima. The drift instability of lattice solitons, which is characterized by humps moving from the lattice maxima to the nearby lattice minima, was previously observed numerically for fundamental lattice solitons by Ablowitz et al. [28] (see also [14–16]).

5. Conclusion

We have numerically investigated dipole and multi-vortex structures associated with both periodic ($N = 4$) and Penrose ($N = 5$) lattices. First, we showed the numerical existence of vortex and dipole solitons on periodic and Penrose lattices. We investigated the nonlinear stability properties by simulations of saturable NLS equation. The simulations of the NLS equation showed that

i. both the dipole and the vortex solitons centered at the periodic lattice maxima are found to be nonlinearly *stable*;

ii. both the dipole and the five-hump vortex solitons centered at the Penrose lattice maxima are found to be nonlinearly *stable*;

iii. 10-hump vortex solitons situated at Penrose lattice maxima are nonlinearly *unstable*. In this case, the nonlinear instability occurs as a result of the amalgamation of two or more vortex humps.

In conclusion, if one compares the cubic nonlinearity case to saturable nonlinearity case, saturation helps to suppress collapse for both dipole and vortex solitons located on lattice maxima. Even for a complex vortex structure such as 10-hump vortex on Penrose lattice maxima, collapse is suppressed and amalgamation of vortex humps and phase entanglement is delayed.

Author details

İlkay Bakırtaş

Address all correspondence to: ilkayb@itu.edu.tr

İstanbul Technical University, Faculty of Science and Letters, Department of Mathematics, Maslak, Istanbul, Turkey

References

[1] M. R. Matthews, B. P. Anderson, P. C. Haljan, D. S. Hall, C. E. Wieman, E. A. Cornell, Vortices in a Bose-Einstein condensate, Phys. Rev. Lett. 83 (13) (1999) 2498–2501.

[2] J. R. Abo-Shaeer, C. Raman, J. M. Vogels, W. Ketterle, Observation of vortex lattices in Bose-Einstein condensates, Science 292 (5516) (2001) 476–479.

[3] B. Luther-Davies, Y. S. Kivshar, Dark optical solitons: physics and applications, Physics Reports 298 (2–3) (1998) 81–197.

[4] O. Manela, O. Cohen, G. Bartal, J. W. Fleischer, M. Segev, Two-dimensional higher-band vortex lattice solitons, Opt. Lett. 29 (2004) 2049–2051.

[5] J. W. Fleischer, G. Bartal, O. Cohen, O. Manela, M. Segev, J. Hudock, D. N. Christodoulides, Observation of vortex-ring "discrete" solitons in 2d photonic lattices, Phys. Rev. Lett. 92 (12) (2004) 123904.

[6] B. Freedman, G. Bartal, M. Segev, R. Lifshitz, D. Christodoulides, J. Fleischer, Wave and defect dynamics in nonlinear photonic quasicrystals, Nature 440 (2006) 1166–1169.

[7] H. Leblond, B. A. Malomed, D. Mihalache, Spatiotemporal vortex solitons in hexagonal arrays of waveguides, Phys. Rev. A 83 (2011) 063825.

[8] Y. V. Kartashov, B. A. Malomed, L. Torner, Solitons in nonlinear lattices, Rev. Mod. Phys. 83, (2011) 247–305.

[9] M. I. Weinstein, Modulational stability of ground states of NLS equations, SIAM J. Math. Anal. 16 (1985) 472–490.

[10] M. I. Weinstein, Lyapunov stability of ground states of nonlinear dispersive evolution equations, Comm. Pure Appl. Math. 39 (1986) 51–68.

[11] M. Weinstein, The nonlinear Schrödinger equation: Singularity formation, stability and dispersion, Contemp. Math. 99 (1989) 213–232.

[12] H. A. Rose, M. I. Weinstein, On the bound states of the nonlinear Schrödinger equation with a linear potential, Phys. D 30 (1988) 207–218.

[13] V. A. Brazhnyi, V. V. Konotop, M. Pérez-Carcía, Driving defect modes of Bose-Einstein condenstates in optical lattices, Phys. Rev. Lett. 96 (2006) 060403.

[14] Y. Sivan, G. Fibich, M. I. Weinstein, Waves in nonlinear lattices: Ultrashort optical pulses and Bose-Einstein condensates, Phys. Rev. Lett. 97 (2006) 193902.

[15] Y. Sivan, G. Fibich, B. Ilan, M. I. Weinstein, Qualitative and quantitative analysis of stability and instability dynamics of positive lattice solitons, Phys. Rev. E 78 (2008) 046602.

[16] B. Ilan, Y. Sivan, G. Fibich, A quantitative approach to soliton instability, Opt. Lett. 36 (3) (2011) 397–399.

[17] J. Yang, Z. H. Musslimani, Fundamental and vortex solitons in a two-dimensional optical lattice, Opt. Lett. 28 (2003) 2094–2096.

[18] H. Sakaguchi, B. A. Malomed, Gap solitons in quasiperiodic optical lattices, Phys. Rev. E 74 (2) (2006) 026601.

[19] M. J. Ablowitz, B. Ilan, E. Schonbrun, R. Piestun, Solitons in two-dimensional lattices possessing defects, dislocations and quasicrystal structures, Phys. Rev. E – Rap. Comm. 74 (2006) 035601.

[20] J. Yang, Stability of vortex solitons in a photorefractive optical lattice, New J. Phys. 6 (1) (2004) 47.

[21] H. Sakaguchi, B. A. Malomed, Higher-order vortex solitons, multipoles and supervortices on a square optical lattice, EPL (Europhys. Lett.) 72 (5) (2005) 698.

[22] J. Wang, J. Yang, Families of vortex solitons in periodic media, Phys. Rev. A 77 (3) (2008) 033834.

[23] M. J. Ablowitz, N. Antar, İ. Bakırtaş, B. Ilan, Vortex and dipole solitons in complex two-dimensional nonlinear lattices, Phys. Rev. A 86 (2012) 033804.

[24] M. Bağcı, İ. Bakırtaş, N. Antar, Vortex and dipole solitons in lattices possessing defects and dislocations, Opt. Commun. 331 (2014) 204–218.

[25] M. Senechal, Quasicrystals and Geometry, Cambridge University Press, New York, 1995.

[26] V. I. Petviashvili, Equation for an extraordinary soliton, Sov. J. Plasma Phys, 2 (1976) 257–258.

[27] M. J. Ablowitz, Z. H. Musslimani, Spectral renormalization method for computing self-localized solutions to nonlinear systems, Opt. Lett. 30 (2005) 2140–2142.

[28] M. J. Ablowitz, N. Antar, İ. Bakırtaş, B. Ilan, Band-gap boundaries and fundamental solitons in complex two-dimensional nonlinear lattices, Phys. Rev. A 81 (3) (2010) 033834.

Chapter 8

Holographic Optical Elements to Generate Achromatic Vortices with Ultra-Short and Ultra-Intense Laser Pulses

María-Victoria Collados, Íñigo J. Sola,
Julia Marín-Sáez, Warein Holgado and Jesús Atencia

Additional information is available at the end of the chapter

http://dx.doi.org/10.5772/66314

Abstract

The requirements for the generation of optical vortices with ultra-short and ultra-intense laser pulses are considered. Several optical vortice generation procedures are analysed, specifically those based on diffractive elements, such as computer generated holograms (CGH). Optical vortices achromatization techniques are studied. Volume phase holographic (VPH) elements are considered for highly efficient, broad spectrum, high damage-threshold generation of vortices. VPH compound systems, including a compact one, for achromatic vortex generation are presented. Experimental results of vortice generation with ultra-short and ultra-intense pulses are shown.

Keywords: optical vortex, pulse shaping, volume phase holograms (VPH), aberration compensation, holographic optical elements

1. Introduction

An optical vortex is a wave that has a phase singularity, so that the intensity figure is ring-shaped, with zero intensity at the centre due to the indeterminacy of phase at that point. The phase varies helically around the singularity, from 0 to $2\pi m$, m being an integer called the *topological charge*. This kind of beams is associated with an amount of angular momentum, which makes them very attractive for certain types of applications, such as the development of optical tweezers [1]. Another application of interest is the vortex coronagraphy, which allows astronomers to create a blind spot that blocks the starlight revealing orbiting bodies such as planets or dust clouds [2], a technique that requires achromatic elements.

In recent years interest in the generation of vortices from ultra-short and ultra-intense pulses has been increased, opening access to the experimental study of phenomenology in vortex propagation in non-linear regime [3, 4] and their possible applications, such as remote laser-induced breakdown spectroscopy (LIBS) [5], or phase control in higher order harmonics generation. The generation of high-energy vortices from ultra-intense and ultra-short laser pulses requires elements that, on the one hand, have a high damage threshold and, on the other, are able to work with a wide bandwidth.

There are several techniques for generating vortices with short laser pulses, all extrapolated from their use in continuous-wave regime, under conditions of monochromaticity and low energy. One of them is the use of modal converters, in which a vortex can be obtained from a HG_{01} mode with a combination of cylindrical lenses [6]. This technique is not feasible in the case of intense laser pulses, since it is difficult in this case to get the laser to emit in HG_{01} mode. There are modal converters based on LCD spatial light modulators, but are not applicable to high-energy lasers because of their damage threshold [7].

To date, there are two vortex generation techniques applicable to femtosecond lasers. The first is the use of spiral phase masks, manufactured either by depositing quartz onto a quartz substrate [8], by lithography in resins [9] and photoresists [10] or directly by carving on fused silica. In all cases, the mask is made with steps; the thickness increases with the azimuth angle and therefore the output phase varies helically. This type of masks has an efficiency of approximately 55% in the case of manufacturing by deposition or up to 80% in the case of photoresist, and they have a high damage threshold, allowing illumination with ultra-intense pulses obtaining high-energy vortices. A disadvantage of these phase masks is that they have some chromatic aberration when the half-width of the pulse is greater than 40 nm, which is manifested in the variation of the topological charge of the vortex with the wavelength, which limits their use to pulses higher than 30 fs [11].

Recently, Swartzlander [12] proposed a solution to try to achromatize this type of elements, by joining two different materials bonded together with a spiral phase (achieved by varying thickness) between them. With this method, achromatization is achieved for two wavelengths. In any case, the manufacturing process is expensive and requires photolithography facilities, vapour deposition or spin coating, the achromatization that can be achieved depends on the materials used and not total achromatization for the entire bandwidth is achieved.

The second technique uses computer generated holograms (CGH). In this case, the transmission gratings have a dislocation that generates the vortex and are usually printed on transparency or recorded in photographic film [13].

The efficiency of such amplitude masks is low, 6%, but may be increased by applying them a bleaching process [14] or generating patterns directly on LCD [15]. The main disadvantage of this type of dislocation gratings is the low damage threshold, which does not allow its use with high intensities. An interesting solution is the one proposed by Sacks et al. [16], which, using the vortex generated with an amplitude mask, records a volume and phase hologram with a reference plane wave, obtaining high efficiency elements.

To prevent the angular chromatic dispersion inherent to diffractive elements, a combination of two gratings with two lenses may be disposed, so that the chromatic dispersion introduced by the grating with dislocation is compensated and the vortex generated is achromatic [17, 18]. Sola et al. [19] performed an assembly similar to that of Mariyenko et al. [17], replacing printed or photographed gratings with volume holographic gratings. A holographic dislocation grating was achieved by the interference of a plane wave with a vortex generated by an amplitude grating, by a procedure similar to that described by Sacks et al. [16]. The material used was dichromated gelatin (Slavich PFG04), with which highly efficient gratings were achieved and with a high damage threshold [20]. In this work, high-energy vortices were generated with a femtosecond laser, so that non-linear vortex propagation effects in air could be observed. The fundamental problem is the complexity of the assembly, which should include a vacuum chamber between the two lenses to avoid non-linear effects in the beam concentration.

There are other solutions, such as those proposed by Martínez-Matos et al. [21], in which the generation of femtosecond paraxial beams with a combination of only two volume gratings separated by some distance is proposed. The disadvantage of this solution is that the separation of gratings introduces a temporal and a spatial chirp at the ends of the Gaussian intensity profile. Atencia et al. [22] have recently developed a compact achromatic holographic vortex generator design based on attached gratings, built from a CGH. The holographic element obtained completely avoids the presence of spatial chirp across the beam intensity profile.

In this chapter the recording of volume holographic elements for the generation of vortices and two achromatization techniques with different features will be explained. In all cases the achromatism condition for a continuous bandwidth is met.

2. Basics on volume holography

2.1. Holographic recording and reconstruction

Holography is a method for recording the amplitude and phase of a wave $U_o(x,y)$ (named *object wave*) along with a reference coherent wave $U_r(x,y)$ on an intensity sensitive medium. The recording of the interference of both waves is called a hologram and the object wave can be recovered from it. The method consists of two stages: recording and reconstruction.

The complex amplitude at each point of the hologram is the sum of the amplitude of the two waves,

$$U(x,y) = U_o(x,y) + U_r(x,y) = A_o e^{-i\phi(x,y)} + A_r e^{-i\varphi(x,y)} \tag{1}$$

so intensity will be given by

$$\begin{aligned} I(x,y) &= U(x,y) \cdot U^*(x,y) = U_o U_o^* + U_r U_r^* + U_o U_r^* + U_r U_o^* \\ &= A_o^2 + A_r^2 + A_o A_r e^{-i\phi} e^{i\varphi} + A_o A_r e^{i\phi} e^{-i\varphi} \\ &= A_o^2 + A_r^2 + 2A_o A_r \cos\left[\phi(x,y) - \varphi(x,y)\right] \end{aligned} \tag{2}$$

It can be seen that the intensity varies harmonically with the phase difference. This intensity pattern is recorded in a photosensitive medium, so its transmittance changes. Assuming the amplitude transmittance is linear with exposure of the recording material and that the intensity of the reference wave is uniform over the holographic plate, the transmittance of an amplitude hologram is given by

$$\tau(x,y) = \tau_0 + \gamma\{A_o^2 + 2A_oA_r\cos[\phi(x,y)-\varphi(x,y)]\} \quad (3)$$

where τ_0 is the uniform transmittance due to constant exposure of the reference wave and γ is the slope of the curve of transmittance versus exposure in the linear region. If the recorded hologram is illuminated with a wave with wavelength equal to the one of the recorded reference wave, the amplitude of the transmitted wave will be

$$\begin{aligned} U_i(x,y) &= U_r \cdot \tau(x,y) \\ &= (\tau_0 + \gamma A_o^2)A_r e^{i\varphi(x,y)} + \gamma A_o A_r^2 e^{-i\phi(x,y)} + \gamma A_o A_r^2 e^{i\phi(x,y)} e^{-i2\varphi(x,y)} \\ &= U_1(x,y) + U_2(x,y) + U_3(x,y) \end{aligned} \quad (4)$$

Due to the linearity of Maxwell's equations each of the addends can be interpreted as an independent wave. $U_1(x,y)$ is the beam transmitted by the hologram; it is a replica of the reference wave and corresponds to the 0 diffracted order. $U_2(x,y)$ is a replica of the object wave and corresponds to the +1 diffraction order. $U_3(x,y)$ is similar to the object wave conjugate and corresponds to the −1 diffraction order.

2.2. Volume phase holograms (VPH)

If a suitable material is used in the recording of the hologram, the intensity variations can be translated into variations of the refractive index, so that a phase hologram is recorded. In phase gratings, modulation of the refractive index is given by

$$n(x,y) = n_0 + n_1 \sin(\vec{K} \cdot \vec{r}) \quad (5)$$

where n_0 is the average index, n_1 is the amplitude modulation of the refractive index and $\vec{K}$ is the grating vector.

Consider a hologram region as shown in **Figure 1**. The holographic grating at this zone is recorded with an object wave with wave vector $\vec{k}_o$ and a reference wave with $\vec{k}_r$. The grating vector is determined with $\vec{K} = \vec{k}_o - \vec{k}_r$, and the distance between planes of constant index is $\Lambda = 2\pi/K$.

According to the relationship between the thickness d of the emulsion and the spatial period Λ of the grating thin holograms ($d \leq \Lambda$) and volume ones ($d \gg \Lambda$) can be distinguished. While in a thin hologram the three diffracted orders of Eq. (4) appear, in a volume hologram only diffracted orders 0 and +1 (U_1 and U_2 terms of Eq. (4)) are obtained.

The criterion to determine whether a hologram is volume type is given by the Q parameter,

$$Q = \frac{2\pi\lambda d}{n_0\Lambda^2} \quad (6)$$

where λ is the wavelength of the reconstruction beam. A volume hologram is considered when $Q \geq 10$.

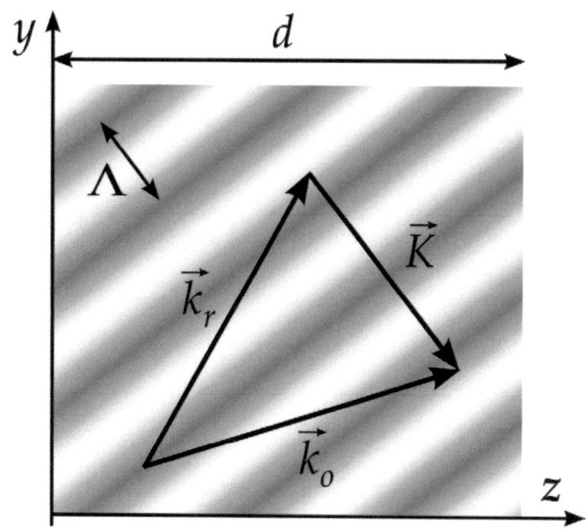

Figure 1. Volume transmission grating with object and reference wavevectors $\vec{k}_o$ and $\vec{k}_r$ and grating vector $\vec{K}$. The planes with constant refractive index, perpendicular to the grating vector, are marked with equal grey levels.

The main advantage of volume phase holograms (VPH) is that, for a given wavelength, it is possible to get 100% of the incident light diffracted to +1 order. For this to happen the so-called *Bragg condition* given by

$$2n_0\Lambda \sin\theta_0 = \lambda \tag{7}$$

must meet, where θ_0 is the half angle between the wave vectors of the incident wave $\vec{\rho}$ and the diffracted wave $\vec{\sigma}$, as shown in **Figure 2(a)**.

While the volume grating determines the energy performance, the surface grating determines the geometric behaviour. So, if in the reconstruction step a wave whose propagation vector is $\vec{\rho}$ reaches the hologram, the propagation vector $\vec{\sigma}$ of the diffracted wave must meet the following relationship:

$$\sigma_y = \rho_y \pm K_y \tag{8}$$

The distribution of energy between $\vec{\sigma}$ and $\vec{\rho}$ waves at the exit of the medium depends on the characteristics of the hologram itself and the conditions of reconstruction. The diffraction efficiency η is defined as the fraction of energy of the incident wave that is diffracted by the holographic grating.

If light hits the surface of the hologram at an angle other than θ_0, or at θ_0 but with a different wavelength, the Bragg condition is not fulfilled. Diffracted wave $\vec{\sigma}$ satisfies the geometric condition (Eq. (8)), as seen in **Figure 2(b)**. According to Kogelnik's Coupled Wave Theory [23], the efficiency when reconstructing transmission VPH near Bragg condition, is given by

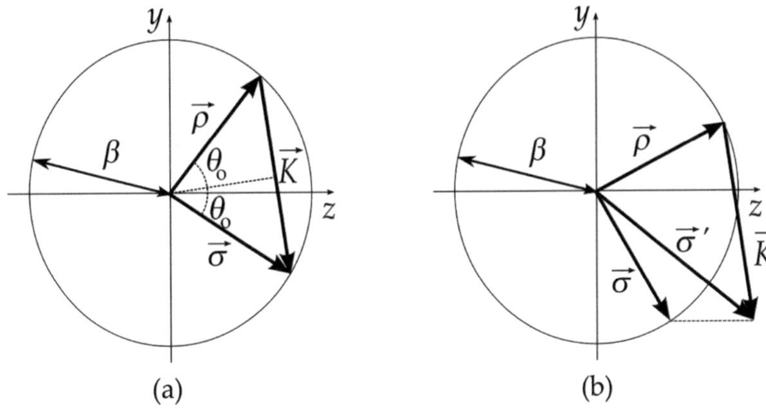

Figure 2. Wave vectors of the incident and diffracted waves and grating vector when the Bragg condition (a) is fulfilled and when (b) it is not fulfilled.

$$\eta = \frac{\sin^2(\sqrt{\nu^2 + \xi^2})}{1 + \xi^2/\nu^2} \tag{9}$$

where ξ, ν are dimensionless parameters defined as

$$\nu = \frac{\pi n_1 d}{\lambda\sqrt{c_r c_s}} \tag{10}$$

$$\xi = \frac{\vartheta d}{2c_s} \tag{11}$$

c_r and c_s are the so called *obliquity coefficients*; they are the z-component of the unit vectors in the direction of $\vec{\rho}$ and $\vec{\sigma}$, respectively. For transmission holograms, $c_s > 0$. The parameter ϑ is called *phase shift* and measures the deviation from the Bragg condition (**Figure 2(b)**). It is defined as

$$\vartheta = \frac{|\vec{\rho}|^2 - |\vec{\sigma}'|^2}{2|\vec{\rho}|} = \frac{\beta^2 - \sigma'^2}{2\beta} \tag{12}$$

with $\vec{\sigma}' = \vec{\rho} + \vec{K}$ and $\beta = 2\pi n_0/\lambda$. If the Bragg condition is strictly fulfilled (monochromatic light of wavelength λ reaches the hologram at an angle θ_0), then $\vartheta = 0$ and $\xi = 0$ and the efficiency is

$$\eta(\vartheta = 0) = \sin^2\nu \tag{13}$$

A 100% efficiency can be achieved if the value of ν for the reconstruction wavelength λ is forced to be $\pi/2$.

When the reconstruction conditions move away from the Bragg condition the ξ parameter increases and the efficiency decreases. For polychromatic illumination (ultra-short pulses, for example), Bragg condition can be fulfilled only for a certain wavelength. To analyse the effect of this in the performance of the holographic element, a polychromatic reconstruction wave of wavelength $\lambda = \lambda_0 + \Delta\lambda$ with an incident direction that meets the Bragg condition for λ_0 is considered. Usually it is assumed that $\Delta\lambda \ll \lambda_0$ [24] but for pulsed waves it is necessary to consider a broad spectrum, so $\Delta\lambda \approx \lambda_0$. Taking into account this fact, a relation between the diffraction efficiency (Eq. (9)) and the reconstruction wavelength λ can be obtained.

In v parameter expression (Eq. (10)) only the wavelength changes, d does not vary and n_1 can be chosen to have 100% efficiency for λ_0. The obliquity factors c_r and c_s depend on the direction of illumination, determined with the Bragg condition for λ_0, so both remain constant. From these conditions,

$$\frac{\pi}{2}\lambda_0 = \frac{\pi n_1 d}{\sqrt{c_r c_s}} = cte. \tag{14}$$

So, for a wavelength λ,

$$v = \frac{\pi}{2}\frac{\lambda_0}{\lambda} \tag{15}$$

For ξ parameter (Eq. (9)) it is necessary to expand ϑ from Eq. (12). Taking into account **Figure 2** for an incident wave with angle θ_0 that fulfils Bragg condition for λ_0,

$$\begin{aligned} \beta^2 - \sigma'^2 &= \beta^2 - (\vec{\rho} - \vec{K})^2 = 2\vec{\rho}\vec{K} - K^2 = \\ &= 2\beta K \cos\left(\frac{\pi}{2} - \theta_0\right) - K^2 = 2\beta K \sin\theta_0 - K^2 \end{aligned} \tag{16}$$

So the phase shift ϑ is

$$\vartheta = \frac{2\beta K \sin\theta_0 - K^2}{2\beta} \tag{17}$$

From the Bragg condition for λ_0 the following relationship is obtained,

$$2n_0\Lambda \sin\theta_0 = \lambda_0 \tag{18}$$

Replacing Eq. (18) in Eq. (17) and substituting the expression of β,

$$\vartheta = \frac{K^2(\lambda_0 - \lambda)}{4\pi n_0} \tag{19}$$

To have the widest possible diffracted spectrum a small angle between beams is used, and wave vectors near the normal direction are chosen, so $c_s \approx 1$, and ξ parameter (Eq. (11)) is

$$\xi = \frac{K^2 d(\lambda_0 - \lambda)}{8\pi n_0} \tag{20}$$

Substituting the obtained values of ξ and v in Eq. (9) the efficiency as a function of the reconstruction wavelength for a broad spectrum ($\Delta\lambda \approx \lambda_0$) is obtained.

A numerical simulation of the spectrum diffracted by a holographic grating has been carried out with typical values for ultra-short pulses (λ_0 = 800 nm) and for dichromated gelatin Slavich PFG04 commercial plates [20] (d = 30 μm and $n_0 = 1.52$). In the volume hologram limit, for Q = 10, a grating period of $\Lambda = 3.15\ \mu$m is obtained. The graph of efficiency vs. reconstruction wavelength is shown in **Figure 3**. It can be seen that the efficiency is higher than 90% for a band with $\Delta\lambda$ = ±100 nm, so this type of VPH could be adequate for its use with ultra-short pulses.

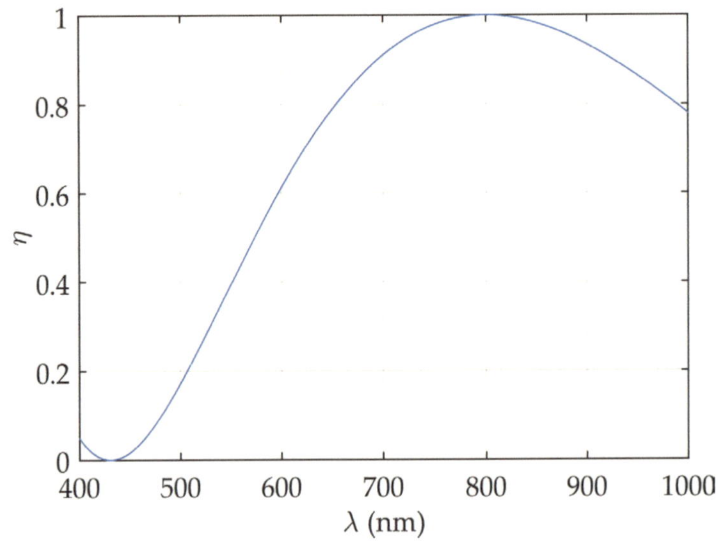

Figure 3. Theoretical efficiency of a transmission VPH as a function of the reconstruction wavelength for $\lambda_0 = 800$ nm, $d = 30\ \mu$m, $n_0 = 1.52$, $Q = 10$.

3. Vortex generation with holographic optical elements

A holographic vortex generation element is recorded by the interference of a plane (or spherical) wave and a vortex beam. Since the recording is with monochromatic light, any of the methods described in Section 1 can be used to generate this vortex object wave. For this work a computer generated hologram (CGH) was chosen.

In order to generate the CGH, the interference between a plane wave (reference wave), with its wave vector forming an angle α_1 with the z-axis, and a vortex-carrying plane wave (object wave), which propagates in the z-axis, is calculated. The reference plane wave at the hologram plane can be expressed as

$$U_r(x, y) = e^{-i\frac{2\pi}{\lambda}x\sin\alpha_1} \tag{21}$$

and the object vortex wave as

$$U_o(x, y) = e^{-im\theta} \tag{22}$$

where $\theta = \tan^{-1}(y/x)$ is the angular coordinate at the hologram plane. The intensity distribution on the CGH is given by

$$I = |U_r + U_o|^2 = 2\left[1 + \cos\left(\frac{2\pi}{\lambda}x\sin\alpha_1 + m\theta\right)\right] \tag{23}$$

λ and α_1 are chosen to obtain the desired spatial period $\Lambda = \lambda/\sin\alpha_1$. The intensity distribution is calculated with MatLab, binarized, and printed on paper with a laser printer to obtain a grating with dislocation. **Figure 4** shows a CGH calculated for $m = 1$.

Figure 4. Dislocated grating obtained with the interference of a plane wave and a vortex wave with topological charge $m = 1$.

The printed pattern is photographically reduced using a reflex camera with Kodak TMax100 film, obtaining an amplitude thin grating of 14.7 lines/mm with a dislocation at the centre. This CGH has high absorption, so it cannot be illuminated with a high-intensity laser, because it could be damaged. Therefore the resulting film is contact-copied onto a commercial Slavich PFG04 dichromated gelatin plate using incoherent light. After this complex procedure a thin phase CGH is obtained. It can be illuminated with an intense laser, although multiple diffracted orders are obtained and their efficiency is low (25%).

The CGH is used to generate the object wave for the recording of a VPH. The recording scheme is shown in **Figure 5**. The beam of a Coherent Verdi 6 W CW laser emitting at 532 nm is divided in two by means of a beam splitter BS. One of the beams is filtered and expanded with a spatial filter and collimated with a lens L_1, and acts as a plane reference beam. The other beam is first filtered and expanded and then collimated with a lens L_2, and illuminates the thin phase CGH. The effects of bitmap resolution in the previous laser-printed CGH require to use spatial filtering techniques to select the desired diffraction order and to avoid bitmap artefacts [16]. The +1 diffracted order, containing the vortex, is selected, filtered and imaged with magnification

$-f_4/f_3$ onto the recording plate to cancel phase propagation terms by means of the 4-f processor formed with lenses L_3 and L_4. Slavich PFG04 dichromated gelatin plates are used, following the process described in [20] to obtain maximum efficiency at 800 nm. The two recording beams form an angle of 10°, the minimum angle required to obtain a volume hologram ($Q = 10$) with the 30 μm-thick PFG04 emulsion.

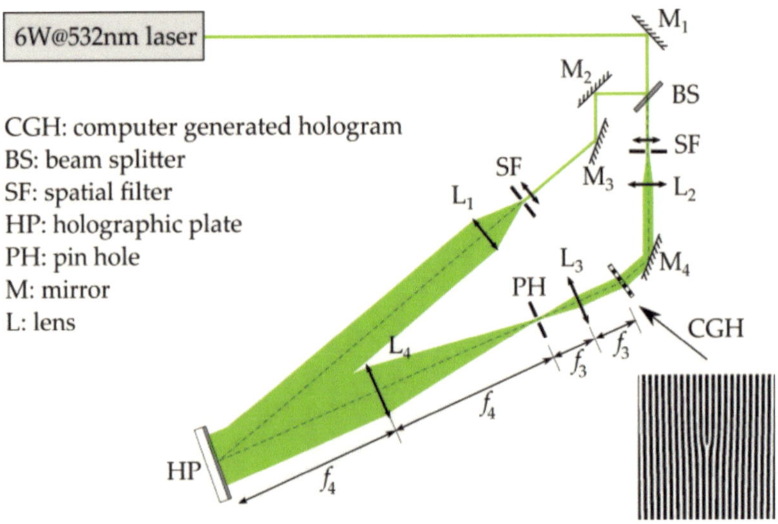

Figure 5. Scheme of the recording setup of a volume holographic optical element with a vortex-carrying plane wave as object wave.

The recorded VPH presents an efficiency of 95% at 800 nm. The spatial period is $\Lambda = 2.93\ \mu$m that corresponds to 340 lines/mm spatial frequency. The diffracted spectrum is shown in **Figure 6**. It can be seen that the maximum efficiency appears for 800 nm, and the wavelength

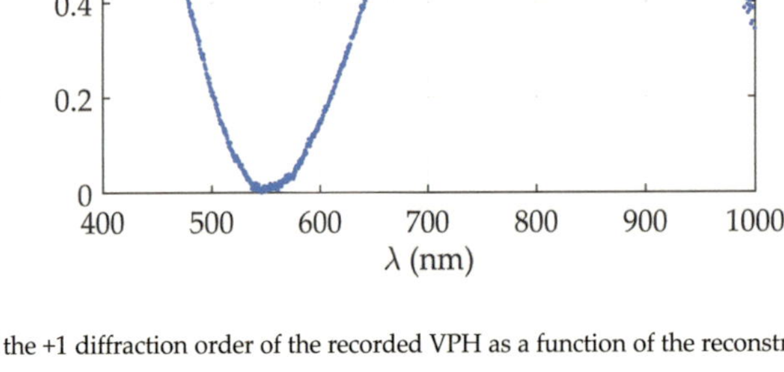

Figure 6. Efficiency of the +1 diffraction order of the recorded VPH as a function of the reconstruction wavelength.

http://dx.doi.org/10.5772/66314

range in which the diffractive efficiency is above 80%, is almost 200 nm. The holographic material Slavich PFG04 has high damage threshold [25], so the recorded VPH can be used with ultra-short and ultra-intense laser pulses.

In **Figure 7**, the Fourier transform of the vortex generated by the VPH when illuminated with a CW laser beam of 800 nm is shown.

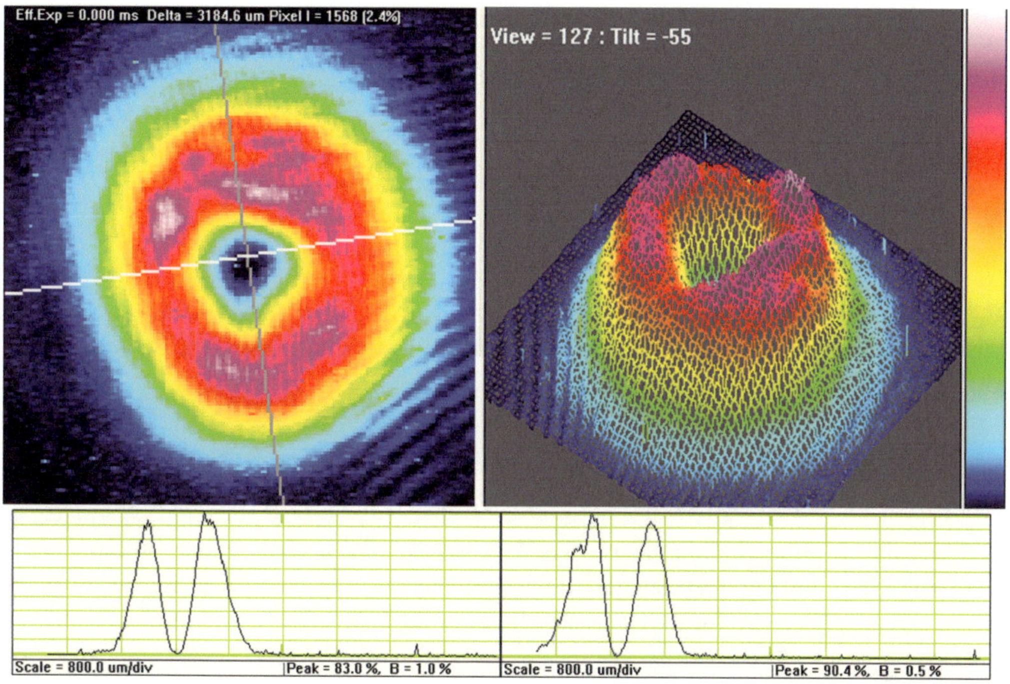

Figure 7. Fourier transform of the vortex generated by the VPH when illuminated with a CW laser beam of 800 nm.

4. Holographic generation of achromatic vortex beams

A disadvantage of using the previous holographic optical element to generate vortex beams with ultra-short pulses is the chromatic dispersion. As we have seen, the VPH can work properly in a spectral range of around 200 nm, but the direction of the diffracted vortex beam depends on the wavelength, so the vortex beam presents an important spatial chirp, as it is shown in **Figure 8**.

This problem can be overcome by designing an achromatic set-up in which the chromatic dispersion of the vortex beam is compensated by using a volume holographic plane grating (H_I) with the same spatial frequency as the vortex generator VPH (H_{II}).

H_I is recorded with two plane waves of wavelength λ' that propagate forming an angle α_o and α_r with z-axis, so the complex transmittance of the grating (Eq. (3)), considering a phase hologram, is proportional to

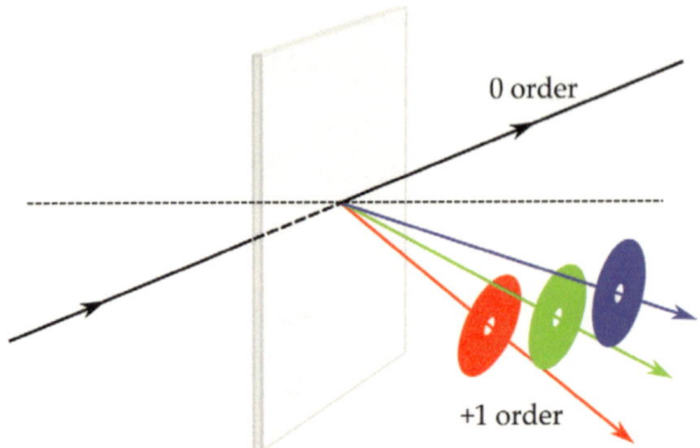

Figure 8. Chromatic dispersion of the vortex generator VPH described in Section 3.

$$\tau_I(x,y) = 1 + \frac{1}{2} e^{i\frac{2\pi}{\lambda'} x\ (\sin\alpha_r - \sin\alpha_o)} + \frac{1}{2} e^{-i\frac{2\pi}{\lambda'} x\ (\sin\alpha_r - \sin\alpha_o)} \tag{24}$$

H_{II} is recorded with the interference of a plane wave, forming an angle α_r with z-axis, and a vortex beam forming an angle α_o with z-axis, so the complex transmittance of H_{II} is given by

$$\tau_{II}(x,y) = 1 + \frac{1}{2} e^{i\left(\frac{2\pi}{\lambda'} x\ (\sin\alpha_r - \sin\alpha_o) + m\theta\right)} + \frac{1}{2} e^{-i\left(\frac{2\pi}{\lambda'} x\ (\sin\alpha_r - \sin\alpha_o) + m\theta\right)} \tag{25}$$

4.1. Achromatization with separated elements

The first approach to combine these two VPH is proposed by Sola et al. [19] based on a solution with printed gratings suggested by Mariyenko et al. [17]. H_{II} is located at the plane where a positive lens (L_1) forms the image of the grating H_I in a 2*f*-2*f* configuration, as shown in **Figure 9**.

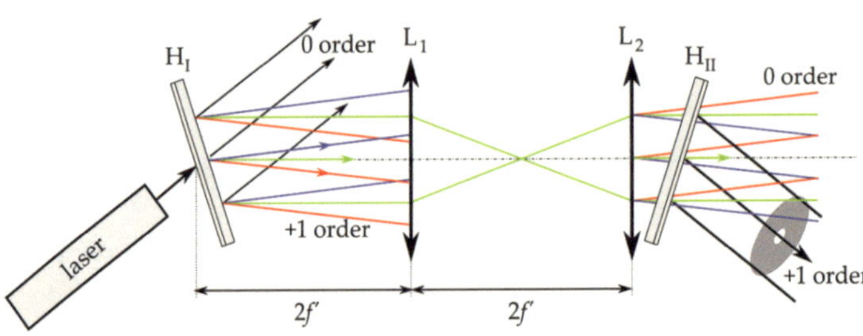

Figure 9. Scheme of the experimental set-up for generating vortices with separated holographic elements.

If we consider that H_I is illuminated with a polychromatic plane wave with a Gaussian spatial profile, $U_c(x,y)$,

$$U_c(x,y) = \int_{-\infty}^{\infty} u(\lambda) e^{-\frac{(x^2+y^2)}{w^2}} e^{-i\frac{2\pi}{\lambda} x\ \sin\alpha_c} d\lambda \tag{26}$$

where $u(\lambda)$ is the wave spectrum and w is the half wide of the Gaussian beam, then the wave emerging from H_I, $U_{i,I}(x,y)$, can be expressed as

$$\begin{aligned} U_{i,I}(x,y) &= U_C(x,y)\cdot\tau_I(x,y) = \\ &= \int_{-\infty}^{\infty} u(\lambda)\mathrm{e}^{-\frac{(x^2+y^2)}{w^2}}\mathrm{e}^{-i\frac{2\pi}{\lambda}x\,\sin\alpha_c}\mathrm{d}\lambda + \\ &+\frac{1}{2}\int_{-\infty}^{\infty} u(\lambda)\mathrm{e}^{-\frac{(x^2+y^2)}{w^2}}\mathrm{e}^{-i\left\{\frac{2\pi}{\lambda}x\left[\sin\alpha_c-\frac{\lambda}{\lambda'}(\sin\alpha_r-\sin\alpha_o)\right]\right\}}\mathrm{d}\lambda + \\ &+\frac{1}{2}\int_{-\infty}^{\infty} u(\lambda)\mathrm{e}^{-\frac{(x^2+y^2)}{w^2}}\mathrm{e}^{-i\left\{\frac{2\pi}{\lambda}x\left[\sin\alpha_c+\frac{\lambda}{\lambda'}(\sin\alpha_r-\sin\alpha_o)\right]\right\}}\mathrm{d}\lambda = \\ &= U_{1,I}(x,y)+U_{2,I}(x,y)+U_{3,I}(x,y) \end{aligned} \quad (27)$$

$U_{1,I}$ is the transmitted order; $U_{2,I}$ and $U_{3,I}$ are the +1 and -1 diffracted orders, respectively. In the case of volume holograms, only one of these orders is diffracted, depending on the α_C value. In this case, the order efficiently diffracted by the VPH is $U_{2,I}$, which is a set of polychromatic plane waves with a Gaussian spatial profile. The direction of propagation of each wavelength, α_i, is a function of λ,

$$\sin\alpha_i = \sin\alpha_c - \frac{\lambda}{\lambda'}(\sin\alpha_r - \sin\alpha_o) \quad (28)$$

Considering L_1 as an ideal lens, the complex amplitude at the image plane $U'_{2,I}(x,y)$ is given by [26]

$$U'_{2,I}(x,y) = \frac{1}{|M|}U_{2,I}\left(\frac{x}{M},\frac{y}{M}\right) \quad (29)$$

where M is the magnification. In this case, as the lens L_1 works in a 2*f*-2*f* configuration, $M = -1$, so

$$U'_{2,I}(x,y) = \frac{1}{2}\int_{-\infty}^{+\infty} u(\lambda)e^{\frac{-(x^2+y^2)}{w^2}}e^{i\left\{\frac{2\pi}{\lambda}x\left[\sin\alpha_c-\frac{\lambda}{\lambda'}(\sin\alpha_r-\sin\alpha_o)\right]\right\}}d\lambda \quad (30)$$

$U'_{2,I}$ illuminates the second hologram H_{II}, so the amplitude distribution that leaves H_{II} is

$$U_{i,II} = \tau_{II}U'_{2,I} = U_{1,II} + U_{2,II} + U_{3,II} \quad (31)$$

The term $U_{2,II}$ is the efficiently diffracted order, and is given by

$$U_{2,II} = \frac{1}{2}\int_{-\infty}^{+\infty} u(\lambda)e^{\frac{-(x^2+y^2)}{w^2}}e^{i\left\{\frac{2\pi}{\lambda}x\sin\alpha_c+m\theta\right\}}d\lambda \quad (32)$$

$U_{3,II}$ is an achromatic vortex without chromatic dispersion, as the direction of the propagation does not depend on the wavelength.

A second lens L_2, with the same focal length as L_1, is placed just in front of or behind H_{II}, restoring the phase of the propagating beam into a plane wavefront. The VPH to generate the vortex beam has been recorded with a plane wave as a reference wave, so placing the lens L_2 in front of H_{II} (as it is shown in **Figure 9**) provides a reconstruction plane wave, which assures the reconstruction of the vortex beam with no geometrical aberrations.

Depending on the intensity of the pulses, the use of a vacuum chamber between the two lenses could be necessary to avoid non-linear effects in the focalization produced by the lens L_1.

4.2. Achromatization with a compact element

The second approach was proposed by Atencia et al. [22]. In this case both holographic elements H_I and H_{II} are placed parallel to each other, as it can be seen in **Figure 10**.

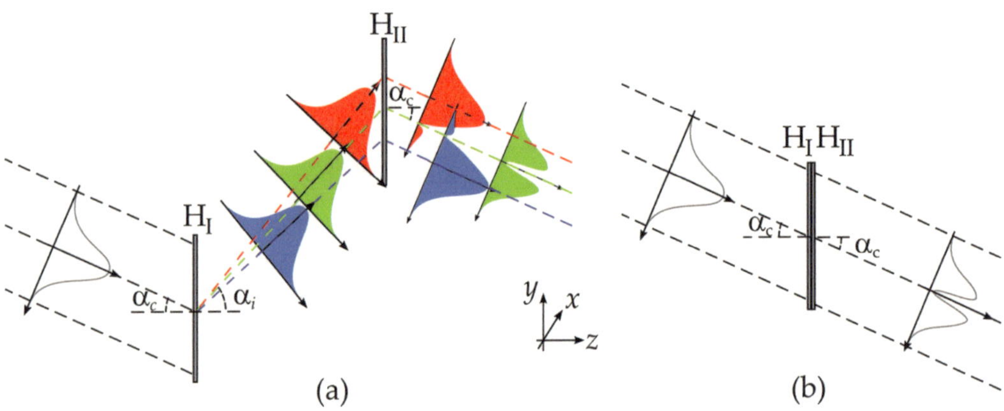

Figure 10. Double-grating vortex generator. (a) Separated VPH, (b) compact VPH.

If H_I and H_{II} are separated (**Figure 10(a)**), the Gaussian profile of the incident beam is kept in the propagation between H_I and H_{II}, but the different wavelengths are spatially separated and de-phased when they reach H_{II}. A set of vortices emerges from H_{II}, all with the same topological charge m forming the same angle with the z-axis. The position of the vortex is the same for all wavelengths, but the Gaussian distribution is spatially shifted for each wavelength.

If the two VPH are placed together, as in **Figure 10(b)**, the amplitude distribution that reaches H_{II} is $U_{2,I}$, so the distribution that leaves H_{II} is

$$U_{i,II} = \tau_{II} U_{2,I} = U_{1,II} + U_{2,II} + U_{3,II} \tag{33}$$

where, in this case, $U_{3,II}$ is the order that is diffracted efficiently and is given by

$$U_{3,II} = \frac{1}{2} \int_{-\infty}^{+\infty} u(\lambda) e^{\frac{-(x^2+y^2)}{w^2}} e^{-i\left\{\frac{2\pi}{\lambda} y \sin \alpha_c + m\theta\right\}} d\lambda \tag{34}$$

http://dx.doi.org/10.5772/66314

$U_{3,II}$ is a set of vortices emerging from H_{II}, all with the same topological charge m and forming an angle of α_c with the z-axis.

In the construction of this compact VPH, the recording scheme for H_I is similar to that shown in **Figure 5** by replacing the vortex wave with a collimated wave. The angle between beams has to be the same for H_I and H_{II} to guarantee the chromatic compensation. For a simpler alignment procedure, the recording geometry for H_I and H_{II} is chosen to give Bragg condition at 800 nm for an angle $\alpha_c = 0$.

After processing, H_I and H_{II} are cemented to each other, using Norland NOA61 optical adhesive between the two emulsions. This ensures that no change in the refraction index between the holograms occurs and preserves the emulsion from degradation effects of the environment.

When illuminating each VPH at Bragg angle for 800 nm, the maximum efficiency obtained is approximately 95%, so the sandwich has a total maximum efficiency of 90%. Reflection on glass surfaces and glass absorption causes losses of about 15%, so the compound holographic element reaches an absolute efficiency of 77%. When λ moves away from 800 nm, this efficiency decreases. For the compound element, **Figure 11** shows the wavelength dependence of the diffractive efficiency. It can be observed that the full width at half maximum (FWHM) covers 250 nm, and the diffracted efficiency is above 80% for a wavelength range of 100 nm.

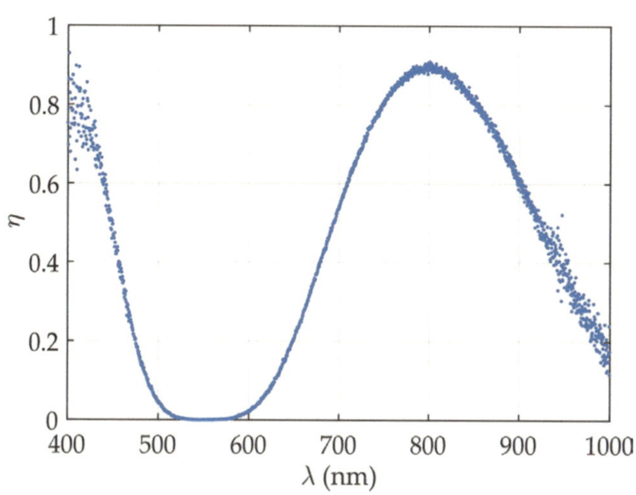

Figure 11. Dependence of the diffractive efficiency of the compact VPH with the wavelength.

For wavelengths different from 800 nm the diffractive efficiency drops and the transmitted light increases, so the transmitted and diffracted light can spatially overlap for some wavelengths. To prevent this, H_I is slightly rotated around the z-axis before joining the two holograms. So, the transmitted and diffracted beams emerge from H_{II} with different directions, but the efficiency is not affected.

Figure 12(a) and **(b)** demonstrate that the compound element generates a broadband achromatic optical vortex. A 5 mm-wide white light vortex can be seen in **Figure 12(a)**, when illuminating the compact VPH with a tungsten light source (Ocean Optics LS-1) with 600 nm FWHM centred at 850 nm. The photograph was taken at 1 m propagation from the element output. **Figure 12(b)** shows the interference of transmitted and diffracted light. To obtain an appreciable intensity on the transmitted beam, the holographic element is illuminated out of Bragg condition. The dislocation of the interference pattern is clearly visible at the centre of the image.

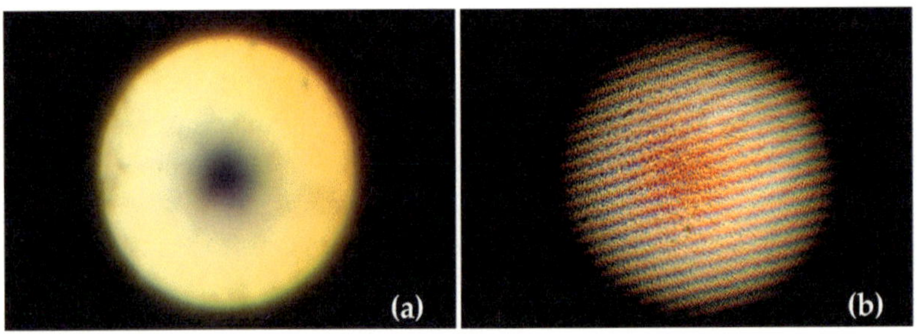

Figure 12. White light vortex. (a) Propagated at 1 m, (b) interference with a plane wave.

This compact vortex generator holographic element is very easy to align and avoids pulse concentration zones, so for ultra-intense pulses the use of a vacuum chamber is not required.

5. Applications of vortex volume phase holograms to ultra-fast optics

Once the recording process of a vortex generating VPH has been reviewed, and the two solutions for the vortex achromatization have been described, in the following section we will comment some of the characteristics of the generated vortex and some possible applications using femtosecond high power pulses. The achromatization of the VPH is very attractive from the point of view of ultra-short pulses, which present broadband spectra, allowing the generation of short vortex pulses with high peak power [19]. The main limitation to obtain shorter pulses is the spectral bandwidth of the hologram; however, the present performance allows the generation of ultra-short vortex beams (e.g., compact vortex VPH is compatible with few tenths of fs pulses).

One of the first questions arising is what the spatiotemporal structure of the vortex looks like. In the literature, fs pulse vortices generated by spiral phase masks have been characterized by using spatially resolved interferometry [27]. In the case of the compact VPH, when studying spatial pattern of interferences in a similar way, a peculiar behaviour was observed. **Figure 13** shows the interference pattern between two replicas of the laser beam (pulses of 100 fs FWHM, central wavelength at 795 nm), obtained at a Mach-Zender interferometer. One of the beams passed through a topological charge $m = 1$ compact VPH, while the other arm acted as a spatio-

http://dx.doi.org/10.5772/66314

temporal reference. As expected, fork-like interference pattern appears when both optical paths of the interferometer match, disappearing otherwise. However, a spatial evolution of the interference region is observed. Starting from a situation of unbalanced optical path length of each interferometer arms, with no visible spatial interference pattern, the time delay between the vortex pulse and the reference one is gradually reduced. When the pulse distance becomes within the coherence length, spatial interferences arise, but localized in a spatial region (**Figure 13(a)**). Reducing the arm difference increases the contrast of the interferences but it also shifts them in the diagonal direction (**Figure 13(b)–(d)**). Finally, when the optical path mismatch increases again, interferences disappear, located on the opposite region of the beam profile (**Figure 13(d)** and (**e**)). This time dependence of interference spatial regions shows that the vortex beam is generated with a pulse tilt. According to the experimental result, the tilt could be quantified around 40 fs/mm in the direction of the VPH incident beam-vortex mismatch (induced, as commented in 4.2 by the small rotation of H_I related to H_{II} in order to avoid overlapping between transmitted and diffracted beams). Similar results were obtained using pulses of 25 fs FWHM.

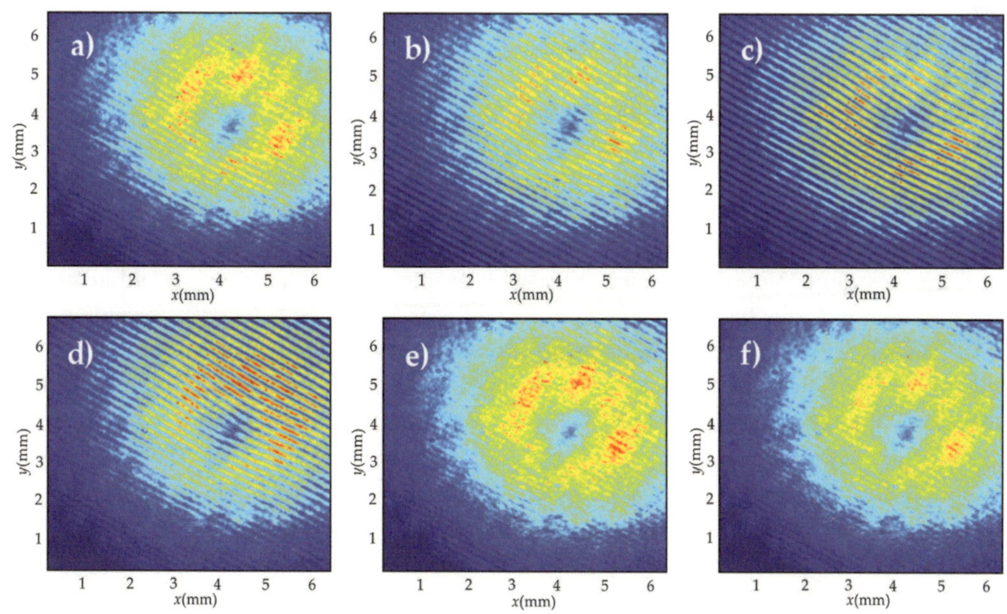

Figure 13. Interference patterns between an $m = 1$ vortex beam generated with the compact VPH described in Section 4.2 and a collimated Gaussian beam, presenting a small angle between them. The interference region varies as the relative interferometer arm delay changes (time shift between arms: 0 fs (reference) (a); 67 fs (b); 133 fs (c); 266 fs (d); 400 fs (e); 533 fs (f)).

Although this pulse tilt is small and for the most of applications it plays no relevant role (e.g., in the following non-linear interaction cases presented in this section no relevant effect has been observed), it is important to be aware of its existence. One way of reducing it would be to better match transmitted-diffracted beam directions. Nevertheless, if some of the incident beam is transmitted through the compact VPH, it may interfere with the vortex beam. On the other hand, it has recently been shown that controlled pulse tilts can be a tool to induce interesting effects, such as the so-called attosecond lighthouses [28]. Therefore, the ability of imprinting controlled spatiotemporal alterations on the VPH may have potential applications.

One of the main advantages of using the compact VPH consists on the possibility of creating a high power vortex quite easily. The set-up for generating the high power vortices is simplified to the point of making the ultra-short pulses pass through it, while orienting the VPH in the proper angle, defined by the recording configuration. Since the element presents a high damage threshold of, at least, hundreds of GW/cm^2 [19], the incident pulse peak power can be high, allowing to generate high-power vortices because of the high efficiency of the VPH. Therefore, it makes it possible to induce non-linear effects on the vortex beams while propagating though a certain medium.

In order to explore the potential of VPH creating high intensity femtosecond vortices, different experiments have been performed. Firstly, using the achromatic set-up described in Section 4.1, the dynamics of vortices at different intensity regimes were studied [19], analysing the evolution of the spatial distribution of the light along propagation after passing through a focusing lens (f = 2.2 m). While working on low-intensity regime (linear regime), the vortex maintains its structure at focus and after it, remains stable. However, for high intensity regime the dynamics varies, as shown in **Figure 14**. The vortex structure, unaltered in the first centimetres of the focusing propagation, begins to be affected by the non-linear propagation. Just before the focus, because of the Kerr effect driven self-focusing, the vortex splits into two fragments. It is intriguing to observe how the two splinters rotate along the optical axis as the beam propagates, following a spiralling path. When the beam diverges, away from the focal region, the splinters stop from rotating. This behaviour can be explained as an effect of angular momentum conservation. Numerical calculations [19] were carried out from (2+1)-dimensional numerical simulations based on solving the non-linear Schrödinger equation including Kerr effect, as described in Ref. [29]. Agreement between theory and experiments shows the major role of the Kerr effect on the dynamics, since other possible players (as ionization) were not included on the calculations.

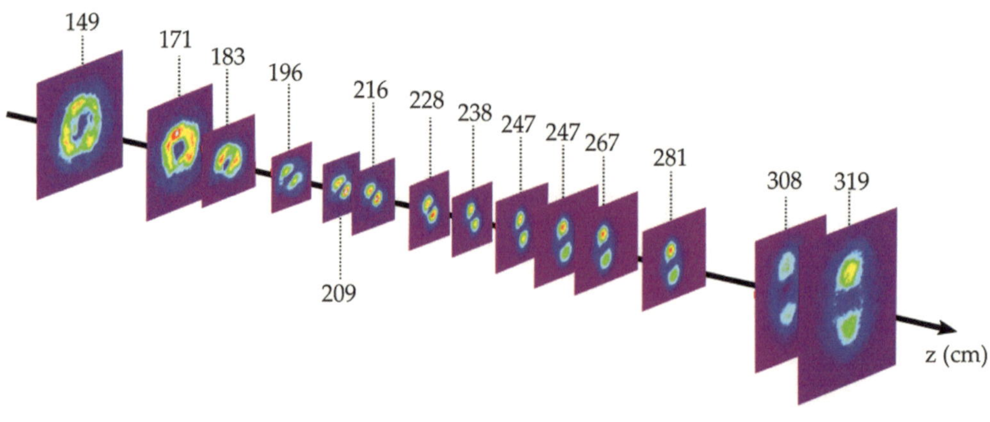

Figure 14. Spatial beam profile evolution of 14 GW vortex focused by an f = 2.2 m convergent lens.

An increase of beam intensity would involve utterly medium ionization. Then, light propagation may become even more complex, being the result of the interplay between effects competing for focusing or unfocusing the beam (e.g. Kerr effect will introduce focusing on the beam, while ionization makes the opposite). This may produce what is called filamentation [30], exhibiting a self-guiding of the beam because of the balance between linear regime light propagation and the different non-linear effects raised from interaction with the medium. Splitting and filamentation of the vortex fragments have been observed for high enough intensities (e.g., 14 GW vortices propagating in a tube filled with nitrogen at a pressure of 1600 mbar [19]).

In the last decades, an extremely non-linear process known as high order harmonic generation (HHG) has become a hot topic and the basis of the new discipline called *Attoscience* [31]. This process rises from the interaction of intense light and matter and enables to obtain harmonic frequencies of the driving radiation well inside the ultraviolet spectral region (known as extreme ultraviolet, or XUV) or even the soft X-ray region [32]. In the case of interaction with gases, it can be interpreted as a three-step process [33, 34], starting with matter ionization, freeing an electron (typically in tunnelling regime). In a second step, the electron is accelerated by the intense electromagnetic field of the pulse. Finally, depending on the electric-field phase, the electron may return towards the parent ion, recombining and emitting the accumulated energy (kinetic energy plus the ionization potential) as a photon. This XUV radiation inherits coherence from the generating beam, while showing a peculiar temporal emission pattern, a train of sub-femtosecond pulses. It can be eventually reduced to a single XUV burst by means of few cycle driving pulses and gating techniques [35–37]. This kind of extremely short light sources is an unprecedented tool for studying fast dynamics, as molecular or electronic [31]. Within this context, during the last years, special attention is being devoted to the HHG of pulses with orbital angular momentum (OAM), as vortices. The first experimental work showing the generation of XUV vortices from IR ones was reported on 2012 [38]. The measurements of the topological charges of the different harmonic yielded unexpectedly a constant charge $m = 1$ (coming from the same charge in the driving field), independent of the harmonic order. The authors explained this discrepancy as caused by vortex decay during the non-linear process. Later, numerical simulations [39] predicted topological charge scaling with the order of the harmonics and subsequent experiments [40] confirmed this point.

The basis of HHG and its use for the generation of XUV beams exhibiting OAM are analysed in depth in another chapter of this book [41]. Here we will focus on the use of vortex generating compact VPH in HHG experiments. In the pioneering experimental studies [38, 40], the OAM driving beam was obtained from a 30 fs pulsed beam and $m =$ 1 topological charge was imprinted by means of a spatial light modulator (SLM) but using two different ways. In the case of Zürch et al. [38], the reflective SLM was used to imprint the AOM in a spiral phase mask way, introducing diphase of the wavefront phases azimuthally dependent. On the other hand, Gariepy et al. [40] used the SLM for creating a fork type diffraction pattern, i.e. presenting dislocation introducing the topological charge. In order to compensate the first order angular dispersion, a long period-grating pattern was used. This way of generating vortices allows to imprint the same topological charge to all the wavelengths, in contrast with the spiralling phase mask procedure that, as shown in Ref. [27], produces topological charges depending on the wavelength, being even not an integer.

Therefore, the case of Gariepy et al. [40] is quite similar to the way the compact VPH generates the vortices, with a dislocated grating and a second non-dislocated grating to suppress angular dispersion. The main difference lays on the fact of using a SLM element or a holographic plate. While the former is more flexible and dynamic, the compact VPH is a more robust and cheaper element, presenting a simpler alignment.

Then, in order to explore its potential, we tested the compact VPH in a HHG experiment. The experimental set-up is shown in **Figure 15**. A 25 fs pulsed beam up to 2 mJ per pulse (1 kHz repetition rate, spectrum centred on 790 nm) is focused by means of a $f = 50$ cm spherical Ag mirror into an Ar gas jet on vacuum (10^{-3} mbar of residual pressure when the gas jet is operating). After light-matter interaction, radiation passes through a 4 mm long slit to a Rowland circle spectrometer (Model 248/310G, McPherson Inc.), using a 1 m radius spherical grating (133 lines/mm). A 150 nm Al filter just after the slit absorbs the fundamental radiation and lower harmonics, up to the 11th. The compact VPH was placed before the focusing element, creating the AOM beam. Because of the physical dimension of the set-up and XUV beam divergence, the size of the harmonic vortices was bigger than the slit dimensions. Therefore, several measurements were performed, shifting the height of IR beam and gas jet referred to the slit position. **Figure 16** shows the combined spectra for different divergence angle. The result shows annular distribution of the harmonics, with a minimum of intensity in the centre (divergence 0 mrad) and a ring around 3-6 mrad. Although no measurement concerning topological charge has been performed, results are compatible with the obtained at [40].

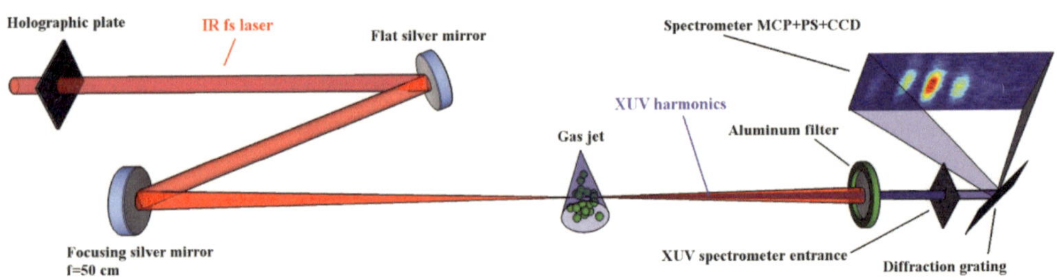

Figure 15. HHG experimental set-up for generating XUV. A 25 fs pulsed beam passes through the compact VPH (Holographic plate), creating an $m = 1$ vortex. The beam is focused by means of a spherical Ag mirror ($f = 0.5$ m) on an Ar gas jet, where HHG occurs. XUV radiation is filtered by an Al filter and spectrally analysed using a Rowland circle spectrometer (Model 248/310G, McPherson Inc.) including an spherical diffraction grating (133 lines/mm), and a detector composed by a micro-channel plate (MCP), a phosphor screen (PS) and a CCD camera.

In summary, the compact VPH [22] is a simple, cheap and robust element for generating high power AOM beams. Their achromaticity allows the generation of pulsed vortices down a few tenths of fs and their high damage threshold permits to obtain high peak powers. High power vortex generated by this way are able to interact non-linearly with material, producing effects from Kerr effect and filamentation to HHG.

http://dx.doi.org/10.5772/66314

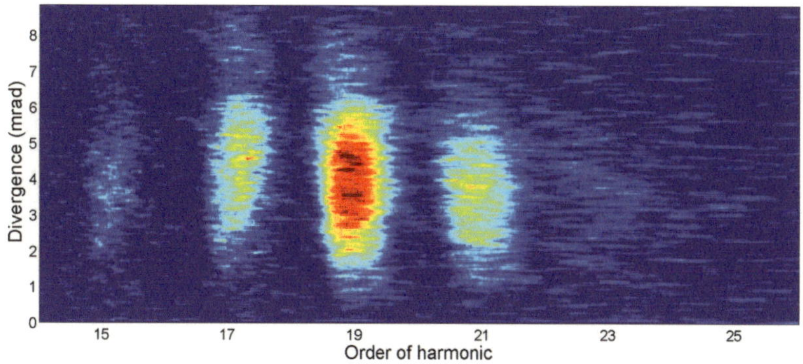

Figure 16. Spectrum of high order harmonic radiation depending on divergence, once the IR beam and lower harmonic have been filtered.

Acknowledgements

This research was supported by 'Ministerio de Economía y Competitividad' of Spain for the funding (projects FIS2013-44174-P and FIS2015-71933-REDT), Diputación General de Aragón-Fondo Social Europeo (TOL research group, T76), Junta de Castilla y León (projects SA116U13, UIC016 and SA046U16), Generalitat de Catalunya (grant 2016 FI_B1 00019), and Centro de Láseres Pulsados (CLPU) for granting access to its facilities.

Author details

María-Victoria Collados[1], Íñigo J. Sola[2], Julia Marín-Sáez[1,3], Warein Holgado[2,4] and Jesús Atencia[1]*

*Address all correspondence to: atencia@unizar.es

1 Applied Physics Department, Aragon Institute of Engineering Research (I3A), University of Zaragoza, Zaragoza, Spain

2 Photonics and Laser Applications Group, Applied Physics Department, University of Salamanca, Salamanca, Spain

3 Enviromental Sciences Departament, Applied Physics Section, University of Lleida, Lleida, Spain

4 Pulsed Lasers Center (CLPU), Villamayor (Salamanca), Spain

References

[1] J. E. Curtis, B. A. Koos, and D. G. Grier. Dynamic holographic optical tweezers. Optics Communications. 2002;**207**(1–6):169–175. DOI: 10.1016/S0030-4018(02)01524-9

[2] G. A. Swartzlander Jr. The Optical Vortex Lens. Optics & Photonics News. 2006;**17** (11):34–41. DOI: 10.1364/OPN.17.11.000039

[3] M. S. Bigelow, P. Zerom, and R. W. Boyd. Breakup of ring beams carrying orbital angular momentum in sodium vapor. Physical Review Letters. 2004;**92**(8):083902. DOI: 10.1103/ PhysRevLett.92.083902

[4] Z. Chen, M. Shih, M. Segev, D. W. Wilson, R. E. Muller, and P. D. Maker. Steady-state vortex-screening solitons formed in biased photorefractive media. Optics Letters. 1997;**22** (23):1751–1753. DOI: 10.1364/OL.22.001751

[5] M. Fisher, C. Siders, E. Johnson, O. Andrusyak, C. Brown, and M. Richardson. Control of filamentation for enhancing remote detection with laser induced breakdown spectroscopy. Proceedings of SPIE. 2006;**6219**:621907. DOI: 10.1117/12.663824

[6] J. Courtial, M. J. Padgett. Performance of a cylindrical lens mode converter for producing Laguerre-Gaussian laser modes. Optics Communications. 1999;**159**(1–3):13–18. DOI: 10.1016/S0030-4018(98)00599-9

[7] J. A. Rodrigo, T. Alieva, and M. L. Calvo. Experimental implementation of the gyrator transform. Journal of the Optical Society of America A. 2007;**24**(10):3135–3139. DOI: 10.1364/JOSAA.24.003135

[8] K. Sueda, G. Miyaji, N. Miyanaga, and M. Nakatsuka. Laguerre-Gaussian beam generated with a multilevel spiral phase plate for high intensity laser pulses. Optics Express. 2004;**12**(15):3548–3553. DOI: 10.1364/OPEX.12.003548

[9] W. C. Cheong, W. M. Lee, X-C Yuan, L-S Zhang, K. Oholakia, and H. Wang. Direct electron-beam writing of continuous spiral phase plates in negative resist with high power efficiency for optical manipulation. Applied Physics Letters. 2004;**85**:5784–5786. DOI: 10.1063/1.1830678

[10] J. W. Sung, H. Hockel, J. D. Brown, and E. G. Johnson. Development of a two-dimensional phase-grating mask for fabrication of an analog-resist profile. Applied Optics. 2006;**45** (1):33–43. DOI: 10.1364/AO.45.000033

[11] K. J. Moh, X-C Yuan, D. Y. Tang, W. C. Cheong, L-S Zhang, D. K. Y. Low, X. Peng, H. B. Niu, and Z. Y. Lin. Generation of femtosecond optical vortices using a single refractive optical element. Applied Physics Letters. 2006;**88**:091103. DOI: 10.1063/1.2178507

[12] G. A. Swartzlander Jr. Achromatic optical vortex lens. Optics Letters. 2006;**31**(13):2042–2044. DOI: 10.1364/OL.31.002042

[13] V. Y. Bazhenov, M. S. Soskin, and M. V. Vasnetsov. Screw dislocations in light wavefronts. Journal of Modern Optics. 1992;**39**(5):985–990. DOI: 10.1080/09500349214551011

[14] H. He, N. R. Heckenberg, and H. Rubinsztein-Dunlop. Optical particle trapping with higher-order doughnut beams produced using high efficiency computer generated holograms. Journal of Modern Optics. 1995;**42**(1):217–223. DOI: 10.1080/09500349514550171

[15] K. Crabtree, J. A. Davis, and I. Moreno. Optical processing with vortex-producing lenses. Applied Optics. 2004;**43**(6):1360–1367. DOI: 10.1364/AO.43.001360

[16] Z. S. Sacks, D. Rozas, and G. A. Swartzlander. Holographic formation of optical-vortex filaments. Journal of the Optical Society of America B. 1998;**15**(8):2226–2234. DOI: 10.1364/JOSAB.15.002226

[17] I. G. Mariyenko, J. Strohaber, and C. J. G. J. Uiterwaal. Creation of optical vortices in femtosecond pulses. Optics Express. 2005;**13**(19):7599–7608. DOI: 10.1364/OPEX.13.007599

[18] K. Bezuhanov, A. Dreischuh, G. G. Paulus, M. G. Schätzel, and H. Walther. Vortices in femtosecond laser fields. Optics Letters. 2004;**29**(16):1942–1944. DOI: 10.1364/OL.29.001942

[19] I.J. Sola, V. Collados, L. Plaja, C. Méndez, J. San Román, C. Ruiz, I. Arias, A. Villamarín, J. Atencia, M. Quintanilla, and L. Roso. High power vortex generation with volume phase holograms and non-linear experiments in gases. Applied Physics B. 2008;**91**(1):115–118. DOI: 10.1007/s00340-008-2967-9

[20] A. Villamarín, J. Atencia, M.V. Collados, and M. Quintanilla. Characterization of transmisión volume holographic gratings recorded in Slavich PFG04 dichromated gelatin plates. Applied Optics. 2009;**48**(22):4348–4353. DOI: 10.1364/AO.48.004348

[21] O. Martínez–Matos, J. A. Rodrigo, M. P. Hernández-Garay, J. G. Izquierdo, R. Weigand, M. L. Calvo, P. Cheben, P. Vaveliuk, and L. Bañares. Generation of femtosecond paraxial beams with arbitrary spatial distribution. Optics Letters. 2010;**35**(5):652–654. DOI: 10.1364/OL.35.000652

[22] J. Atencia, M. V. Collados, M. Quintanilla, J. Marín-Sáez, and I. J. Sola. Holographic optical element to generate achromatic vortices. Optics Express. 2013;**21**(18):21056–21061. DOI: 10.1364/OE.21.021056

[23] H. Kogelnik. Coupled wave theory for thick hologram gratings. Bell System Technical Journal. 1969;**48**(9):2909–2947. DOI: 10.1002/j.1538-7305.1969.tb01198.x

[24] R. J. Collier, C. H. Burckhardt, and L. H. Lin. Optical Holography. 1st ed. New York: Academic Press; 1971. 605 p.

[25] A. Villamarín, I. J. Sola, M. V. Collados, J. Atencia, O. Varela, B. Alonso, C. Méndez, J. San Román, I. Arias, L. Roso, and M. Quintanilla. Compensation of second-order dispersion in femtosecond pulses after filamentation using volume holographic transmission gratings recorded in dichromated gelatin. Applied Physics B. 2012;**106**(1):135–141. DOI: 10.1007/s00340-011-4770-2

[26] J. W. Goodman. Introduction to Fourier Optics. 2nd ed. Singapore: McGraw-Hill; 1996. 441 p.

[27] M. Miranda, M. Kotur, P. Rudawski, C. Guo, A. Harth, A. L'Huillier and C. L. Arnold. Spatiotemporal characterization of ultrashort optical vortex pulses. Journal of Modern Optics. 2016. DOI: 10.1080/09500340.2016.1257751

[28] K. T. Kim, C. Zhang, A. D. Shiner, B. E. Schmidt, F. Légaré, D. M. Villeneuve, and P. B. Corkum. Petahertz optical oscilloscope. Nature Photonics. 2007;**13**:958–962. DOI: 10.1038/nphoton.2013.286

[29] C. Ruiz, J. San Román, C. Méndez, V. Díaz, L. Plaja, I. Arias, and L. Roso. Observation of Spontaneous Self-Channeling of Light in Air below the Collapse Threshold. Physical Review Letters. 2005;**95**:053905. DOI: 10.1103/PhysRevLett.95.053905

[30] A. Couairon and A. Mysyrowicz. Femtosecond filamentation in transparent media. Physics Reports. 2007;**441**(2–4):47–189. DOI: 10.1016/j.physrep.2006.12.005

[31] P. B. Corkum and F. Krausz. Attosecond science. Nature Physics. 2007;**3**:381–387. DOI: 10.1038/nphys620

[32] T. Popmintchev, M.-C. Chen, D. Popmintchev, P. Arpin, S. Brown, S. Alisauskas, G. Andriukaitis, T. Balciunas, O. D. Muecke, A. Pugzlys, A. Baltuska, B. Shim, S. E. Schrauth, A. Gaeta, C. Hernandez-Garcia, L. Plaja, A. Becker, A. Jaron-Becker, M. M. Murnane, and H. C. Kapteyn. Bright coherent ultrahigh harmonics in the keV X-ray regime from mid-infrared femtosecond lasers. Science. 2012;**336**(6086):1287–1291. DOI: 10.1126/science.1218497

[33] P. B. Corkum. Plasma perspective on strong field multiphoton ionization. Physical Review Letters. 1993;**71**(13):1994-1997. DOI: 10.1103/PhysRevLett.71.1994

[34] K. Schafer, B. Yang, L.F. DiMauro, and K.C. Kulander. Above threshold ionization beyond the high harmonic cutoff. Physical Review Letters. 1993;**70**:1599. DOI: 10.1103/PhysRevLett.70.1599

[35] M. Hentschel, R. Kienberger, C. Spielmann, G. A. Reider, N. Milosevic, T. Brabec, P. Corkum, U. Heinzmann, M. Drescher, and F. Krausz. Attosecond metrology. Nature. 2001;**414**:509–513. DOI: 10.1038/35107000

[36] G. Sansone, E. Benedetti, F. Calegari, C. Vozzi, L. Avaldi, R. Flammini, L. Poletto, P. Villoresi, C. Altucci, R. Velotta, S. Stagira, S. De Silvestri, and M. Nisoli. Isolated single-cycle attosecond pulses. Science. 2006;**314**(5798):443–446. DOI: 10.1126/science.1132838

[37] I. J. Sola, E. Mevel, L. Elouga, E. Constant, V. Strelkov, L. Poletto, P. Villoresi, E. Benedetti, J. P. Caumes, S. Stagira, C. Vozzi, G. Sansone, and M. Nisoli. Controlling attosecond electron dynamics by phase-stabilized polarization gating. Nature Physics. 2006;**2**:319–322. DOI: 10.1038/nphys281

[38] M. Zürch, C. Kern, P. Hansinger, A. Dreischuh, and C. Spielmann. Strong-field physics with singular light beams. Nature Physics. 2012;**8**:743–746. DOI: 10.1038/nphys2397

[39] C. Hernandez-Garcia, A. Picon, J. San Roman, and L. Plaja. Attosecond extreme ultraviolet vortices from high-order harmonic generation. Physics Review Letters. 2013;**111**:083602. DOI: 10.1103/PhysRevLett.111.083602

[40] G. Gariepy, J. Leach, K. T. Kim, T. J. Hammond, E. Frumker, R. W. Boyd, and P. B. Corkum. Creating high-harmonic beams with controlled orbital angular momentum. Physics Review Letters. 2014;**113**:153901. DOI: 10.1103/PhysRevLett.113.153901

[41] L. Rego, J. San-Román, L. Plaja, A. Picón, and C. Hernández-García. Ultrashort extreme-ultraviolet vortices. In: H. Pérez-de-Tejada, editor. Vortex Dynamics. 1st ed. Rijeka: Intech; 2016.

Chapter 9

Ultrashort Extreme Ultraviolet Vortices

Laura Rego, Julio San Román, Luis Plaja,

Antonio Picón and Carlos Hernández-García

Additional information is available at the end of the chapter

http://dx.doi.org/10.5772/64908

Abstract

Optical vortices are very attractive because they transport a well-defined orbital angular momentum (OAM) associated with the singularity of the beam. These singular beams, commonly generated in the optical regime, are used in a wide range of applications: communication, micromanipulation, microscopy, among others. The production of OAM beams in the extreme ultraviolet (XUV) and X-ray regimes is of great interest as it allows to extend the applications of optical vortices down to the nanometric scale. Several proposals have been explored in order to generate XUV vortices in synchrotrons and FEL facilities. Here, we study the generation of XUV vortices through high-order harmonic generation (HHG). HHG is a unique source of coherent radiation extending from the XUV to the soft X-ray regime, emitted in the form of attosecond pulses. When driving HHG by OAM beams, highly charged XUV vortices with unprecedented spatiotemporal properties are emitted in the form of helical attosecond beams. In this chapter, we revise our theoretical work in the generation of XUV vortices by HHG. In particular, we illustrate in detail the role of macroscopic phase matching of high-order harmonics when driven by OAM beams, which allows to control the production of attosecond beams carrying OAM.

Keywords: high harmonic generation, attosecond pulses, orbital angular momentum, nonlinear optics, ultrafast phenomena, attosecond helical beams, extreme ultraviolet vortices, phase matching

1. Introduction

Light beams are known to carry spin angular momentum (SAM), related to their polarization, or orbital angular momentum (OAM), related to their spatial phase profile [1–3]. Helical phase beams, also called optical vortices, exhibit a transversal spiral phase structure around the beam axis, thus having a well-defined OAM that is characterized by the topological charge, ℓ, i.e., the number of 2π-phase shifts along the azimuthal coordinate of the light beams. These beams are typically produced in the optical and infrared (IR) regimes, being used in a broad range of applications, such as optical communication [4, 5], micromanipulation [6, 7], and phase contrast microscopy [8, 9], among others [10]. There is a wide interest to bring helical phase beams into the extreme ultraviolet (XUV) and X-ray regimes, in order to extend their microscopy and spectroscopy applications down to the nanometric scale [11–17].

Schemes for the generation of helical phase beams in the XUV/X-ray regime have been reported in accelerator-based sources, such as synchrotrons and X-ray free-electron lasers (XFELs) [18–22], and in laser-based sources, such as plasma induced in solid surfaces [23, 24] or underdense plasmas [25] and high-order harmonic generation (HHG) in gas-phase media [16, 17, 26–30]. In general, accelerator-based sources produce more intense pulses but they are in the femtosecond timescale (1 fs = 10^{-15} s) and their spatiotemporal coherence is not always ideal. On the other hand, laser-based sources produce shorter pulses, in the attosecond timescale (1 as = 10^{-18} s), with higher spatiotemporal coherence. In this chapter, we explore the unique properties of XUV vortices generated via HHG.

HHG is an extreme nonlinear process, driven in an atomic gas-phase medium interacting with an intense IR laser beam. This highly nonlinear interaction results in the emission of coherent harmonic radiation in the XUV and soft X-ray spectral regimes [31, 32]. The HHG radiation presents unique properties, the higher orders being emitted in the form of pulse trains at the attosecond timescale [33–37] or even at the zeptosecond scale if driven by mid-IR lasers [38]. Harmonics with well-defined OAM (OAM-HHG) can be generated using helical phase beams as driving fields. In such scenario, the phase twist is imprinted in the driving IR field—for which a single setup (diffractive mask, for instance) is required [39–41]—and is subsequently transferred to the short-wavelength radiation by nonlinear conversion. The first experiment of OAM-HHG [26] reported the generation of harmonic vortices with topological charge nearly equal to that of the fundamental field, i.e., the spiral phase structures of the harmonics and the fundamental driver beam are similar. Those experimental observations were attributed by the authors to nonlinear propagation effects and not to the OAM-HHG generation process itself. For instance, this finding was unexpected in terms of the present understanding of HHG, in which the harmonic phase scales roughly with the harmonic order [42]. Indeed, the spiral phase structure of the harmonics is expected to be that of the fundamental but multiplied by the harmonic order. In other words, if the driving OAM beam presents a topological charge ℓ, the qth-order harmonic should be emitted with a higher topological charge of $q\ell$. Shortly afterward, this expectation was demonstrated theoretically [27] and proven experimentally [16]. Interestingly, the highly-charged XUV vortices generated through HHG can by synthesized into helical attosecond beams [27–29]. Hence, HHG can be considered a direct route to generate XUV vortices, and nowadays, stands for a promising technique to produce very short (attosecond) vortices even in the soft X-ray regime.

Besides the promising applications of generating vortices in the XUV regime via HHG, there are still some fundamental questions that are very interesting to explore. High-harmonic radiation is a macroscopic output resulting from the emission of many atoms, but this

http://dx.doi.org/10.5772/64908

observable has its origin in a quantum mechanical (single-atom) mechanism and has indeed a strong correspondence, i.e., what occurs at the single-atom level in a particular spatiotemporal position will affect the macroscopic emission. OAM-HHG is indeed a clear example of this striking effect. Each atom in the target perceives a different transversal phase of the incoming IR field, and even though the atoms do not know the phases of their corresponding neighbors (they are not entangled), this information will be imprinted in the superposition of the photons emitted by the whole sample, giving rise to a transverse phase in the emitted radiation. This fascinating property raises questions such as how this process occurs and which are the significant mechanisms behind the imprinted (OAM) phase in the XUV harmonics. It is in fact a much unexplored field and more research is needed to understand the formation of XUV vortices via HHG.

In this chapter, we revise our theoretical work in the generation of XUV vortices by HHG. We provide a self-explanatory introduction of our theoretical models and we emphasize the most prominent results. In particular, we illustrate in detail the role of macroscopic phase matching of high-order harmonics when driven by OAM beams, leading to helical attosecond beams with unprecedented spatiotemporal properties.

The chapter is organized as follows. First, in Section 2 we introduce HHG, paying special attention to the semiclassical understanding of the process in terms of quantum paths. Second, in Section 3, we present the two theoretical methods that we have developed to simulate the HHG process driven by OAM beams. On one hand, we have developed advanced HHG simulations including quantum single-atom harmonic generation and macroscopic propagation. Our code, which is especially suited for beams without cylindrical symmetry, has been widely contrasted against experiments. On the other hand, we have further developed a simple semiclassical model that allows us to unveil the contribution from different quantum paths in the HHG process. This model is of special interest for the optical community using OAM-HHG, as it allows for qualitative description of future experiments. In Section 4, we present our results of OAM-HHG. We show the main properties (OAM content and divergence) of the generated XUV vortices. In addition, we analyze the phase-matching conditions of HHG driven by OAM, showing that different quantum path contributions to the HHG spectrum can be naturally selected by modifying the target position with respect to the driving beam focus. Finally, we analyze the properties of the temporal emission of the XUV vortices. In particular, we show that different XUV harmonic vortices can be synthesized to produce attosecond helical beams, whose spatiotemporal properties (such as divergence and/or temporal chirp) can be controlled through phase matching.

2. High-order harmonic generation

One of the major challenges since the invention of the laser [43] has been to extend coherent radiation to the extreme regions of the optical electromagnetic spectrum, and, in particular, to higher frequencies. For that purpose, harmonic generation offers a direct route. Although perturbative harmonics were obtained shortly afterwards the invention of the laser [44], nonperturbative harmonics, i.e., those whose efficiency does not decrease exponentially with their order, were not reported until the late 1980s [45, 46]. At the very beginning, this behavior was attributed to multiphoton excitation of atomic subshells, but later, a nice classical interpretation was proposed, relating the harmonic generation with tunneling ionization in the so-called three-step model [47, 48]. Since then, the close interplay between theory and experiments has boosted the development of the field, providing a unique mechanism for the generation of extreme ultraviolet/soft X-ray radiation in the form of attosecond pulses [49–55].

HHG is achieved by focusing an intense femtosecond IR field into a gas target, as depicted in **Figure 1**. The target, either atomic or molecular, is commonly found in experiments as a gas jet, a gas cell, or a gas-filled capillary. The high nonlinear interaction between the IR field and each atom results in the emission of harmonics of the fundamental IR field, whose frequency extends into the XUV or even the soft X-ray regime. Notably, radiation emitted by the atoms is coherent. Hence, the harmonic signal reaching the detector is strongly affected by the phase matching of the high-order harmonics emitted by each atom of the target. Thus, HHG radiation results from the interplay between the microscopic single-atom emission and the macroscopic superposition of the contributions of all the atoms of the target.

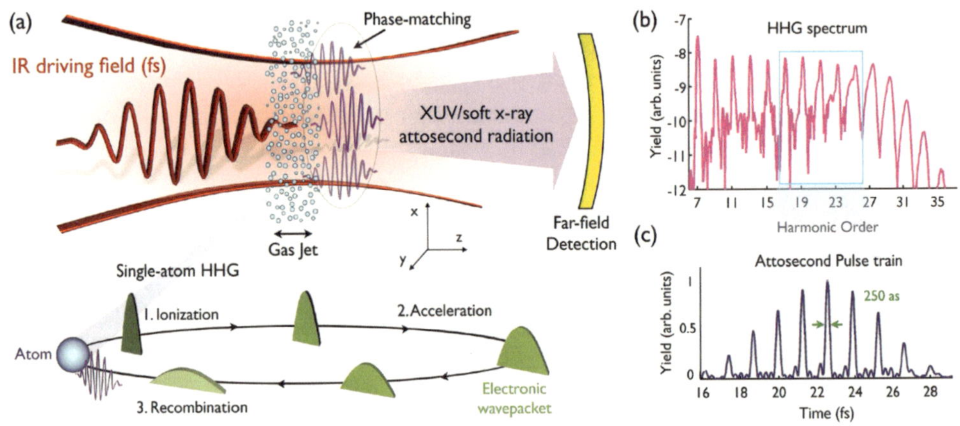

Figure 1. (a) (Top) Schematic view of a HHG experiment: an intense fs IR field is focused into a gas jet. Each atom interacting with the IR field emits XUV radiation in the form of higher-order harmonics. The harmonic signal at the detector is strongly affected by the phase matching of the radiation from all the emitters. At the bottom, we present a scheme of the single-atom HHG mechanism explained through the semiclassical three-step model: (i) ionization of the electronic wave packet through tunnel ionization; (ii) acceleration by the laser field; and (iii) recombination with the parent ion, resulting in the emission of higher-frequency radiation. (b) HHG spectrum obtained from a hydrogen atom driven by an 800 nm, 5.8 cycles (15 fs) FWHM, and 1.57×10^{14} W/cm^2 peak intensity laser pulse. The HHG spectrum is calculated with the quantum SFA+ theory in hydrogen (see Section 3.1). (c) Attosecond pulse train envelope obtained after performing the Fourier transform of the HHG spectrum presented in (b), windowed by a rectangular mask (blue) at the high-energy *plateau* region. Two attosecond pulses per cycle are obtained, of $\simeq$250 as FWHM pulse duration each one.

From the microscopic point of view, the single-atom HHG process is well described by the three-step model [47, 48], explained at the bottom of **Figure 1**: (i) an electronic wave packet is ionized by an intense laser via tunneling; (ii) once in the continuum, it is accelerated by the field and driven back to the parent ion; and (iii) finally, upon recollision, it recombines, releasing high-frequency radiation whose energy corresponds to the kinetic energy acquired from the laser field plus the ionization potential. The dynamics of the electronic wave packet in the continuum can be described accurately using classical trajectories, from which it is simple to associate a particular recombination energy with an initial (ionization) time, as we will see in Section 2.2. Interestingly enough, for every half-cycle, there are two different electron paths leading to the same recombination energy, the so-called short and long trajectories. These short and long trajectories appear naturally as path contributions in the quantum framework within the strong-field approximation (SFA) [56, 57]. The phase of the HHG emission depends on the particular path followed by the electron, thus the interference of the emission from different paths may affect the structure of the harmonic spectrum. In addition, the time ordering of the short and long quantum paths gives rise to a positive or a negative chirp, respectively, in the temporal structure of the associated attosecond pulses [58]. We should also mention the possibility of longer quantum paths, evolving over more than a laser cycle, which give rise to higher-order rescatterings [59, 60] (providing zeptosecond waveforms in the high-order harmonic radiation driven by mid-IR pulses [38]). The HHG interferences from short and long quantum paths have been experimentally observed in standard HHG experiments [61]. In the OAM-HHG context, we started the first studies exploring the contributions of the different quantum paths to the harmonics, see Ref. [28]. In this chapter, we will revise the importance of the quantum paths in OAM-HHG.

Macroscopically, HHG in extensive targets can be intricate. Atoms located at different positions in the target emit harmonic radiation whose phase depends on the amplitude and phase of the local driving field. As a consequence, harmonic phase matching plays an essential role, limiting the spatial regions in which the harmonics contribute efficiently. Typically, harmonic phase matching is described in terms of the longitudinal coherence length that corresponds to the distance between two atoms whose emitted radiation interferes destructively, being a critical parameter for the optimization of HHG in macroscopic targets [62]. In addition, transverse phase matching is defined in terms of the transverse coherence length that describes the phase matching of the radiation coming from atoms placed in a plane perpendicular to the propagation axis [63]. Phase matching is essential to understand most of the macroscopic features of HHG radiation and to develop unique sources such as soft X-ray harmonics driven by mid-IR lasers [31]. Harmonic phase matching has been extensively studied in different macroscopic geometries, such as Gaussian beams focused into gas jets or gas cells [62, 64, 65] or transversal Bessel beams propagating in waveguides [66], among others. In Section 4.2, we discuss in detail some of the phase-matching properties of high-order harmonics driven by beams carrying OAM.

2.1. HHG radiation: XUV to soft X-ray radiation emitted in the form of attosecond pulses

Let us now analyze the main properties of the harmonic spectrum generated by an atom irradiated by an intense laser pulse. The harmonic spectrum presents certain peculiar features that have been observed since the earliest experiments and theoretical works [45, 46, 67, 68]. A typical HHG spectrum is composed of few, odd, low-order harmonics whose intensities decrease exponentially, in accordance with the perturbative scaling, followed by a wide region of odd harmonics (the so-called *plateau*) with similar intensities. The *plateau* ends sharply at a cutoff frequency given by the law $\hbar\omega_{\text{cut-off}} \simeq I_p + 3.17U_p$, where I_p is the ionization potential and U_p is the ponderomotive energy. The appearance of a *plateau* is a remarkable property, since it conveys the efficient emission of shorter wavelengths. Beyond the cutoff frequency, the intensity of the harmonics decreases again quickly and higher-order harmonics are hardly visible. The scaling of the harmonic q with the driving field intensity follows the law I^p, with $p < q$ at the *plateau* and cutoff regions, revealing their nonperturbative origins [69]. In **Figure 1(b)**, we present an example of an HHG spectrum. In particular, we show the simulated HHG spectrum obtained from the interaction of a hydrogen atom with a 800 nm, 5.8 cycles full width at half maximum (FWHM), and 1.57×10^{14} W/cm^2 peak intensity laser pulse.

There is a fundamental interest in extending the cutoff frequency to higher energies. From the dependences of the ponderomotive energy, $U_p \propto \lambda_0^2 I$, and the cutoff law, the HHG spectrum can be extended toward higher energies by increasing the intensity (I) or/and the wavelength (λ_0) of the driving field. However, there are physical restrictions that limit the maximum photon energy of the harmonic spectra. First, if we increase the laser intensity beyond a critical point, we move to the barrier-suppression regime for ionization, where the spectral *plateau* is degraded [70]. Second, the efficiency for the generation of the highest harmonic orders has been shown to decrease drastically with the driving wavelength [59, 71]. In addition, phase-matching conditions may decrease the efficiency of the harmonics when both increasing the intensity and the wavelength of the driving laser field. Nevertheless, recently, it has been possible to define experimental conditions in which favorable phase matching is naturally achieved, leading to the efficient emission of soft X-ray HHG radiation, if driven by mid-IR [31], or vacuum ultraviolet fields [32].

On the other hand, HHG offers the exciting perspective of synthesizing XUV pulses of attosecond duration [33, 34]. An attosecond pulse train is obtained by the selection of the higher frequency part of the HHG spectrum, i.e., the *plateau*. This synthesis is possible because the *plateau* is composed of harmonics with similar intensities (like a frequency comb) and phases [33]. In **Figure 1(c)**, we present the attosecond pulse train envelope computed from the inverse Fourier transform of the complex harmonic spectrum presented in **Figure 1(b)**, windowed by a rectangular mask (blue) of ten photons bandwidth ($10\omega_0$, where ω_0 is the fundamental frequency) at the high-energy *plateau*. An attosecond pulse is emitted in each half-cycle of the incident laser pulse. Note that the obtention of odd-order harmonics results from the integration of attosecond pulses delayed by half a cycle, and with an overall phase shift of π, coming from the sign of the driving field.

The first experimental measurement of an attosecond pulse train was performed by selecting five consecutive harmonics generated in argon, obtaining 250 as pulses [35]. In addition, isolated pulses with duration of 650 as were produced by spectrally filtering few cutoff harmonics produced by an ultrashort laser pulse [36]. At present, after postcompression, isolated pulses with temporal durations < 100 attoseconds have been measured experimentally [37, 72]. Moreover, there is a great interest to extend the production of isolated attosecond pulses to the soft X-ray regime [73, 74], and there are proposals to produce subattosecond waveforms by using mid-IR driving pulses [38].

2.2. Quantum path contributions to HHG

A beautifully simple picture to understand the single-atom HHG process is given by the three-step model [47, 48], based on the so-called simple man's model. In the tunneling regime, one can assume that the ionization process depends only on the instantaneous value of the electromagnetic field, and right after ionization, the electron is located at the coordinate origin with zero velocity. Another assumption of this model consists in considering the dynamics subsequent to ionization as corresponding to a classical free electron in the electromagnetic field, thus neglecting the influence of the Coulomb potential. As a consequence, HHG can be understood in terms of simple semiclassical arguments, as explained at the bottom of **Figure 1(a)**.

We can study in detail the electronic dynamics in the three-step model by integrating the classical equations of motion given by Newton's law and using the conditions given by the simple man's model. In **Figure 2(a)**, we have depicted some electron trajectories for different ionization times, in the presence of a monochromatic laser field of $\lambda_0 = 800$ nm and peak intensity of 1.57×10^{14} W/cm^2. The gray-dashed line represents the electric field in arbitrary units, whereas the green line represents the nucleus position, located at the coordinate origin. Note that we have represented some of the trajectories that lead to recollision, as those that do not recollide will not contribute to the HHG process. In **Figure 2(b)**, we have plotted the recollision kinetic energy of the particles as a function of the recollision (green points) and ionization (red points) times. Note that there is one particular trajectory, represented in blue, whose recollision energy is maximum, taking the well-known value of $3.17U_p$. This trajectory leads to the emission of the cutoff energy, and presents an excursion time of approximately 0.63 times the period of the laser pulse. Further recollisions do not raise this maximum energy. Noticeably, in every half-cycle of the laser pulse, there are two possible electron trajectories leading to the same kinetic energy at recollision and therefore two possible paths for the generation of the same harmonic (each named accordingly to the excursion time as *short* and *long* trajectories). We have represented in **Figure 2(a)** three pairs of short and long trajectories with energies at recollision of $3.0U_p$ (purple), $2.5U_p$ (dark pink), and $1.5U_p$ (light pink).

The short and long path contributions also emerge from the quantum framework within the strong-field approximation [56, 57, 75]. The phase of the harmonic emission not only depends on the phase of the fundamental field but also on the particular path followed by the electron. This additional nonperturbative term, the so-called *intrinsic* phase, is proportional to the ponderomotive potential times the excursion time of the quantum path [75]. As a consequence, the structure of the HHG spectrum may be intricate, as it results from the interference of the emissions of the different quantum paths [61]. In addition, since the different electronic trajectories rescatter at different times (see **Figure 2**), a chirp is imprinted in the emitted radiation [58, 76, 77]. The emission from the short trajectories is positively chirped, i.e., the lower harmonics are emitted before the higher, as depicted by the positive slope of the green curve in **Figure 2(b)**. On the other hand, the emission from the long trajectories exhibits a negative chirp, as depicted by the negative slope of the green curve in **Figure 2(b)**. As a result, the HHG mechanism itself prevents the harmonics to be emitted in the Fourier limit.

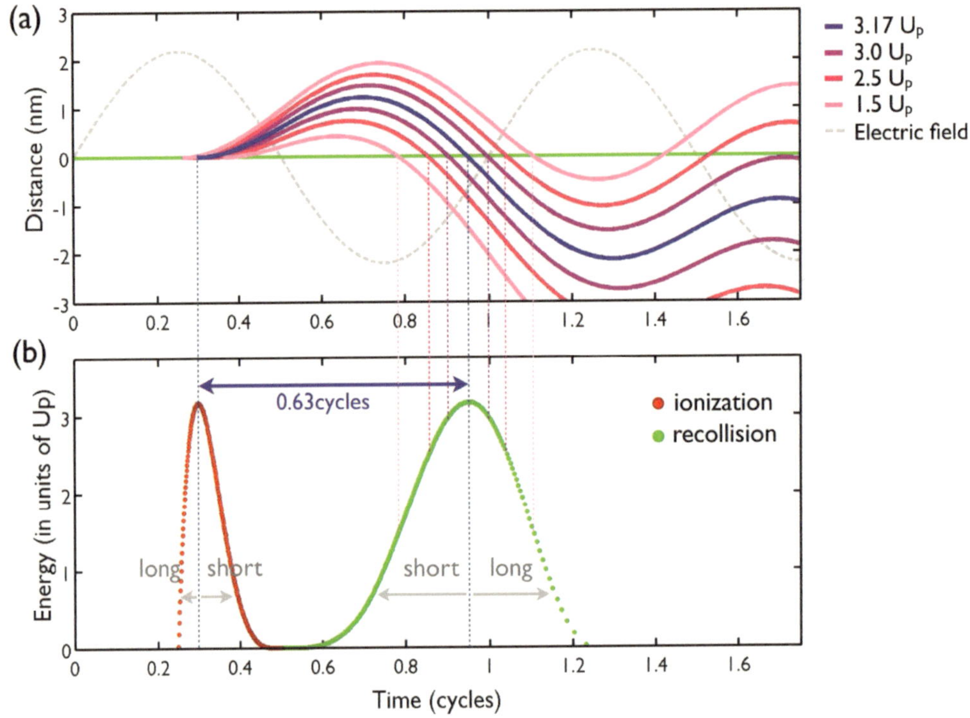

Figure 2. (a) Sample of electronic trajectories in a monochromatic laser field of λ_0 = 800 nm and peak intensity of 1.57×10^{14} W/cm^2. The gray-dashed line represents the electric field (arb. units), whereas the green line, the nucleus position. Three pairs of short and long trajectories are represented for energies at recollision of 3.0U_p (purple), 2.5U_p (dark pink), and 1.5U_p (light pink), whereas the most energetic trajectory, raising 3.17$_p$ at recollision, is represented in blue. The vertical axis represents the distance from the nucleus. (b) Returning kinetic energy of the electrons at the instant of the first recollision in (a). The green points represent the recollision time, whereas the red points, the ionization time.

3. Theoretical treatment of OAM-HHG

The theoretical treatment of HHG admits different levels of description, ranging from the classical to semiclassical and full quantum. For instance, in the previous section, we have shown how HHG can be studied through classical electron trajectories. In this section, we build a theoretical model to describe HHG driven by OAM beams, which takes into account both microscopic and macroscopic physics. We have developed two alternative, full quantum and semiclassical, methods that allow us to describe and have an insight into the microscopic quantum paths involved in the formation of OAM vortices.

http://dx.doi.org/10.5772/64908

Let's start putting in context the physical scenario of HHG driven by OAM beams. In **Figure 3**, we present an schematic view of OAM-HHG. A pulsed vortex beam centered at $\lambda_0 = 800$ nm, a typical wavelength used for HHG, is focused into a gas jet. In this work, we use an argon gas, and the amplitude of the field E_0 is chosen to give a peak intensity of 1.4×10^{14} W/cm^2 at the focus. High-order harmonics are emitted at each atom in the target, and, upon propagation, their coherent addition results in the far-field emission of XUV vortices, represented along the divergence β and azimuth φ coordinates.

The spatial structure of the IR vortex beam is represented by a monochromatic Laguerre-Gaussian beam propagating in the z-direction, with wavelength $\lambda_0 (k_0 = 2\pi/\lambda_0)$ expressed as $LG_{\ell,p}(r,\phi,z;k_0)e^{ik_0 z}$, where

$$LG_{\ell,p}(r,\phi,z;k_0) = E_0 \frac{w_0}{w(z)} \left(\frac{\sqrt{2}r}{w(z)} \right)^{|\ell|} L_p^{|\ell|} \left(\frac{2r^2}{w^2(z)} \right) \exp\left(-\frac{r^2}{w^2(z)} \right) \exp\left[i\ell\phi + i\frac{k_0 r^2}{2R(z)} + i\Phi_G(z) \right], \quad (1)$$

and $w(z) = w_0\sqrt{1 + (z/z_0)^2}$, with w_0 being the waist of the mode, $z_0 = k_0 w_0^2/2$ the Rayleigh range, $R(z) = z\left[1 + (z_0/z)^2\right]$ the phase-front radius, $\Phi_\ell(z) = -(2p + |\ell| + 1)\arctan(z/z_0)$ the Gouy phase, and $L_p^{|\ell|}(x)$ the associated Laguerre polynomials

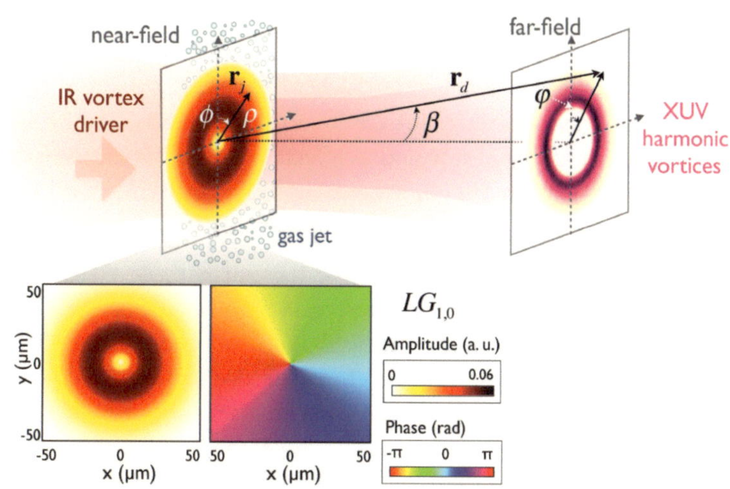

Figure 3. Scheme of HHG driven by OAM beams. An intense IR vortex beam carrying OAM (with $\ell = 1$ in this case) is focused into an argon gas jet. The near-field coordinates are (ρ,ϕ). Each atom emits HHG radiation that, upon propagation, results in the far-field emission of XUV vortices with some divergence and azimuth (β,φ). In the bottom, we show the near-field amplitude (left) and phase (right) of the $LG_{1,0}$ IR mode, with beam waist of 30 μm.

$$L_p^{|\ell|}(x) = \sum_{m=0}^{p}(-1)^m \frac{(|\ell|+p)!}{(p-m)!(|\ell|+m)!m!}x^m. \quad (2)$$

The indices $\ell = 0, \pm 1, \pm 2,...$ and $p = 0,1,2,...$ correspond to the topological charge and the number of nonaxial radial nodes of the mode, respectively. At the bottom of **Figure 3**, we present the amplitude (left) and phase (right) of the $LG_{1,0}$ IR mode. In the simulations presented in this work, we have considered a well-defined OAM of $\ell = 1$ and $p = 0$ as a driving field, with a beam waist of $w_0 = 30$ μm and, therefore, a Rayleigh range of $z_0 = 3.5$ mm.

3.1. 3D quantum SFA theory

We have developed a quantum method to compute HHG including both single-atom (microscopic) and phase matching (macroscopic) physics. In order to take into account macroscopic phase matching, we compute harmonic propagation using the electromagnetic field propagator [78]. To this end, we discretize the target (gas jet) into a set of *j* elementary radiating volumes (typically > 10^5), as sketched in **Figure 3**, and propagate the emitted field $\mathbf{E}_j(\mathbf{r}_j, t)$ to the far-field detector,

$$\mathbf{E}_j(\mathbf{r}_d,t) = \frac{q_j \mathbf{s}_d}{c^2 |\mathbf{r}_d - \mathbf{r}_j|} \times \left[\mathbf{s}_d \times \mathbf{a}_j\left(t - \frac{|\mathbf{r}_d - \mathbf{r}_j|}{c}\right)\right] \quad (3)$$

where s_d is the unitary vector pointing to the detector, $\mathbf{r}_d$ and $\mathbf{r}_j$ are the position vectors of the far-field detector and of the near-field elementary radiator *j*, respectively, and $\mathbf{a}_j$ is the dipole acceleration of each elementary source. Note that in Eq. (3), the harmonic radiation is assumed to propagate with the vacuum phase velocity, which is a reasonable assumption for high-order harmonics. The harmonic field reaching the detector is thus calculated as the coherent addition of these elementary contributions. Propagation effects in the driving field, such as the production of plasma, the refractive index of the neutrals, the group-velocity walkoff [79], as well as the absorption in the propagation of the harmonics are also taken into account.

In the case of intense fields, the computation of the microscopic HHG dynamics of the elementary radiators is not trivial, as the interaction is nonperturbative. Due to the large number of radiators, the use of exact numerical integration of the time-dependent Schrödinger equation becomes extremely expensive. Therefore, the use of simplified models is almost mandatory. In the case of intense fields, S-matrix approaches combined with the strong-field approximation [80–82] are demonstrated to retain most of the features of the HHG process [75, 83]. We use an extension of the standard SFA, hence we will refer it as SFA+, where the total dipole acceleration of the *j* radiator ($\mathbf{a}_j$) is found from two contributions, a_b and a_d, the first being the standard SFA expression and the latter being a correction due to the instantaneous dressing of the ground state. This method computes the dipole acceleration directly from the superposition of the contributions of each Volkov wave, each can be integrated separately as an ordinary 1D equation, leading to a very efficient algorithm [71, 84]. The single-atom HHG spectrum can be found after performing the Fourier transform of the dipole acceleration.

One of the advantages of this method, which takes into account both microscopic and macroscopic HHG, is that it is well-fitted to nonsymmetric geometries, therefore, it is especially suited for computing HHG driven by singular beams, such as those carrying OAM. The method has been successfully used for describing regular HHG with near- and mid-IR lasers, in good agreement with several experiments [31, 63, 73, 85–88].

For the simulation results presented in this work, we have considered a laser pulse with a well-defined OAM of $\ell = 1$ as a driving field. The laser pulse is assumed to have a temporal envelope of $\sin^2(\pi t/NT)$, where N=16 cycles and T is the laser period (2.67 fs). The FWHM of the pulse is 15.4 fs and the peak intensity is 1.4×10^{14} W/cm^2. The detector is assumed to be far enough from the interaction area in order to comply with the far-field condition. The target is an argon gas jet, directed along the x-axis, and modeled by a Gaussian distribution along the y and z dimensions (whose FWHM is 500 μm), and a constant profile along its axial dimension, x. The peak density is 10^{17} atoms/cm^3.

3.2. The thin slab model (TSM)

We have developed a simple model, the *thin slab model* (referred as quantum path OAM model in [28]), to gain insight on the physical origin of the structure of the XUV vortices generated via HHG. In this simplified description, the target is represented by a thin (2D) slab placed perpendicular to the propagation axis of the fundamental beam. As a consequence, our approach does not take into account longitudinal phase matching, allowing us to focus the discussion on transverse phase matching [63] that is expected to be relevant due to the transverse structure of OAM beams.

The complex beam profile of the fundamental field in the thin slab, placed at a propagation distance z_t, can be written as

$$A(r) = U(\rho,\phi,z_t)e^{i\Phi(\rho,\phi,z_t)}, \tag{4}$$

where $\Phi(\mathbf{r})$ contains the phase terms presented in the exponential of Eq. (1) and the amplitude terms are grouped in $U(\mathbf{r})$.

We consider the high-order harmonics to be emitted at the thin slab. Let us see how their intensity and phase profiles are related to that of the driving field. In the perturbative regime, the amplitude of the qth harmonic is proportional to the power of the fundamental beam amplitude $\propto U(\mathbf{r})^q$. However, since HHG is highly nonperturbative, the amplitude of the generated harmonic scales with a lower power [69]. As a consequence, the harmonic field at the slab can be approximated by an analytic SFA representation, first used in [42]. This representation gives a different scaling of the amplitude and harmonic phase with the driving field. First, the amplitude of the qth harmonic is proportional to the pth power of the fundamental amplitude $U(\mathbf{r})^p$, where $p < q$ reflects the nonperturbative HHG behavior, being approximately constant for the harmonics in the *plateau* region. In order to extract the value of the p exponent, and thus the scaling of the harmonic amplitude with the driving field, we perform single-atom HHG calculations using the SFA+ approach. We present in **Figure 4(a)**, the HHG spectra (log scale) driven by an 800 nm laser field for three different peak intensities. The driving laser pulse is assumed to have a temporal envelope of $\sin^2(\pi t/NT)$, where N= 64 cycles and T is the laser period (2.67 fs). The FWHM of the pulse is 61.4 fs. The amplitude of the field is chosen to give peak intensities of 0.875×10^{14} W/cm^2 (red), 1.177×10^{14} W/cm^2 (green), and 1.524×10^{14} W/cm^2 (blue). In **Figure 4(b)–(e)**, we show the scaling for the peak of the 17th, 19th, 21st, and 23rd harmonics, respectively, with the driving field amplitude (log scale). The oscillations of the harmonic peak amplitudes with the driving amplitude are a known consequence of the interference between the short and long quantum path contributions [61]. By performing a linear fit out of the logarithmic plot, we extract the p exponent, being 3.7, 3.9, 3.4, and 3.6 for the 17th, 19th, 21st, and 23rd harmonics, respectively. As commented above, the value obtained for p is very similar for the harmonics in the *plateau* region of the spectrum.

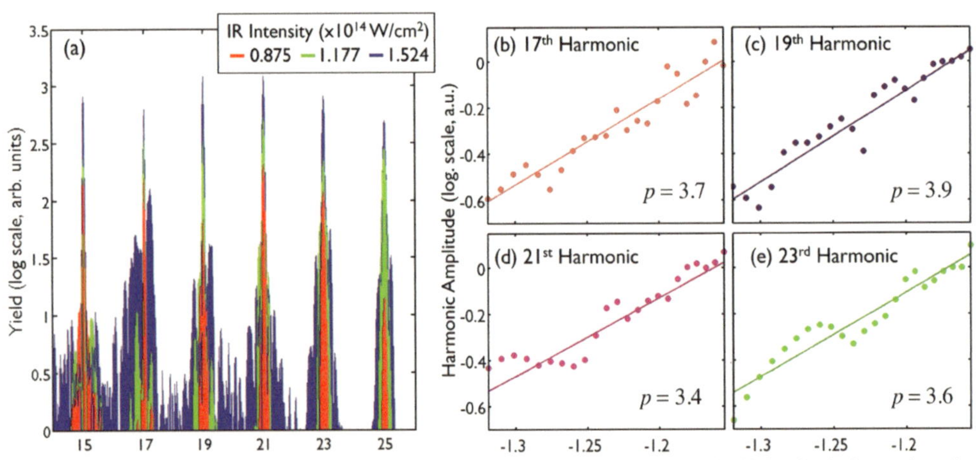

Figure 4. Scaling of the harmonic amplitude with the driver field. (a) Single-atom HHG spectrum (log scale) calculated using the SFA+ approach, driven by a 800 nm laser pulse modeled by a $\sin^2$ envelope with 61.4 fs FWHM, and peak intensities of 0.875×10^{14} W/cm^2 (red), 1.117×10^{14} W/cm^2 (green), and 1.524×10^{14} W/cm^2 (blue). The scaling of the 17th (b), 19th (c), 21st (d), and 23rd (e) harmonics with the driver amplitude is fitted to a power p of 3.7, 3.9, 3.4, and 3.6, respectively.

On the other hand, the phase of the harmonics scales with q times the phase of the driving field $q\Phi(\mathbf{r})$, and has an additional nonperturbative term, the *intrinsic* phase, which is proportional to the intensity of the fundamental field $\alpha_q^j|U(\mathbf{r})|^2$. The strong-field parameters α_q^j

depend on the electronic quantum path (j) (short or long) followed by the semiclassical description of the HHG process (see Section 2.2). They can be calculated from the action of classical trajectories with recollision energies leading to the considered harmonic [62]. For the simulations considered in this chapter, the strong-field parameters for the 19th harmonic for the short (S) and long (L) quantum path contributions are found to be $\alpha_{19}^{S} = 2.75 \times 10^{-14}$ cm^2 W^{-1} and $\alpha_{19}^{L} = 2.220 \times 10^{-13}$ cm^2 W^{-1}.

Taking into account the previous arguments for the description of the high-order harmonics in the analytic SFA representation, the contribution of the j quantum path (short or long) to the qth harmonic emitted at a slab located at z_t can be written as

$$A_q^j(\rho,\phi,z_t) = \left(\frac{C}{\tau^j}\right)^{3/2} |U(\rho,\phi,z_t)|^p \, e^{iq\Phi(\rho,\phi,z_t)} e^{i\alpha_q^j |U(\rho,\phi,z_t)|^2} \quad (5)$$

where C is a constant and τ^j is the excursion time associated with the j quantum path [75], affecting the efficiency of HHG due to the quantum diffusion of the electron wave packet.

Once we have the description of the qth harmonic field at the slab, we use Fraunhofer diffraction theory to calculate the far-field signal. The qth harmonic field at the divergence β and azimuthal ϕ coordinates is given by the Fraunhofer integral

$$F_q^j(\beta,\varphi) \propto \int_0^\infty \int_0^{2\pi} \rho' d\rho' d\phi' A_q^j(\mathbf{r}') e^{-i\frac{2\pi}{\lambda_q}\rho' \tan\beta \cos(\varphi-\phi')}, \quad (6)$$

$$F_q^j(\beta,\varphi) \propto \left(\frac{C}{\tau^j}\right)^{3/2} \int_0^\infty \int_0^{2\pi} \rho' d\rho' d\phi' |U(\rho',\phi',z_t)|^p \, e^{iq\Phi(\rho',\phi',z_t)} e^{i\alpha_q^j |U(\rho',\phi',z_t)|^2} e^{-i\frac{2\pi}{\lambda_q}\rho' \tan\beta \cos(\varphi-\phi')} \quad (7)$$

where (ρ',φ') represents the near-field polar coordinates and λ_q is the wavelength of the qth harmonic related to the fundamental wavelength as $\lambda_q = \lambda_0/q$. Now, using (5), we obtain

While this equation is valid for a driving field composed of any combination of Laguerre-Gaussian modes, let us now consider a driving field composed of a single OAM mode. In this case, the intensity of the driving does not vary along the azimuthal coordinate, i.e., $|U(\rho',\varphi', z_t)| = |U(\rho',z_t)|$, and we can perform the analytic integral over φ'. By considering the phase part of the fundamental Laguerre-Gaussian beam in Eq. (1), we finally obtain a compact formula that includes the dependence on both the fundamental OAM (ℓ) and the quantum path contribution (j) [28],

$$F_q^j(\beta) \propto \left(\frac{C}{\tau^j}\right)^{3/2} \int \rho' d\rho' \, |U(\rho', z_t)|^p \, e^{iq\frac{k_0 \rho'^2}{2R(z_t)}} e^{iq\Phi_G(z_t)} e^{i\alpha_q^j |U(\rho', z_t)|^2} J_{q\ell}\left(\frac{2\pi}{\lambda_q} \beta \, \rho'\right). \tag{8}$$

Note that the detected harmonic signal will result from the superposition of short and long quantum path contributions. We remark that Eq. (8) is valid when the fundamental beam is composed by a single OAM mode. For any combination of different OAM modes, one should use Eq. (7) [30].

4. Results: XUV harmonic vortices

Once we have introduced the theoretical methods, we can proceed with the discussion of the main results of OAM-HHG. First, in Section 4.1, we describe the main properties of the XUV vortices that are generated. We will concentrate on their OAM content and divergence. Second, in subsection 4.2, we will analyze how macroscopic phase-matching conditions affect the process of OAM-HHG. We will observe how short and long quantum path contributions can be naturally isolated by adjusting the relative position between the gas jet and the driving beam focus. Finally, in Section 4.3, we will describe the rich spatiotemporal structure of the helical attosecond beams that emerges when several XUV vortices are synthesized.

4.1. XUV vortices: OAM content and divergence

Let us now investigate the main properties of the XUV vortices generated via HHG. In **Figure 5**, we present the intensity (top) and phase (bottom) far-field angular profiles of the 17th (a), 19th (b), 21st (c), and 23rd (d) harmonic vortices. These results are obtained using the 3D quantum SFA simulations, where we have placed the argon jet at the focus position of the $LG_{1,0}$ mode shown in **Figure 3**. Two main conclusions arise from these plots. First, the radius of the annular intensity distribution is similar for all the harmonics, and thus, they are emitted with similar divergence. Note that some of the harmonics exhibit secondary rings due to phase matching, as we discuss in detail in subsection 4.2. Second, the phase-plots show that the topological charge of the qth-order harmonic is q, leading to a simple scaling of the topological charge with the harmonic order $\ell_q = q\ell$ (ℓ_q being the topological charge of the qth harmonic). This scaling of OAM in HHG was first predicted theoretically in Ref. [27] and later confirmed experimentally in Ref. [16]. Note that this simple selection rule follows the energy conservation law for the harmonic conversion process $\omega_q = q\omega$ (ω being the frequency of the fundamental field), which was previously found in perturbative OAM harmonic generation [89–92]. However, this simple rule may result unexpected in the nonperturbative regime, since the intrinsic phase is not directly related to a specific multiphoton channel. Recently, we have shown that the nonperturbative nature of HHG modifies the selection rule for the OAM build-up only if the driving beam is composed of different $LG_{\ell,p}$ modes [30]. If pure OAM modes are used, the OAM build-up in HHG is thus governed by the simple selection rule $\ell_q = q\ell$.

http://dx.doi.org/10.5772/64908

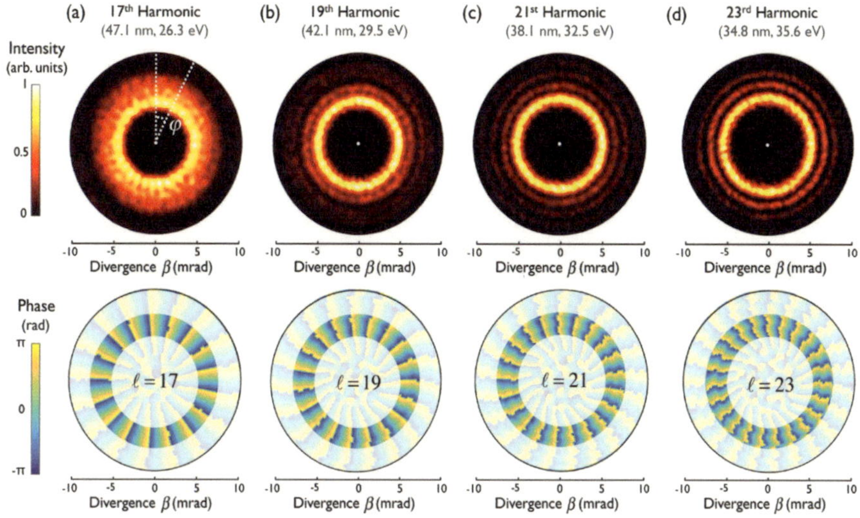

Figure 5. Intensity (top) and phase (bottom) far-field angular profiles of the 17th (a), 19th (b), 21st (c), and 23th (d) harmonics, obtained with the 3D quantum SFA simulations. The argon gas jet is placed at the focus of a $LG_{1,0}$ mode, with a beam waist of $w_0 = 30$ μm and $\lambda_0 = 800$ nm. The amplitude is chosen to give a peak intensity of 1.14×10^{14} W/cm^2 at the focus. Note that the resulting topological charge (obtained from the phase plots) follows the scaling law $l_q = ql$ and thus, resulting in $l_{17} = 17$ to $l_{23} = 23$.

It is interesting to note that in the OAM-HHG experiments [16, 26, 29] and theoretical works [27, 28] performed to date, the divergence of all the harmonics was found to be similar, in contrast to standard HHG driven by Gaussian beams, where the divergence decreases with the harmonic order. In fact, this result is a consequence of the OAM build-up law $\ell_q = q\ell$ [29]. We can disentangle it from the final expression of our TSM approach, given by Eq. (8). The dependence on the divergence in the TSM appears in the Bessel function, which is a part of the amplitude of the integrand of the equation. The order of this Bessel function is directly related to the OAM of the qth harmonic generated in the process $q\ell$. The argument, which is proportional to the divergence, is inversely proportional to the wavelength of the qth harmonic λ_q. Consequently, both, the order and the argument of the Bessel function, are proportional to the harmonic order q, which is the reason why the position of the maximum of the amplitude of the integrand of Eq. (8) is almost invariant with respect to the harmonic order. To prove this similarity, we show in **Figure 6** this amplitude as a function of the divergence for different harmonic orders [19th (blue), 21st (green), 23rd (red), 25th (cyan), and 27th (purple)]. The amplitudes have been calculated at the radius at which the function $rU^p(r,z_t = 0)$ takes the maximum value. As it can be observed, the signal of all the harmonics rises around $\beta \simeq 6$ mrad, where all of them present a maximum amplitude.

HHG leads to a perfect vortex generation process in terms of its applicability [93], as all these XUV vortices of topological charge $q\ell$ are emitted with similar size. First, HHG offers a unique possibility to generate energy-tunable OAM beams, by selecting different harmonic orders (for example, in **Figure 5**, from 26.3 to 35.6 eV), which, in principle, could be extended to the soft X-ray regime if mid-IR drivers were used [31]. Second, HHG provides a unique tool to produce

OAM-tunable XUV vortices, by directly changing the topological charge of the driving beam ℓ, or, for example, by using a crossed beam geometry, where the noncollinear beams present different OAM and wavelength properties [16]. Recently, the use of driving beams carrying a superposition of different OAM contributions revealed a rich scenario in which XUV vortices present multiple OAM contributions arising from the nonperturbative behavior of HHG [30], offering a new route to generate OAM-tunable vortex beams.

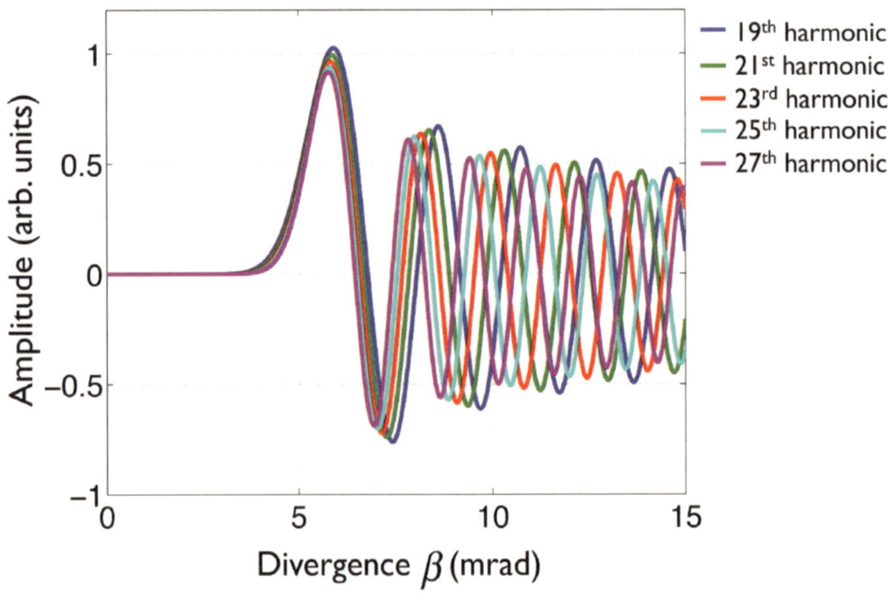

Figure 6. Amplitude of the integrand of Eq. (8), $r_0U^p(r_0,z_t = 0)J_{q\ell}(2\pi\beta r_0/\lambda_q)$, for different harmonic orders (q going from 19th to 27th), and r_0 being the radius where we obtain the maximum of the function $rU^p(r,z_t = 0)$. The divergence where the integrand is maximized is very similar for all harmonic orders ($\beta \simeq 6$ mrad). Note that in the order of the Bessel function, we have considered the OAM build-up law, $\ell_q = q\ell$, which appears to be a necessary condition for emitting XUV harmonic vortices with similar divergences.

4.2. Phase-matching effects in OAM-HHG

Once we have presented the main properties of the XUV vortices generated via HHG (divergence and OAM content), we study the effect of macroscopic phase matching on their generation. Harmonic phase-matching conditions are known to depend strongly on the position of the target with respect to the focus of the driving field [62]. In particular, in standard HHG experiments with Gaussian beams, short quantum path contributions dominate the detected HHG emission if the gas target is placed after the focus position. However, if the target is placed before the focus, long quantum paths dominate for low divergence angles while the short ones dominate at larger angles [64, 65]. In this section, we discuss the effects of the relative position between the target and the beam focus in HHG driven by helical-phase beams carrying OAM. First, we make use of our TSM approach to disentangle the short and long quantum path

http://dx.doi.org/10.5772/64908

contributions under different phase-matching conditions. Second, we perform a time-frequency analysis (TFA) to identify the quantum path contributions in the 3D quantum SFA simulations. More information about the phase matching in OAM-HHG can be found in Ref. [28].

4.2.1. Disentangling quantum path contributions with the thin slab model

In **Figure 7**, we show the intensity profile (log scale) of the 19th harmonic as a function of the target position along the propagation distance. To this end, the intensity angular profile along the divergence β for a fixed azimuth $\varphi = 0$ is shown as a function of the target position z_t, where $z_t < 0$ ($z_t > 0$) stands for a target placed before (after) the focus position. Note that the horizontal axis can be read as the radial axis of the emitted OAM ring. First, we show in panel (a), the results obtained with the 3D quantum SFA model, where we have considered the target as a thin slab of 1 μm. We observe that both the divergence and the thickness of the harmonic vortex ring vary substantially with respect to the slab position. Moreover, in addition to the main intensity structure emitted at 5–6 mrad, we observe additional structures with larger divergence if $z_t > 0$ and with lower divergence if $z_t < 0$.

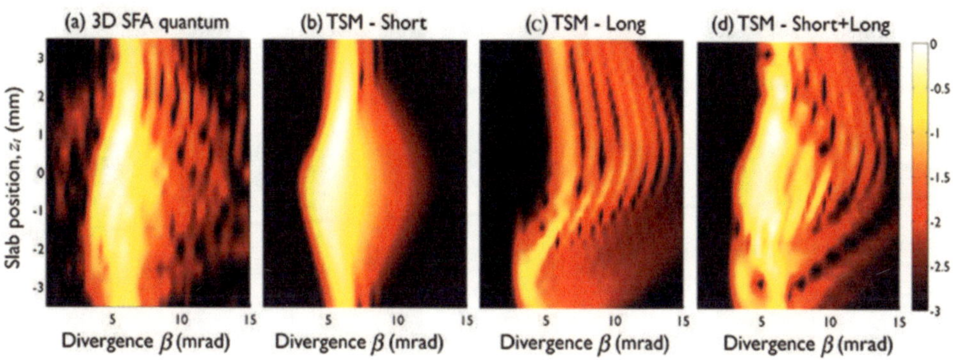

Figure 7. Angular dependence of the 19th harmonic as a function of the slab position z_t. Panel (a) shows the result obtained from the full 3D quantum SFA model. Panels (b–d) are obtained from the TSM, using the same parameters as in (a), but considering the short quantum paths (b), the long ones (c) and the coherent addition of both of them (d).

In order to get a qualitative explanation of the divergence profiles observed in panel (a) of **Figure 7**, we use the *thin slab model*. To this end, we calculate the 19th harmonic far-field signal by integrating Eq. (8) with the same beam parameters as used in panel (a), where the pth power of the harmonic amplitude is $p = 3.9$ and the strong-field parameters are $\alpha_{19}^{S} = 2.75 \times 10^{-14}$ cm²/W and $\alpha_{19}^{L} = 2.220 \times 10^{-13}$ cm²/W. We can separately obtain the harmonic contributions from different quantum paths with the TSM. Thus, we present the intensity profile of the short (b), long (c) and short + long (d) quantum path contributions to the 19th harmonic as a function of z_t. Note that in panel (d), short and long contributions are added coherently taking into account the semiclassical excursion times [see Eq. (8)]. The excellent agreement between the 3D SFA quantum simulations in panel (a) and the short + long TSM results in panel (d) serves as a validation of the TSM. As a result, we can use the results in panels (b) and (c) to identify the main features of panel (a) in terms of short and long quantum paths.

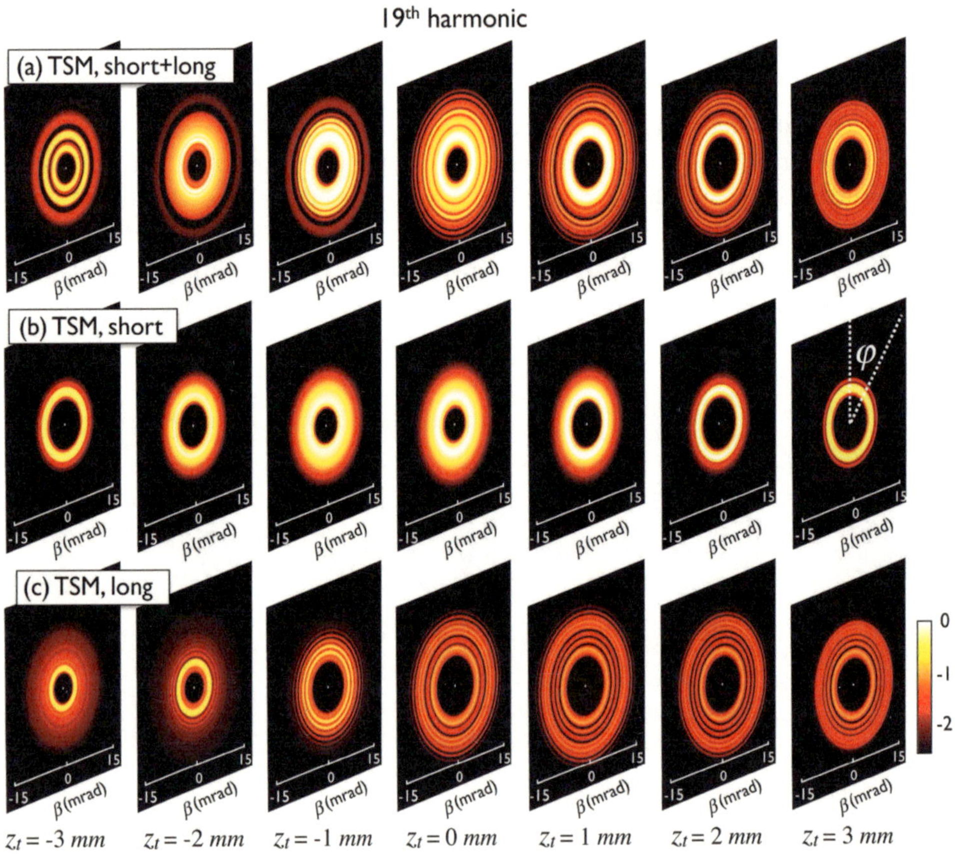

Figure 8. Far-field spatial intensity profile of the 19th harmonic for a slab placed at seven near-field positions, from z_t = − 3 mm (left) to z_t = 3 mm (right), calculated with the TSM considering (a) short + long, (b) short, and (c) long quantum path contributions. Whereas short quantum path contributions exhibit similar intensity and structure independently of the near-field slab location, long ones are more intense if the slab is placed before the focus position. As a consequence, a rich vortex structure profile is obtained depending on the near-field slab position.

For better interpretation of the results that can be extracted from **Figure 7**, we represent in **Figure 8** the far-field divergence intensity profile of the 19th harmonic for seven slab positions z_t, from $z_t = -3$ mm to $z_t = 3$ mm, using the TSM and considering (a) short + long, (b) short, and (c) long quantum path contributions. From **Figures 7** and **8**, we observe that the emission dependence on the slab position is completely different for short and long quantum paths, being the short quantum path contribution the dominant, as expected from the shorter excursion time [see Eq. (8)]. On the other hand, the intensity profile obtained from the short quantum path contributions is symmetric with z_t and that of the long ones is asymmetric, being more intense if $z_t < 0$. This behavior allows us to identify two different phase-matching regimes depending on the slab position with respect to the focus position. First, when the slab is placed before the focus ($z_t < 0$), long quantum paths are emitted with smaller divergence than the short ones. In fact, for $z_t = < -2$ mm, in **Figure 7(a)** and **(d)**, we can identify two well-separated contributions, with similar weight, corresponding to long quantum paths at low divergence and to short paths at higher divergence. This result becomes evident when analyzing the first column of **Figure 8**: two separated rings, from long (inner) and short (outer) quantum path contributions are obtained. Second, when the slab is placed after the focus position ($z_t > 0$), the short quantum path contributions dominate at smaller divergence angles while the long ones generate a less intense wriggle structure at large divergence angles.

As the results presented in **Figures 7** and **8** are performed for a thin slab, longitudinal phase-matching effects are neglected, and we can confirm that the behavior presented for different target positions is a direct consequence of transverse phase matching [28, 63]. Note that although in **Figure 7(a)** we have used an unrealistic 2D gas jet (1 μm -thick), when using a realistic 3D gas jet (500 μm-thick) the results are very similar [28].

4.2.2. Quantum path contributions in the 3D quantum SFA simulations: time-frequency analysis

In order to confirm the qualitative picture given by the TSM for the separation of short and long quantum paths, we perform a time-frequency analysis of the harmonic emission based on the 3D quantum SFA simulations using a realistic 500 μm thick 3D gas jet. The 3D quantum SFA model gives a full quantum description of the HHG emission but it does not provide an insight of the semiclassical picture in terms of quantum trajectories. However, with the help of the TFA, we can resolve the temporal order in which high-order harmonics are emitted, and thus, it allows us to extract relevant information about the quantum path contributions.

In the TFA of the harmonic emission [94], we select a spectral window in the harmonic spectrum and take its Fourier transform. By shifting the window to cover the entire harmonic spectrum, it is possible to resolve the time in which the different harmonics are generated. Nevertheless, due to the uncertainty principle, we have to be careful when interpreting the results of the TFA, as the width of the spectral window determines the resolution in time; the narrower the spectral window is, the less resolution we obtain in the temporal domain. The TFA has served to reveal unique features of HHG, like for example, the quantum path interferences due to multiple rescatterings in the HHG process [38, 95]. As it is usually observed in HHG calculations, the time evolution of the harmonic emission follows faithfully the distribution of rescattering energies of classical trajectories. As a consequence, from the TFA analysis, we can identify the short (long) quantum paths, as the TFA structures with positive (negative) slope, which give rise to a positive (negative) chirp in the harmonic emission, as commented in Section 2.2.

In **Figure 9**, we present the HHG spectra as a function of the divergence β for a 500 μm-thick gas jet placed 2 mm before (a) and after (b) the focus position, accordingly to the two regimes identified in **Figures 7** and **8**. As commented before, we observe that the harmonics exhibit similar divergence but their structure changes with the gas jet position. In the lower panels of **Figure 9**, we present the TFA of the HHG spectra emitted at different divergence β, where a Gaussian spectral window with FWHM $3\omega_0$ has been used to perform the TFA.

When the gas jet is placed before the focus (a), two ring structures can be identified, centered at divergences 3.5 and 6.2 mrad, whose TFA analysis are shown in panels (a1) and (a2), respectively. Whereas the TFA at 3.5 mrad (a1) presents structures with negative slope, at 6.2 mrad (a2), it presents structures with positive slope. As a consequence, long quantum paths are emitted with lower divergence and short ones with higher divergence, in agreement with the results previously obtained with the TSM in **Figures 7** and **8**. In addition, note that within each half-cycle, the long quantum path structures in (a1) are delayed in time with respect to the short ones in (a2), as expected from the semiclassical HHG theory [96].

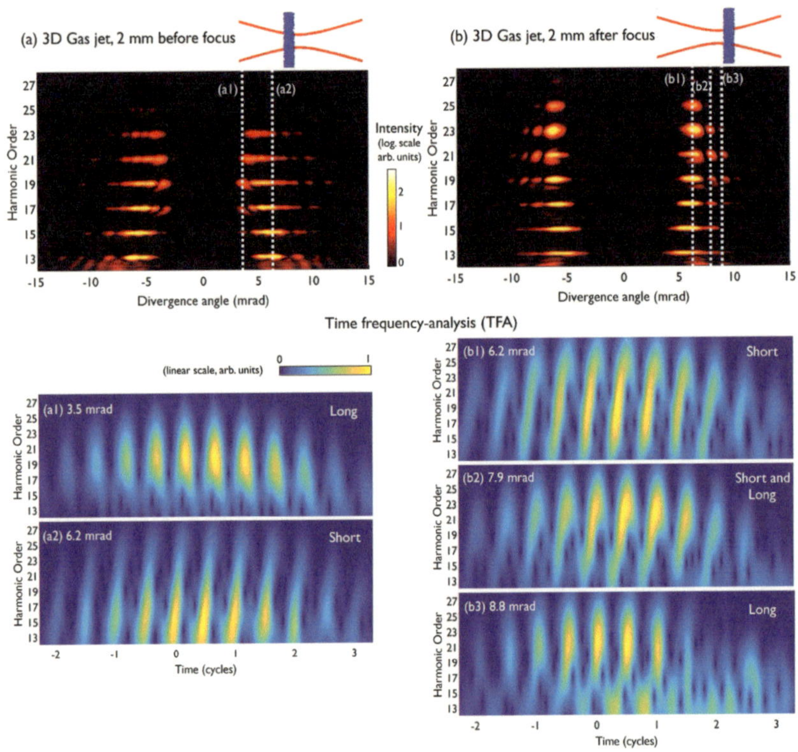

Figure 9. Time-frequency analysis based on the 3D quantum SFA model, for the XUV vortices generated in a 3D gas jet placed 2 mm before (a) and after (b) the focus position of the $LG_{1,0}$ beam. The top panels present the angular dependence of the harmonic radiation. The TFA of some particular angles (indicated by the dashed-white lines) are shown at the bottom panels, revealing the contribution of the short and long quantum paths, through the positive or negative slopes of the TFA structures, respectively.

On the other hand, when the gas jet is located after the focus (b), a prominent ring centered at 6.2 mrad and two secondary ones centered at 7.9 and 8.8 mrad are observed. The TFA analyses at those three divergences are shown in plots (b1) to (b3), respectively. We note that, in agreement with the TSM of **Figures 7** and **8**, (b1) is dominated by short quantum path contributions, whereas (b3) by long ones. Interestingly, the annular structure conformed around 7.9 mrad, (b2) presents contributions from both short and long quantum paths.

As a result, longitudinal phase matching, included through the 500 μm thickness of the gas jet, does not modify our previous conclusions in realistic gas jets, and thus the angular separation of highly-charged HHG vortices from different quantum paths is due to the transverse phase matching. In addition, this analysis through the TFA corroborates the adequacy of the TSM approach developed in Section 3.2. A rich scenario of XUV harmonic vortices is obtained, with different intensity structures and temporal properties due to the phase matching of short and long quantum path contributions. Note that long quantum path contributions are hard to observe experimentally due to their lower weight in the overall harmonic emission. Whereas with standard Gaussian beams, long paths have been successfully observed and characterized, the observation of XUV vortices obtained from long quantum path contributions remains unobserved.

4.3. Helical attosecond pulse trains

Finally, we proceed to study how the XUV vortices obtained from HHG are emitted in the temporal domain. As introduced in Section 2, one of the most exciting perspectives of HHG is the possibility of synthesizing XUV pulses of attosecond duration. For the correct synthesis, the spectrum should approximately satisfy two conditions: its structure should approach to that of a frequency comb, in which the harmonic intensities are similar, and the relative phase between the harmonics should be nearly constant. Fortunately, HHG satisfies these two conditions, and as XUV vortices are emitted with similar divergence in OAM-HHG, the radiation of several high-order harmonics can be synthesized to produce attosecond pulse trains.

In **Figure 10**, we present our simulation results of the temporal evolution of the high-order harmonics produced when the 500 μm-thick gas jet is centered at 2 mm after the focus position. Note that, at this position, short quantum path contributions dominate, and the HHG signal is maximized, as discussed in Section 4.2. We have filtered the low-order harmonics (below 11th) by simulating their transmission through an aluminum filter. First, in panel (a), we present five snapshots of the transverse intensity distribution of the integrated harmonic signal within half a cycle of the fundamental laser pulse, i.e., in a time interval of 1.33 fs. We observe that two well-defined intensity structures rotate in time, with a period of half a cycle.

In order to visualize the spatial structure of the HHG emission, we plot in panel (b) the spatiotemporal evolution of a given HHG intensity. We observe that a helical attosecond pulse train is obtained [27], i.e., an attosecond pulse train delayed along the azimuth φ according to the phase variation of the fundamental $LG_{1,0}$ beam. In the right part of panel (b), we show the attosecond pulse train obtained at four different azimuths (φ=0, $\pi/2$, π, and $3\pi/2$). Note that this particular spatiotemporal structure can also be identified as two interwinded helices that result from the addition of harmonic vortices with different topological charges [97]. The number of interwinded helices is given by $\delta n\ell$, i.e., the order difference between successive harmonics in the spectrum (δn) times the topological charge of the driving beam (ℓ) [97]. In our case, $\ell = 1$ and $\delta n = 2$ as odd-order harmonics are produced, leading to two interwinded helices. As a consequence, the spatiotemporal structure of the helical attosecond beam can be sculpted by driving HHG with different topological charges (i.e., modifying ℓ) or wavelength combinations (and thus changing δn).

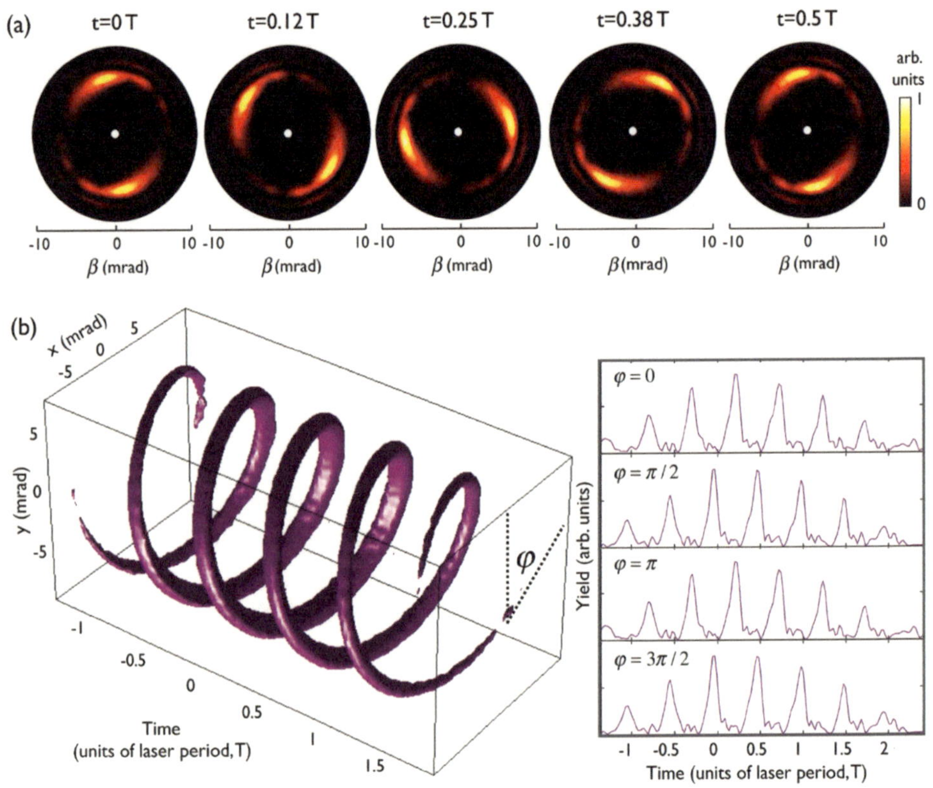

Figure 10. Attosecond helical beam structures of the emitted harmonic radiation obtained with the 3D quantum SFA model, for a 500 μm gas jet placed 2 mm after the focus position. In panel (a), we represent five snapshots of the transversal intensity structure at times equal to 0, 0.12, 0.25, 0.38, and 0.5 times the laser period (T), starting from the center of the IR pulse. We note that two well-defined intensity structures rotate in time, with a period of half a cycle. In panel (b), we represent the spatiotemporal structure of the XUV emission, exhibiting an helical attosecond beam composed of two interwinded helices. The right panels in (b) show the attosecond pulse train at four different azimuths ($\varphi = 0$, $\pi/2$, π, and $3\pi/2$). From this panel, we observe that the helical attosecond beam is composed by an attosecond pulse train that is delayed along the azimuthal coordinate.

The attosecond helical beams obtained from OAM-HHG were theoretically predicted in [27] and have been recently measured experimentally using the RABBITT technique [29]. Note that the carrier-envelope phase of the driving field is imprinted in the attosecond helical beam along

http://dx.doi.org/10.5772/64908

the azimuth. If the pulse duration of the driving field is restricted to few cycles, the number of pulses of the attosecond helical beam changes along the azimuthal direction, from an isolated pulse, to two pulses [27].

Finally, note that helical attosecond pulse trains with different divergence and chirp can be obtained by properly selecting the phase-matching properties to maximize the emission from short or long quantum path contributions [28]. Thus, our results show the possibility of generating helical attosecond pulse trains with different spatiotemporal structures that could be selected depending on the application. In these types of helical beams, which can also be generated using relativistic laser-matter interaction [23, 98], not only the phase but also the intensity profile possess a helical structure. As a consequence, these beams are able to exchange OAM in situations where standard Laguerre-Gauss beams cannot, opening an entirely new light-matter interaction regime [97].

5. Conclusions and outlook

Extreme ultraviolet OAM beams with high spatiotemporal coherence are now produced via HHG. When focusing an intense infrared OAM beam into a gas target, the HHG process combines the microscopic quantum mechanics with the macroscopic physics to imprint OAM into higher-order harmonics of the driving field. For instance, HHG maps the OAM properties of the driver into the harmonic vortices. When driven by pure vortices, the qth-order harmonic vortex is produced with topological charge $\ell_q = q\ell$ (ℓ being the topological charge of the driving vortex beam). In addition, the different harmonic vortices are emitted with similar divergence, thus allowing their synthesis into attosecond helical beams, unprecedented beams whose phase and intensity profiles exhibit a helical structure. Thus, the combination of the properties of OAM beams with the spatiotemporal characteristics of HHG opens an exciting perspective in ultrafast science.

Harmonic phase matching in the generation of XUV vortices allows for the selection of their spatiotemporal properties. Since beams carrying OAM have a very unique transverse structure, transverse phase matching plays a fundamental role in the macroscopic emission. With the help of a thin slab model, we unveil the role of different quantum path contributions in the generation of XUV vortices. For instance, the relative position between the gas jet and the beam focus allows for the spatial selection of XUV vortices produced from short and/or long quantum path contributions in the HHG process.

It has been recently shown that if OAM-HHG is driven by a combination of vortex beams with different topological charge, the generated XUV vortices exhibit a rich OAM content, arising from the nonperturbative behavior of HHG [30]. For instance, experimental methods to produce vortex beams with well-defined OAM may be imperfect, and such beams with several OAM contributions are naturally produced. Even when those imperfections are small, the sensibility of the nonperturbative nature of the HHG makes them visible. Another example where nonpure OAM beams are present is in the case of beams with fractional OAM, which can be generated with spiral phase plates or spatial light modulators [99], or using conical refraction [100, 101].

In all these situations, the OAM dynamics are much more complex than those presented in this review. On the other hand, it has been recently shown that spin angular momentum can be transferred to the XUV harmonics in the HHG process [102], producing circularly polarized harmonics and attosecond pulses [87, 88, 103, 104]. New scenarios, in which orbital and spin angular momenta add a new degree of freedom to light-matter interaction, open an exciting route for the next generation of high-resolution, ultrafast, XUV/X-ray diagnostic tools for fundamental studies and applications.

6. Acknowledgements

C. Hernández-García acknowledges support from the Marie Curie International Outgoing Fellowship within the EU's Seventh Framework Programme for Research and Technological Development (2007–2013), under REA Grant Agreement No. 328334. C. Hernández-García, J. San Román, and L. Plaja acknowledge support from Junta de Castilla y León (Project SA116U13, SA046U16) and MINECO (FIS2013-44174-P, FIS2016-75652-P). A. Picón acknowledges financial support of the U.S. Department of Energy, Basic Energy Sciences, Office of Science, under Contract No. DE-AC02-06CH11357.

Author details

Laura Rego[1], Julio San Román[1], Luis Plaja[1], Antonio Picón[2] and Carlos Hernández-García[1*]

*Address all correspondence to: carloshergar@usal.es

1 Grupo de Investigación en Aplicaciones del Láser y Fotónica, University of Salamanca, Salamanca, Spain

2 Argonne National Laboratory, Argonne, Illinois, USA

References

[1] L. Allen, M. W. Beijersbergen, R. J. C. Spreeuw, and J. P. Woerdman, Phys. Rev. A 45, 8185 (1992).

[2] M. S. Soskin and M. V. Vasnetsov, Progress Optics 42, 219–276 (2001).

[3] G.F. Calvo, A. Picón, and E. Bagan, Phys. Rev. A 73, 013805 (2006).

[4] J. Wang, J. Y. Yang, I. M. Fazal, N. Ahmed, Y. Yan, H. Huang, Y. Ren, Y. Yue, S. Dolinar, M. Tur, and A. E. Willner, Nat. Photon. 6, 488–496 (2012).

[5] X. Cai, J. Wang, M. J. Strain, B. Johnson-Morris, J. Zhu, M. Sorel, J. L. O'Brien, M. G. Thompson, and S. Yu, Science 338, 363 (2012).

[6] N. Simpson, L. Allen, and M. Padgett, J. of Mod. Opt. 43, 2485–2491 (1996).

[7] D. G. Grier, Nature 424, 810 (2003).

[8] S. Fürhapter, A. Jesacher, S. Bernet, and M. Ritsch-Marte, Opt. Lett. 30, 1953 (2005).

[9] A. Jesacher, S. Fürhapter, S. Bernet, and M. Ritsch-Marte, Phys. Rev. Lett. 94, 233902 (2005).

[10] J. P. Torres and L. Torner (Eds.), *Twisted Photons*. Wiley-VCH, Weinheim (2011).

[11] A. Sakdinawat and Y. Liu, Opt. Lett. 32, 2635 (2007).

[12] M. van Veenendaal and I. McNulty, Phys. Rev. Lett. 98, 157401 (2007).

[13] K.A. Nugent, Adv. Phys. 59, 1 (2010).

[14] A. Picón, J. Mompart, J. R. Vázquez de Aldana, L. Plaja, G. F. Calvo, and L. Roso, Opt. Exp. 18, 3660 (2010).

[15] A. Picón, A. Benseny, J. Mompart, J. R. Vázquez de Aldana, L. Plaja, G. F. Calvo, and L. Roso, New J. of Phys. 12, 083053 (2010).

[16] G. Gariepy, J. Leach, K. T. Kim, T. J. Hammond, E. Frumker, R. W. Boyd, and P. B. Corkum, Phys. Rev. Lett. 113, 153901 (2014).

[17] M. Z. Zürch, Optical vortices in the XUV chapter, in book *High-Resolution Extreme Ultraviolet Microscopy*. Springer Theses, Heidelberg (2015).

[18] A. G. Peele, P. J. McMahon, D. Paterson, C. Q. Tran, A. P. Mancuso, K. A. Nugent, J. P. Hayes, E. Harvey, B. Lai, and I. McNulty, Opt. Lett. 27, 1752 (2002).

[19] S. Sasaki and I. McNulty, Phys. Rev. Lett. 100, 124801 (2008).

[20] E. Hemsing, A. Marinelli, and J. B. Rosenzweig, Phys. Rev. Lett. 106, 164803 (2011).

[21] P. R. Ribič, D. Gauthier, and G. De Ninno, Phys. Rev. Lett. 112, 203602 (2014).

[22] E. Hemsing, G. Stupakov, D. Xiang, and A. Zholents, Rev. Mod. Phys. 86, 897 (2014).

[23] Y. Shi, B. Shen, L. Zhang, X. Zhang, W. Wang, and Z. Xu, Phys. Rev. Lett. 112, 235001 (2014).

[24] X. Zhang, B. Shen, Y. Shi, X. Wang, L. Zhang, W. Wang, J. Xu, L. Yi, and Z. Xu, Phys. Rev. Lett. 114, 173901 (2015).

[25] J. T. Mendonça and J. Viera, Phys. Plasmas 22, 123106 (2015).

[26] M. Zürch, C. Kern, P. Hansinger, A. Dreischuh, and Ch. Spielmann, Nat. Phys. 8, 743–746 (2012).

[27] C. Hernández-García, A. Picón, J. San Román, and L. Plaja, Phys. Rev. Lett. 111, 083602 (2013).

[28] C. Hernández-García, J. San Román, L. Plaja, and A. Picón, New J. Phys. 17, 093029 (2015).

[29] R. Géneaux, A. Camper, T. Auguste, O. Gobert, J. Caillat, R. Taeib, and T. Ruchon, Nat. Commun. 7, 12583 (2016).

[30] L. Rego, J. San Román, A. Picón, L. Plaja, and C. Hernández-García, "Nonperturbative Twist in the Generation of Extreme-Ultraviolet Vortex Beams", Phys. Rev. Lett. 117, 163202 (2016).

[31] T. Popmintchev, M. C. Chen, D. Popmintchev, P. Arpin, S. Brown, S. Ališauskas, G. Andriukaitis, T. Balčiunas, O. Mücke, A. Pugzlys, A. Baltuška, B. Shim, S. E. Schrauth, A. Gaeta, C. Hernández-García, L. Plaja, A. Becker, A. Jaron-Becker, M. M. Murnane, and H. C. Kapteyn, Science 336, 1287 (2012).

[32] D. Popmintchev, C. Hernández-García, F. Dollar, C. Mancuso, J. A. Pérez-Hernández, M. C. Chen, A. Hankla, X. Gao, B. Shim, A. L. Gaeta, M. Tarazkar, D. A. Romanov, R. J. Levis, J. A. Gaffney, M. Foord, S. B. Libby, A. Jaron-Becker, A. Becker, L. Plaja, M. M. Murnane, H. C. Kapteyn, and T. Popmintchev, Science 350, 1225 (2015).

[33] G. Farkas and C. Toth, Phys. Lett. A 168, 447 (1992).

[34] I. P. Christov, M. M. Murnane, and H. C. Kapteyn, Phys. Rev. Lett. 78, 1251–1254 (1997).

[35] P. M. Paul, E. S. Toma, P. Breger, G. Mullot, F. Augé, Ph. Balcou, H. G. Muller, and P. Agostini, Science 292, 1689 (2001).

[36] M. Hentschel, R. Kienberger, C. Spielmann, G. A. Reider, N. Milosevic, U. Heinzmann, M. Drescher, and F. Krausz, Nature 414, 509 (2001).

[37] E. Goulielmakis, M. Schultze, M. Hofstetter, V. S. Yakovlev, J. Gagnon, M. Uiberacker, A. L. Aquila, E. M. Gullikson, D. T. Attwood, R. Kienberger, F. Krausz, and U. Kleineberg, Science 320, 1614 (2008).

[38] C. Hernández-García, J. A. Pérez-Hernández, T. Popmintchev, M. M. Murnane, H. C. Kapteyn, A. Jaroń-Becker, A. Becker, and L. Plaja, Phys. Rev. Lett. 111, 033002 (2013).

[39] I. J. Sola, V. Collados, L. Plaja, C. Méndez, J. San Román, C. Ruiz, I. Arias, A. Villamarín, J. Atencia, M. Quintanilla, and L. Roso, Appl. Phys. B 91, 115 (2008).

[40] K. Yamane, Y. Toda, and R. Morita, Opt. Exp. 20, 18986 (2012).

[41] J. Atencia, M. V. Collados, M. Quintanilla, J. Marin-Saez, and I. J. Sola, Opt. Exp. 21, 21056–21061 (2013).

[42] A. L'Huillier, Ph. Balcou, S. Candel, K. J. Schafer, and K. C. Kulander, Phys. Rev. A 46, 2778 (1992).

[43] Th. Maiman, Nature 187, 493–494 (1960).

[44] P. A. Franken, A. E. Hill, C. W. Peters, and G. Weinreich, Phys. Rev. Lett. 7, 118–119 (1961).

[45] A. McPherson, G. Gibson, H. Jara, U. Johann, T. S. Luk, I. A. McIntyre, K. Boyer, and C. K. Rhode, J. Opt. Soc. Am. B 21, 595–601 (1987).

[46] M. Ferray, A. L'Huillier, X. F. Li, L. A. Lompré, G. Mainfray, and C. Manus, J. Phys. B: At. Mol. Opt. Phys. 21, L31–L35 (1988).

[47] P. B. Corkum, Phys. Rev. Lett. 71, 1994–1997 (1993).

[48] K. Schafer, B. Yang, L. F. DiMauro, and K. C. Kulander, Phys. Rev. Lett. 70, 1599–1602 (1993).

[49] P. Agostini and L. F. DiMauro, Rep. Prog. Phys. 67, 813–855 (2004).

[50] M. F. Kling and M. J. J. Vrakking, Annu. Rev. Phys. Chem. 59, 463–492 (2008).

[51] F. Krausz and M. Ivanov, Rev. Mod. Phys. 81, 163–234 (2009).

[52] M. Nisoli and G. Sansone, Prog. Quant. Electron. 33, 17–59 (2009).

[53] T. Popmintchev, M. Chen, P. Arpin, M. M. Murnane, and H. C. Kapteyn, Nat. Photon. 4, 822 (2010).

[54] L. Plaja, R. Torres, and A. Zaïr (Eds.), Attosecond Physics, Springer Series in Optical Sciences. Springer (Heidelberg, 2013).

[55] J. Miao, T. Ishikawa, I. K. Robinson, and M. M. Murnane, Science 348, 530 (2015).

[56] M. Lewenstein, P. Salieres, and A. L'Huillier, Phys. Rev. A 52, 4747 (1995).

[57] M. Bellini, C. Lynga, A. Tozzi, M. Gaarde, T. Hänsch, A. L'Huillier, and C. Wahlström, Phys. Rev. Lett. 81, 297–300 (1998).

[58] Y. Mairesse, A. de Bohan, L. J. Frasinski, H. Merdji, L. C. Dinu, P. Monchicourt, P. Breger, M. Kovacev, R. Taïeb, B. Carré, H. G. Muller, P. Agostini, and P. Saliéres, Science 302, 1540 (2003).

[59] J. Tate, T. Auguste, H. G. Muller, P. Salières, P. Agostini, and L. F. Di Mauro, *Scaling of Wave-Packet Dynamics in an Intense Midinfrared Field*. Phys. Rev. Lett. 98, 013901-1-4 (2007).

[60] D. D. Hickstein, P. Ranitovic, S. Witte, X. Tong, Y. Huismans, P. Arpin, X. Zhou, K. E. Keister, C. W. Hogle, B. Zhang, C. Ding, P. Johnsson, N. Toshima, M. J. J. Vrakking, M. M. Murnane, and H. C. Kapteyn, Phys. Rev. Lett. 109, 073004 (2012).

[61] A. Zaïr, M. Holler, A. Guandalini, F. Schapper, J. Biegert, L. Gallmann, U. Keller, A. S. Wyatt, A. Monmayrant, I. A. Walmsley, E. Cormier, T. Auguste, J. P. Caumes, and P. Salières, Phys. Rev. Lett. 100, 143902 (2008).

[62] M. B. Gaarde, J. L. Tate, and K. J. Schafer, J. Phys. B: At. Mol. Opt. Phys. 41, 132001 (2008).

[63] C. Hernández-García, I. J. Sola, and L. Plaja, Phys. Rev. A 88, 043848 (2013).

[64] P. Salières, A. L'Huillier, and M. Lewenstein, Phys. Rev. Lett. 74, 3776 (1995).

[65] P. Balcou, P. Salieres, A. L'Huillier, and M. Lewenstein, Phys. Rev. A, 55, 3204 (1997).

[66] T. Popmintchev, M. C. Chen, A. Bahabad, M. Gerrity, P. Sidorenko, O. Cohen, I. P. Christov, M. M. Murnane, and H. C. Kapteyn, Proc. Natl. Acad. Sci. USA 106, 10516 (2009).

[67] X. F. Li, A. L'Huillier, M. Ferray, L. A. Lompré, and G. Mainfray, Phys. Rev. A 39, 5751 (1989).

[68] J. L. Krause, K. J. Schafer, and K. C. Kulander, Phys. Rev. A 45, 4998 (1992).

[69] L. Lompré, A. L'Huillier, M. Ferray, P. Monot, G. Mainfray, and C. Manus, J. Opt. Soc. Am. B 7, 754–761 (1990).

[70] J. Seres, P. Wobrauschek, C. Streli, V. S. Yakovlev, E. Seres, F. Krausz, and C. Spielmann, New J. Phys. 8, 251–251 (2006).

[71] J. A. Pérez-Hernández, L. Roso, and L. Plaja, Opt. Exp. 17, 9891 (2009).

[72] K. Zhao, Q. Zhang, M. Chini, Y. Wu, X. Wang, and Z. Chang, Opt. Lett. 37, 3891 (2012).

[73] M. C. Chen, C. Mancuso, C. Hernández-García, F. Dollar, B. Galloway, D. Popmintchev, P. C. Huang, B. Walker, L. Plaja, A. A. Jaron-Becker, A. Becker, M. M. Murnane, H. C. Kapteyn, and T. Popmintchev, Proc. Natl. Acad. Sci. USA 111, E2361–E2367 (2014).

[74] S. M. Teichmann, F. Silva, S. L. Cousin, M. Hemmer, and J. Biegert, Nat. Comm. 7, 11493 (2016).

[75] M. Lewenstein, Ph. Balcou, M. Yu. Ivanov, A. L'Huillier, and P. B. Corkum, Phys. Rev. A 49, 2117 (1994).

[76] K. Varjú, Y. Mairesse, B. Carré, M. B. Gaarde, P. Johnsson, S. Kazamias, R. López-Martens, J. Mauritsson, K. J. Schafer, Ph. Balcou, A. L'Huillier, and P. Saliéres, J. of Mod. Opt. 52, 379–394 (2005).

[77] C. Hernández-García and L. Plaja, J. of Phys. B 45, 074021 (2012).

[78] C. Hernández-García, J. A. Pérez-Hernández, J. Ramos, E. Conejero Jarque, L. Roso, and L. Plaja, Phys. Rev. A 82, 022432 (2010).

http://dx.doi.org/10.5772/64908

[79] C. Hernández-García, T. Popmintchev, H. C. Kapteyn, M. M. Murnane, L. Plaja, A. Becker, and A. Jaron-Becker, New J. Phys. 18, 073031 (2016).

[80] L. V. Keldysh, Zh. Eksp. Teor. Fiz. 47, 1945 (1964) [Sov. Phys. JETP 20, 1307 (1965)].

[81] F. H. M. Faisal, J. Phys. B 6, L89 (1973).

[82] H. R. Reiss, Phys. Rev. A 22, 1786 (1980).

[83] W. Becker, A. Lohr, M. Kleber, and M. Lewenstein, Phys. Rev. A 56, 645 (1997).

[84] J. A. Pérez-Hernández, C. Hernández-García, J. Ramos, E. Conejero, L. Plaja, and L. Roso, New Methods For Computing High-Order Harmonic Generation and Propagation, Springer Series in Chemical Physics Vol. 100 (Springer, New York, 2011), Chap. 7, p. 145.

[85] C. Hernández-García, W. Holgado, L. Plaja, B. Alonso, F. Silva, M. Miranda, H. Crespo, and I. J. Sola, Opt. Exp. 16, 21497–21508 (2015).

[86] M. Kretschmar, C. Hernández-García, D. S. Steingrube, L. Plaja, U. Morgner, and M. Kovacev, Phys. Rev. A 88, 013805 (2013).

[87] T. Fan, P. Grychtol, R. Knut, C. Hernández-García, D. D. Hickstein, D. Zusin, C. Gentry, F. J. Dollar, C. A. Mancuso, C. W. Hogle, O. Kfir, D. Legut, K. Carva, J. L. Ellis, K. M. Dorney, C. Chen, O. G. Shpyrko, E. E. Fullerton, O. Cohen, P. M. Oppeneer, D. B. Milosevic, A. Becker, A. A. Jaron-Becker, T. Popmintchev, M. M. Murnane, and H. C. Kapteyn, Proc. Natl. Acad. Sci. USA 112, 14206 (2015).

[88] D. D. Hickstein, F. J. Dollar, P. Grychtol, J. L. Ellis, R. Knut, C. Hernández-García, C. Gentry, D. Zusin, J. M. Shaw, T. Fan, K. M. Dorney, A. Becker, A. Jaroń-Becker, H. C. Kapteyn, M. M. Murnane, and Ch. G. Durfee, Nat. Photon. 9, 743 (2015).

[89] I. V. Basistiy, V. Yu. Bazhenov, M. S. Soskin, and M. V. Vasnetsov, Opt. Comm. 103, 422–428 (1993).

[90] K. Dholakia, N. B. Simpson, M. J. Padgett, and L. Allen, Phys. Rev. A. 54, R3742 (1996).

[91] J. Courtial, K. Dholakia, L. Allen, and M. J. Padgett, Phys. Rev. A. 56, 4193 (1997).

[92] A. Berzanskis, A. Matijosius, A. Piskarskas, V. Smilgevicius, and A. Stabinis, Opt. Comm. 140, 273–276 (1997).

[93] A.S. Ostrovsky, C. Rickenstorff-Parrao, and V. Arrizón, Opt. Lett. 38, 534 (2013).

[94] P. Antoine, B. Piraux, and A. Maquet, Phys. Rev. A 51, R1750 (1995).

[95] C. Hernández-García and L. Plaja, Phys. Rev. A 93, 023402 (2016).

[96] P. Antoine, A. L'Huillier, and M. Lewenstein, Phys. Rev. Lett. 77, 1234–1237 (1996).

[97] G. Pariente and F. Quéré, Opt. Lett. 40, 2037–40 (2015).

[98] J. Leach, E. Yao, and M. J. Padgett, New J. Phys. 6, 71 (2004).

[99] A. Turpin, Yu. V. Loiko, A. Peinado, A. Lizana, T. K. Kalkandjiev, J. Campos, and J. Mompart, Opt. Exp. 23, 5704–5715 (2015).

[100] K. E. Ballantine, J. F. Donegan, and P. R. Eastham, Sci. Adv. 2, e1501748 (2016).

[101] Zs. Lécz and A. Andreev, Phys. Rev. E 93, 013207 (2016).

[102] A. Fleischer, O. Kfir, T. Diskin, P. Sidorenko, and O. Cohen, Nat. Photon. 8, 543 (2014).

[103] C. Chen, Z. Tao, C. Hernández-García, P. Matyba, A. Carr, R. Knut, O. Kfir, D. Zusin, C. Gentry, P. Grychtol, O. Cohen, L. Plaja, A. Becker, A. Jaron-Becker, H. Kapteyn, and M. Murnane, Sci. Adv. 2, e1501333 (2016).

[104] C. Hernández-García, C. G. Durfee, D. D. Hickstein, T. Popmintchev, A. Meier, M. M. Murnane, H. C. Kapteyn, I. J. Sola, A. Jaron-Becker, and A. Becker, Phys. Rev. A 93, 043855 (2016).

Chapter 10

Fractal Light Vortices

Federico J. Machado, Juan A. Monsoriu and

Walter D. Furlan

Additional information is available at the end of the chapter

http://dx.doi.org/10.5772/66343

Abstract

Vortex lenses produce special wavefronts with zero-axial intensity, and helical phase structure. The variations of the phase and amplitude of the vortex produce a circular flow of energy that allows transmitting orbital angular momentum. This property is especially in optical trapping, because due to the orbital angular momentum of light, they have the ability to set the trapped particles into rotation. Vortex lenses engraved in diffractive optical elements have been proposed in the last few years. These lenses can be described mathematically as a two-dimensional (2D) function, which expressed in polar coordinates are the product of two different separable one-dimensional (1D) functions: One, depends only on the square of radial coordinate, and the other one depends linearly on the azimuthal coordinate and includes the topological charge. The 1D function that depends on the radial coordinate is known as a zone plate. Here, vortex lenses, constructed using different aperiodic zone plates, are reviewed. Their optical properties are studied numerically by computing the intensity distribution along the optical axis and the transverse diffraction patterns along the propagation direction. It is shown that these elements are able to create a chain of optical traps with a tunable separation, strength and transverse section.

Keywords: optical vortex, optical traps, diffractive optical elements (DOEs), fractal zone plates, devil's lenses, Thue-Morse sequence

1. Introduction

Vortex lenses (VLs) produce special wavefronts with zero-axial intensity, and helical structure with undefined phase in the vortex centre. These structures, called optical vortices [1–3], constitute an intriguing and growing area of research that combines fundamental theoretical aspects and novel applied technologies. A fundamental property of vortices is the conservation of the topological charge which is defined as the 2π module of the total change of the phase along a closed curve surrounding the vortex centre. The variations of the phase and amplitude of the vortex produce a circular flow of energy that allows to carry orbital angular momentum [4, 5]. This property is useful for several applications, such as in astrophysics [6, 7], transmission of information [8, 9], microscopy [10], laser engraving [11] and especially in optical traps [12–16], because due to the orbital angular momentum of light, they have the ability to set the trapped particles into rotation.

Several methods have been proposed to obtain optical vortices [3, 4, 17–20] being spiral phase plates (i.e. lenses with a linear phase dependence on the azimuthal angle) and diffractive optical elements (DOEs) probably the most frequent approach. In fact, due to its simplicity, they can be employed in multiple applications. In particular, zone plates (ZPs) have found a great number of new applications in the last few years [21]. A standard amplitude ZP consists of a series of concentric circular rings of equal area, with alternating transmitting and absorbing zones. This means that along the square of the radial coordinate, a ZP can be thought as a periodic structure. The focusing effect is created by the constructive interference of waves passing through this structure.

On the other hand, in recent years different non-periodic and quasi-periodic sequences [22] have been also employed to design new types of ZPs with curious physical properties. Fractal zone plates (FrZPs) [23, 24], Fibonacci zone plates (FiZPs) [25, 26], Thue-Morse zone plates (TMZPs) [27] and some variations of these basic designs are representative examples [28–30].

FrZPs are characterized by its fractal structure along the square of the radial coordinate that produce multiple foci along the optical axis which are defined by the self-similar Fourier spectrum of the fractal pupil function [23]. These lenses produce a main focus surrounded by numerous secondary foci, which together result in a compound focal volume. It has been demonstrated that these self-similar foci produce reduction of the chromatic aberration under wideband illumination and increase of depth of field [24].

FiZPs are bifocal ZPs with their foci located at certain axial positions given by the Fibonacci numbers, being the 'golden mean', the ratio of the two focal distances [25]. As the name indicates, these lenses are designed following an aperiodic structure generated by the Fibonacci sequence. The focusing and imaging capabilities of Fibonacci lenses have been experimentally demonstrated under monochromatic illumination [26]; however, these lenses are affected by the same limitations of conventional ZPs when broadband illumination is considered, since the twin foci are not self-similar.

TMZPs are based on the deterministic Thue-Morse sequence, which results in a combination of the FrZP and FiZP. For this reason, this new family of aperiodic ZPs combines the advantages of fractal ZPs (reduction of the chromatic aberration) and Fibonacci ZPs (bifocusing along the optical axis) [27].

In this chapter, the combination of a vortex lens and ZPs, constructed using different aperiodic sequences, in a single element is considered. Their optical properties are investigated numerically and experimentally. The focusing properties of different combinations of FZPs and vortex lenses are studied by computing the intensity distribution along the optical axis and the

http://dx.doi.org/10.5772/66343

transverse diffraction patterns along the propagation direction. The diffracted field of these vortex lenses was obtained numerically within the Fresnel approximation.

We emphasize that these elements are able to create a chain of optical traps along the optical axis with a tunable separation, strength and transverse section. We discuss the influence of the topological charge on the irradiance propagation and also we investigate the variation of the angular momentum provided by the doughnut-shaped foci.

An optical set-up was implemented to obtain experimental results in which the VLs were registered in a spatial light modulator (SLM).

We have shown that our VLs are able to generate multiple-plane optical trappings with a volumetric extension. In this sense, they are superior to conventional vortices, whose extension is limited to the depth of focus of the beam.

2. Basic theory

The transmittance of a VL can be expressed as the product of two factors being the first one associated to a given ZP, which has only a radial dependence, and the other one corresponding to a vortex lens with a linear phase dependence on the azimuthal angle θ.

A ZP can be realized from a one-dimensional (1D) compact-supported periodic function $q(\zeta)$ as shown in **Figure 1**, where $\zeta = \left(\frac{r}{a}\right)^2$ is the normalized square radial coordinate and a is the external radius of the outermost ring. Therefore, in a binary ZP, every pair of opaque and transparent zones conforms a period. The area of each period is constant over all the ZPs.

In a similar way, the aperiodic zone plates that we consider from now on can be constructed by replacing the periodic function $q(\zeta)$ by either the Cantor function, the Fibonacci sequence or the Thue-Morse sequence. In fact, when designing VL, each of these sequences can be used to

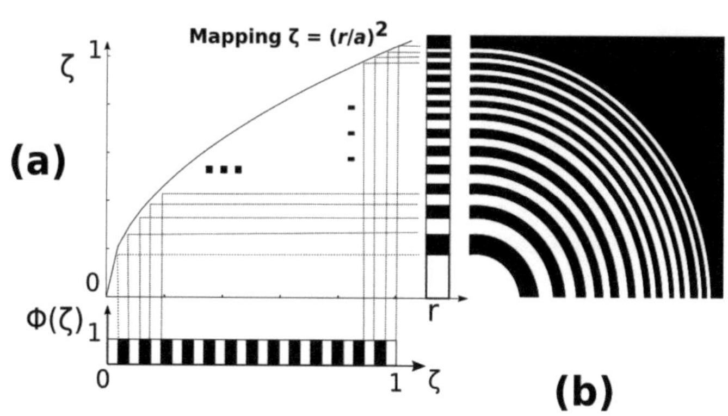

Figure 1. Schematic representation of the geometrical construction of a ZP from a periodic binary function. (a) Variable ζ mapped as function of r^2. (b) Layout of zones on ZP lens.

define the transmission generating function $q(\zeta)$ with compact support on the interval $\zeta\in[0,1]$. This interval is partitioned in 2S sub-intervals of length $dS = \frac{1}{2S}$, and the transmittance value $t_{S,j}$, which takes at the *jth* sub-interval, is associated to the value of the element $D_{S,j}$, being $t_{S,j} = 1$ for the transparent zones and $t_{S,j} = 0$ for the opaque zones.

To study the focusing properties of VLs, we compute the irradiance provided by the transmittance of each lens described in general terms as

$$t(\zeta, \theta_0) = \mathrm{q}(\zeta)\exp[im\theta_0], \tag{1}$$

where the azimuthal dependence is characterized by the topological charge m

We consider monochromatic plane wave illumination of wavelength λ. Thus, within the Fresnel approximation the irradiance function is given by

$$I(u, v; \theta) = u^2 \left| \int_0^1 \int_0^{2\pi} t(\zeta, \theta_0)\exp(-i2\pi u\zeta)\exp[i4\pi uv\zeta^{1/2}\cos(\theta - \theta_0)]d\zeta d\theta_0 \right|^2 \tag{2}$$

where $u = \frac{a^2}{2\lambda z}$ is the dimensionless reduced axial coordinate and $v = \frac{r}{a}$ is the normalized transverse coordinate [23]. By using Eq. (**1**) and taking into account that

$$\begin{aligned} &\int_0^{2\pi} t(\zeta, \theta_0) \exp(-i2\pi u\zeta) \exp[i4\pi uv\zeta^{1/2}\cos(\theta - \theta_{0)}]d\theta_0 \\ &= 2\pi\exp\left[im\left(\theta + \frac{\pi}{2}\right)\right]J_m(4\pi uv\zeta^{1/2}) \end{aligned} \tag{3}$$

Equation 2 is reduced to

$$I(u, v) = 4\pi^2u^2 \left| \int_0^1 q(\zeta, \theta_0)\exp(-i2\pi u\zeta)J_m(i4\pi uv\zeta^{1/2})d\zeta \right|^2 \tag{4}$$

$J_m()$ being the Bessel function of the first kind of order m.

3. Fractal vortex lenses

Two different results were independently obtained by combining spiral phase plates with FrZP to produce a sequence of focused optical vortices along the propagation direction. In Ref. [31], the spiral fractal zone plate is generated as a phase-only FrZP modulated by helical phase structure. Another design of spiral phase plate based on a *blazed* FrZP, the devil's lens [32], was proposed to improve diffraction efficiency of a spiral FrZP. A devil's lens has a characteristic surface relief along the radial coordinate which is obtained using the devil's staircase function

[33]. It is because of its blazed profile that devil vortex lens has improved diffraction efficiency with respect to the spiral fractal zone.

The focal volume generated by Fresnel vortex lenses (FrVLs) results in a chain of vortices that could be used as versatile and very efficient optical tweezers because, in addition to rotating the trapped particles with high refractive index, other particles with a lower refractive index can be trapped in the vortex centre. The relative angular velocity of the particles at the different traps can be changed with different topological charges; the distances between the links of the chain can be modified with different values of S-parameter.

Following the same approach employed in **Figure 1**, an FrVL is mathematically obtained by replacing the periodic sequence by a Cantor structure developed up to certain stage. Let us consider, for example, the triadic regular Cantor sequence. The construction procedure is shown in **Figure 1a**. In the first stage, $S = 1$, the segment is divided into three parts, and the middle one is removed. In the second stage, this slicing-and-removing process is repeated in each one of the remaining two segments from the first stage. This process is repeated in the following stages. In mathematical terms, the FrVL transmittance function, for a given stage S, can be expressed by replacing the following generating function of the FrZP in Eq. (**1**) [23]:

$$q(\zeta, S; N) = \prod_{i=1}^{S} \text{rect}[\zeta^2 - 0.5] \text{rect}\left\{ \text{mod}\left[\zeta^2 - 1 + \frac{1}{(2N-1)^i}, \frac{2}{(2N-1)^i} \right] \frac{(2N-1)^i}{2} \right\} \quad (5)$$

In which the transparent and opaque zones are replaced by pure phase zones differing in π for the design wavelength. In this equation, the function $\text{mod}(x, y)$ gives the remainder on division of x by y.

Typical results are shown in **Figure 2b** for a fractal vortex lens with topological charge $m = 2$ and in **Figure 2c** for an FVL with topological charge$m = 3$.

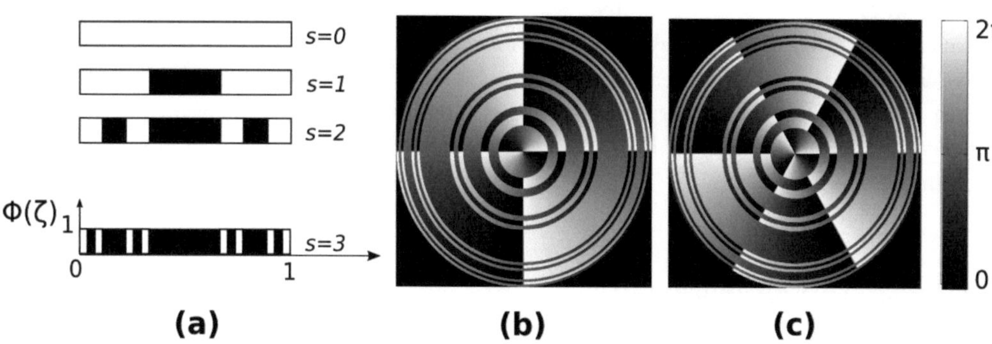

Figure 2. FVL construction: (a) Diagrams of the generation of binary function q(ζ) for an FZP for N = 2 and several values of S . (b) FVL with m=2. (c) FVL with m=3. In this representation, open and filled segments correspond to phase values differing in π of the generating radially binary function.

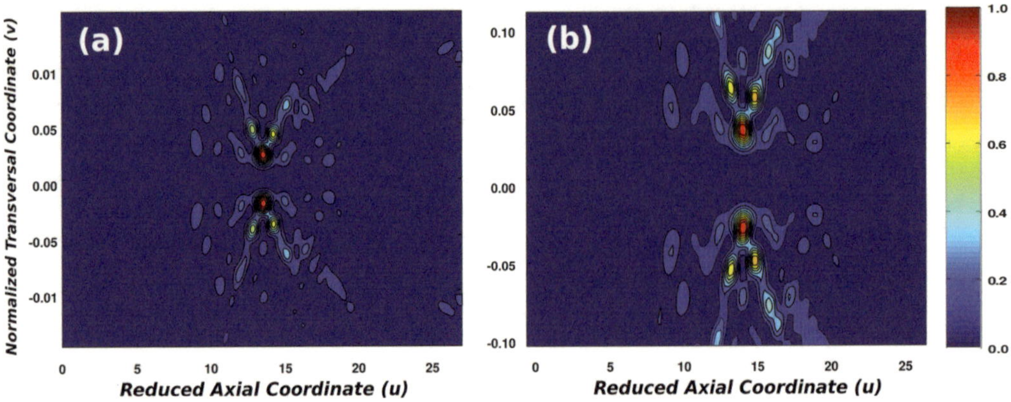

Figure 3. Normalized irradiance contours for the FVL in Figure 1: (a) $m = 2$ and (b) $m = 3$.

By using expression (4), we have computed the irradiance provided by these FrVLs at transverse planes along the optical axis. The result is shown in **Figure 3**.

As we mentioned, the diffraction efficiency of these FVLs can be improved by imposing a blazed profile to the radial coordinate. This can be done by using the Cantor function, or devil's staircase, as a generating $q(\zeta)$ function [32, 34]. The focusing properties of these FVL were experimentally analysed in [35]. It has been demonstrated that for multiple-plane optical trappings, they can generate a light beam with axially distributed optical vortices. The transverse patterns appearing along the propagation distance present several concatenated doughnut modes as represented in **Figure 3**.

Besides, the generation of multiple vortex distributions is of interest as demonstrated by the new methods that have been recently proposed to generate different two-dimensional (2D) and three-dimensional (3D) arrays of vortices. Different methods have been employed, such as interferometric techniques using Michelson or Mach-Zehnder interferometers [36] and Dammann gratings [37]. A simple method to obtain special arrays of vortices is possible by means of a reconfigurable spatial light modulator (SLM). In fact, the use of an SLM allowed the possibility to change in a simple way the characteristics of diffractive lenses, such as their focal length or their topological charge. The implementation of compound 3D optical vortex structures by means of an array of DVLs was reported in [38] with numerical simulations and experimental results.

4. Fibonacci vortex lenses

Fibonacci vortex lenses (FiVLs) are constructed using the Fibonacci sequence [22]. This sequence has been also employed in the development of different photonic devices and applications [39], such as multilayers and gratings [40], cryptography [41] and photonic crystals [42].

http://dx.doi.org/10.5772/66343

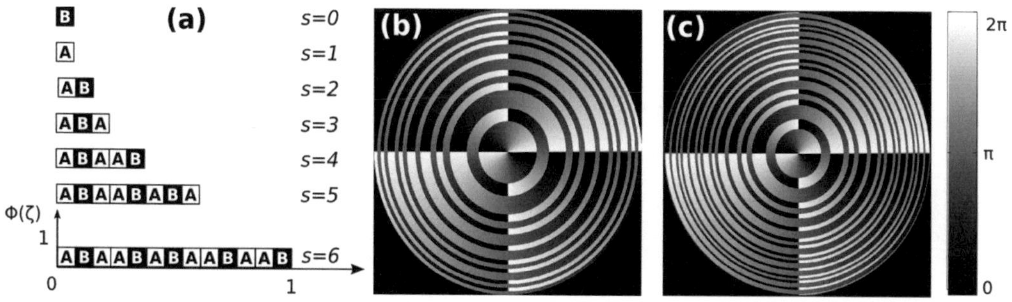

Figure 4. FiVL construction: (a) Diagrams of the generation of binary function $q(\zeta)$for an FiZP and several values of S. (b) FiVL with$S = 7$ and $m = 2$. (c) FiVL with $S = 8$ and $m - 2$. In this representation, open and filled segments correspond to phase values differing in π of the generating radially binary function.

The Fibonacci sequence is a set of ordered numbers, $F_j = \{0, 1, 1, 2, 3, 5, 8, 13, 21, \ldots\}$, that are obtained following a simple rule: Starting with two elements (called seeds) $F_0 = 0$ and $F_1 = 1$, the following numbers of the sequence are obtained as $F_j + 1 = F_j + F_{j\text{-}1}, (j = 0, 1, 2, \ldots)$. The ratio of two consecutive elements of the Fibonacci sequence approaches asymptotically an irrational number known as the golden mean: $\varphi = \lim j {\to} \infty F_j/F_{j\text{-}1} = (1 + \sqrt{5})/2$. Fibonacci series and the golden mean have been ubiquitously observed in nature and on different scientific areas [43, 44].

Based on the Fibonacci numbers, a binary Fibonacci sequence can also be generated with two seed elements, $S_1 = \{A\}$ and $S_0 = \{B\}$, as shown in **Figure 4a**. Then, the next order of sequence is obtained simply as the concatenation of the two previous ones: $S_{j+1} = \{S_j S_{j\text{-}1}\}$ for $j{\geq}1$. Consequently, $S_2 = \{AB\}, S_3 = \{ABA\}, S_4 = \{ABAAB\}, S_5 = \{ABAABABA\}$, and so on. It should be noted that the total number of elements of the order j sequence is F_{j+1} and that, for each S, two consecutive 'B' are separated by either one or two 'A'. Each sequence can be used to define the binary generating function for the radial phase distribution of the FiVL.

In our case, the function, $\Phi_j(\zeta)$, is defined in the domain $[0, 1]$, which is partitioned in F_{j+1} sub-intervals of length $d = 1/F_{j+1}$. Therefore, the function $\Phi_j(\zeta)$ at the k_{th} sub-interval is either 0 or π if the value of the k_{th} element of the S_j sequence, S_{jk}, is 'A' or 'B', respectively. Finally, after performing the coordinate transformation, $\zeta = \left(\frac{r}{a}\right)^2$, the radial part of the transmittance is obtained as $q(\zeta) = \exp[i\Phi_j(\zeta)]$ as shown in **Figure 1**.

An FiVL is defined as a pure phase diffractive element whose phase distribution is given by $\Phi_{\text{FVL}}(\zeta, \theta_0) = \text{mod}2\pi[\Phi_j(\zeta) + m\theta_0]$. Thus, it combines the azimuthal phase variation that characterizes a vortex lens, with the radial phase distribution that is generated through the Fibonacci sequence. **Figure 4b** and **c** shows the results for $S = 7$ and $S = 8$ for the topological charge $m = 2$.

Compared with a Fresnel zone plates (which, we recall, can be considered periodic structures along the square radial coordinate ζ), it can be verified that if both have the same

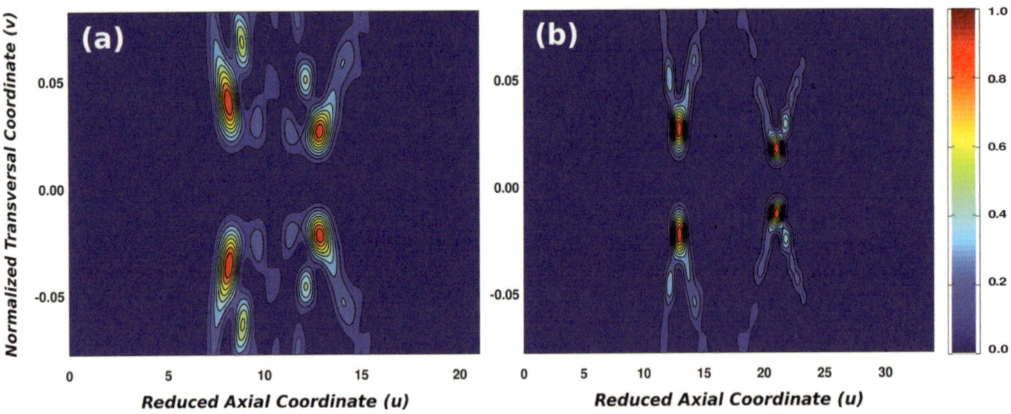

Figure 5. Normalized irradiance contours for the FVL in Figure 4: (a) FiVL $S = 7$ with $m = 2$. (b) FiVL with $S = 8, m = 2$.

number of elements, Fj, but the position of some zones with different phase has been interchanged, the FiZP produces a focal splitting of the main focus of the Fresnel zone plate along the axis. Thus, an FVL with $m = 0$ can be understood as two Fresnel zone plates interlaced [45].

The irradiances provided by the lenses are shown in **Figure 4**. The result is shown in **Figure 5**. The integrals were computed using Eq. (**4**) applying Simpson's rule with a step length 1/2000. Note that FiVLs produce twin foci whose locations are coincident with the Fibonacci numbers. In fact, for S_8 FVLs, the first focus is located at $u_1 = 13 = F_{j-1}$ and the other one at $u_2 = 21 = F_j$. Moreover, the ratio of the focal distances satisfies $\frac{u2}{u1} \approx \varphi$. The diameter of these 'twin' vortices provided by FiVLs is related by the golden mean [46], and the diameter of the rings increases proportionally with the topological charge.

5. Thue-Morse vortex lens

The Thue-Morse sequence [22] is also a binary sequence in which each element is obtained with the previous one by appending to it its Boolean complement. This sequence has been applied in several branches of Physics, as, for example, in the context of photonic crystals [47], quantum wells [48], metamaterials [49] and graphene superlattices [50].

The characteristic function q(ζ) corresponding to the Thue-Morse sequence is constructed defining a seed $D_0 = A$ from which the following elements in the sequence are obtained by replacing A by AB and B by BA. In this way, $D_1 = AB$, $D_2 = ABB$, $D_3 = BBABAAB$, $D4 = ABBABAABBAABABBA$, and so on. **Figure 6a** shows the geometrical construction of the TM sequence up to order $S = 4$. When designing Thue-Morse vortex lens (TMVL), each D_i can be used to define the transmission function $q(\zeta)$ with compact support on the interval $\zeta \in [0, 1]$. This interval is divided in 2S sub-intervals of length $dS = 1/2S$, where the transmittance, $t_{S,j}$, of the jth sub-interval is associated to the element $D_{S,j}$, as $t_{S,j} = 1$ when $D_{S,j\text{is}}$ 'A', and

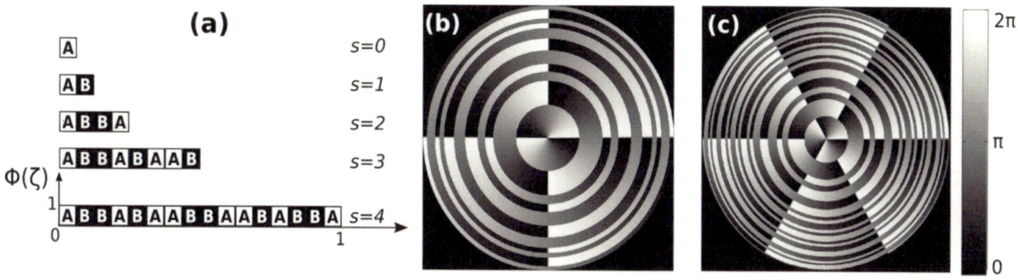

Figure 6. TMVL construction: (a) Diagrams of the generation of binary function $q(\zeta)$ for TMVL and several values of S. (b) TMVL $S = 4$ with $m = 2$. (c) TMVL $S = 5$ with $m = 3$. In this representation, open and filled segments correspond to phase values differing in $q(\zeta)$ of the generating radially binary function.

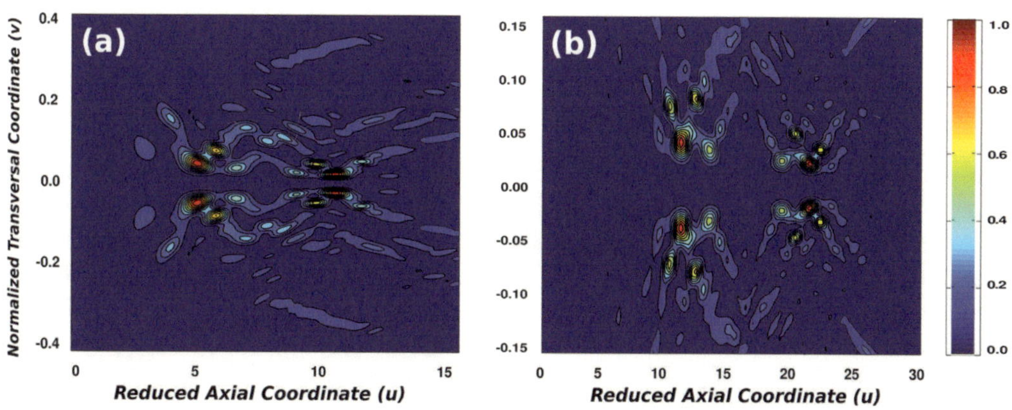

Figure 7. Normalized irradiance contours for the TMVL in Figure 7: (a) TMVL $S = 4$ with $m = 2$. (b) TMVL $S = 5$ with $m = 3$.

$t_{S,j} = 0$ when $D_{S,j}$ *is* $'B'$. **Figure 1**(b) shows the transmittance pupil function of a TMZP of order $S = 6$ and its equivalent periodic ZP. Note that like a conventional ZP the period of a TMZP is $pS = 2dS$, where the position of transparent/opaque zones has been interchanged. In mathematical terms, the transmittance function, $q(\zeta)$, can be written as

$$q(\zeta) = \sum_{j=1}^{2^S} t_{S,j} \mathrm{rect}\left[\frac{\zeta-(j-0.5)d_S}{d_S}\right] \tag{6}$$

Figure 7 shows the axial irradiance provided by TMVLs of orders $S = 4$ and $S = 5$. Note that the Thue-Morse ZP produces a symmetrical splitting of the first-order focus. This zero irradiance in the middle is due to the destructive interference generated by the two conjugated parts of the lens. In this way, like FiVLs, TMVLs are bifocals, but they produce a sequence of secondary foci around each main focus that have a fractal structure. In fact, it has been shown that irradiance provided by these lenses with topological charge $m = 0$ is self-similar [51], that is, the irradiance distribution corresponding to a TMVL of order S is a modulated version of the irradiance distribution corresponding to the previous stage, $S-1$, magnified by a factor 2.

In **Figure 7**, we represent the axial irradiances provided by the TMVLs represented in **Figure 6**. It can be seen that TMVLs produce a bifocal structure with fractal characteristic. It can also be observed that the diameters of these 'twin' vortices provided are proportional to the topological charge.

6. Experimental results

For the experimental study of the properties of FVLs, we implemented the experimental set-up shown in **Figure 8**, where the aperiodic VLs were experimentally implemented in a programmable spatial light modulator. The static aberrations caused by the SLM display were characterized with a Shack-Hartmann wavefront sensor, and then compensated as detailed elsewhere [35]. The vortex-lens performance was studied computing the diffraction patterns along different planes along the optical axis.

The proposed ZPs were recorded on a spatial light modulator (Holoeye PLUTO, eight-bit grey level, pixel size 8 μm and resolution equal to 1920 × 1080 pixels), calibrated for a 2π phase shift at $\lambda = 633$ nm operating in phase-only modulation mode. A linear phase grating was superposed to the diffractive lenses on the SLM to avoid the specular reflection and the pixelated structure of the SLM. This linear phase was compensated by an appropriate tilt of the SLM. A pin hole (PH) was used to filter high diffraction orders. A scaled image of the lens was achieved at the L3 lens focal plane (exit pupil). A collimated He-Ne laser beam ($\lambda = 633$ nm) was sent to the SLM and the diffracted field was registered with an eight-bit grey-level, charge-coupled device (CCD) camera (EO-1312M 1/2″ CCD Monochrome USB Camera, pixel pitch of

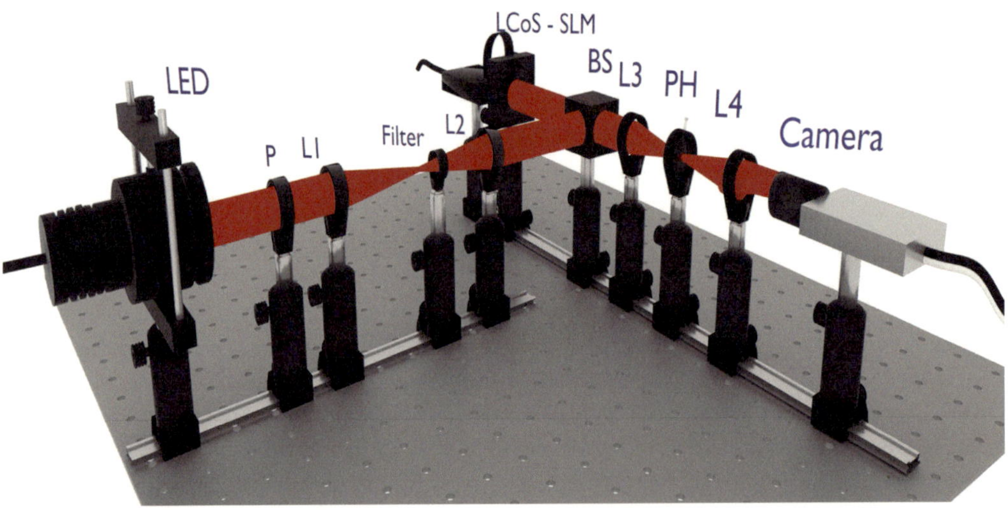

Figure 8. Experimental set-up for characterization and measuring vortices for FVLs.

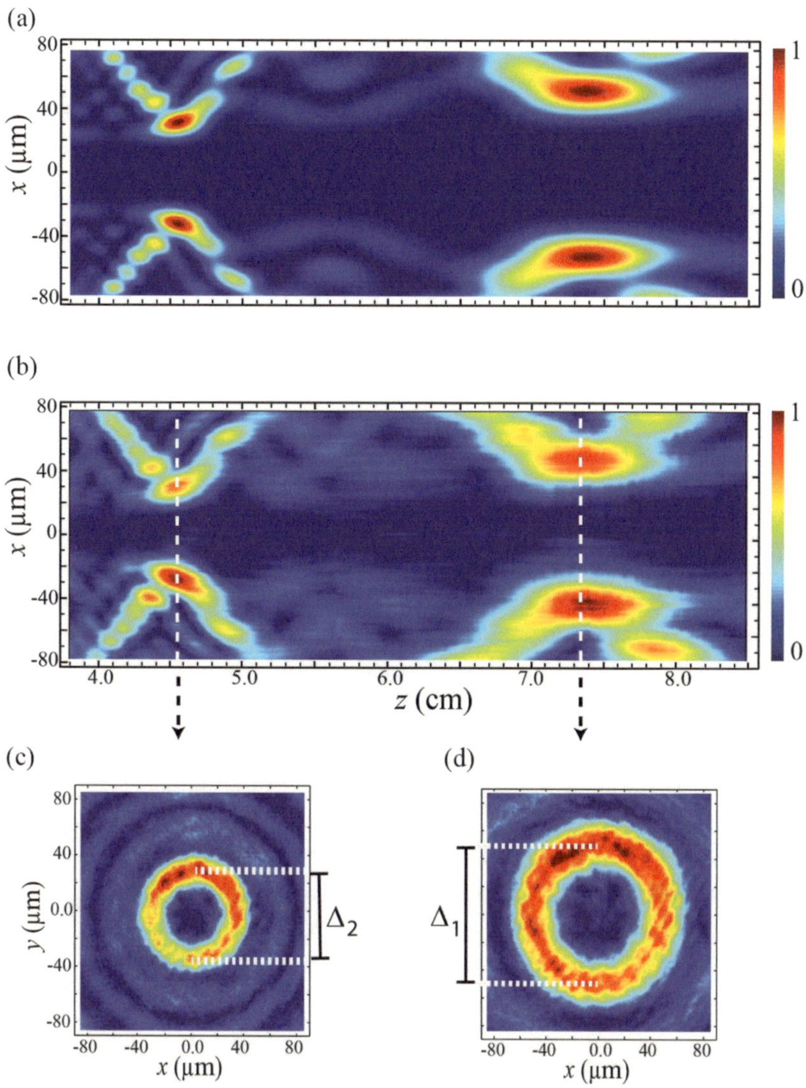

Figure 9. Longitudinal irradiance of different planes produced by $S = 8$ FiVL with $m = 6$ (a) theoretical results, (b) experimental results. (c and d) Transverse irradiance at the planes indicated by the arrows.

4.65 μm, 1280 × 1024 pixels) through a microscope objective (10× Zeiss Plan-Apochromat). The microscope and the CCD were mounted on a translation stage (Thorlabs LTS 300, range: 300 mm and 5 μm precision). The computed and experimental irradiances produced by an S8 FiVL with $m = 6$ are shown in **Figure 9**.

As predicted by the theoretical analysis (**Figure 9a**), the axial localization of the focal rings depends on the Fibonacci numbers F_j and F_{j-1}, and such focal distances satisfy the following relationship: $\frac{f1}{f2} = \frac{F_j}{F_{j-1}} \approx \Phi$. The diameter of the rings also satisfies $\frac{\Delta_1}{\Delta_2} \approx \Phi$.

7. Conclusions

A new family of fractal aperiodic VLs with interesting focusing and imaging capabilities has been presented. The transverse patterns appearing along the propagation distance present several concatenated doughnut modes. The ability of these VLs to produce multiple vortex tweezers has been demonstrated. In fact, it was found that contrary to conventional spiral zone plate, which produces a single vortex, each member of this family generates a delimited chain of vortices that are axially distributed. The distances between the links of the chain depend on the level S of the generating function and the radii of the doughnuts increase with the topological charge. Additionally, the multifocal nature of the lens, resulting from its fractal structure along the radial coordinate, could provide a high depth of field, especially with wideband sources.

In the case of FrVL (i.e. Cantor and Devil's VLs), the evolution of the axial irradiance replicates the fractality of the pupil. The orbital angular momentum in each link on the chain also depends on the topological charge but it is nearly independent of its axial location.

Our analysis demonstrated the possibility of simple design procedure arrays of VLs with any desired range of topological charge. As each individual VL can be understood as a light gear capable of driving microstructures around its circumference, applications involving particle transfer and manipulation could be foreseen.

On the other hand, FiVLs and TMVLs are intrinsically bifocal vortex lenses. It was shown that they produce twin optical vortices along the axial coordinate. The positions of both foci depend on the two incommensurable periods of the Fibonacci sequence in which the lenses are based. Moreover, the evolution of the irradiance along the propagation axis reproduces the fractality of the lens [34]. The diameter of these chains of vortices is proportional to the topological charge, and their ratio is close to the golden mean. The volumetric distribution of the diffracted field generated by FiVLs was assessed experimentally using an SLM. An excellent agreement was found between the experimental results and the theoretical predictions.

The peculiar focal volume obtained with DVLs could be exploited as versatile and efficient optical tweezer, because it can also trap low-index particles in the zero-intensity axial zone of the doughnut and at the same time can exert a torque on small objects having a high refraction index. The distances between the links of the vortex chain can be modified with the level S of the generating function and relative angular velocity of the particles at the trap depends on the topological charge of the vortex.

Another potential application of these optical vortices arises in X-ray microscopy where its azimuthal component can be used for detecting the phase component of objects with complex index of refraction since it acts as a Hilbert phase filter. This feature is especially useful in the case of biological specimens to provide enhanced contrast.

TMVL produces a couple of self-similar vortex located symmetrically one at each side of the focus of the equivalent periodic VL. Therefore, under broadband illumination, a TMVL produces a pair of images with an extended depth of field and a strong reduction in the chromatic

http://dx.doi.org/10.5772/66343

aberration. In this way, TMVLs combine the characteristics of FrVLs and FiVLs and could be used in multiple applications including spectral domain optical coherence tomography (OCT) and X-ray microscopy [35].

Finally, it should be mentioned that all the aperiodic VLs here presented admit fractional topological charges that break down the symmetry of the foci and produce chains of anisotropic vortex foci.

Acknowledgements

This work was founded by the Ministerio de Economía y Competitividad FEDER (Grant DPI2015-71256-R), and by Generalitat Valenciana (Grant PROMETEOII-2014-072), Spain. Federico Machado acknowledges support from the MayaNet - Erasmus Mundus Partnership 552061-EM-1-2014-1-IT-ERA MUNDUS-EMA21 (Grant 2014-0872/001-001).

Author details

Federico J. Machado[1], Juan A. Monsoriu[1] and Walter D. Furlan[2]*

*Address all correspondence to: walter.furlan@uv.es

1 Centre for Physics Technologies, Polytechnic University of Valencia, Valencia, Spain

2 Department of Optics and Optometry and Vision Sciences, University of Valencia, Burjassot, Spain

References

[1] Desyatnikov A, Kivshar Y, Torner L. Optical vortices and vortex solitons. Progress in Optics. 2005;**47**:291–391. DOI: 10.1016/S0079-6638(05)47006-7

[2] Roux FS. Distribution of angular momentum and vortex morphology in optical beams. Optical Communications. 2004;**242**(1-3):45–55. DOI: 10.1016/j.optcom.2004.08.006

[3] Grier DG. A revolution in optical manipulation. Nature. 2003;**424**:810–816. DOI: 10.1038/nature01935

[4] Allen L, Padgett MJ, Babiker M. The orbital angular momentum of light. Progress in Optics. 1999;**39**:291–372. DOI: 10.1016/S0079-6638(08)70391-3

[5] Bekshaev A, Soskin M, Vasnetsov M. Paraxial Light Beams with Angular Momentum. New York, NY: Nova Publishers, 2008. arXiv:0801.2309

[6] Lee JH, Foo G, Johnson EG, Swartzlander GA. Experimental verification of an optical vortex coronagraph. Physics Review Letters. 2006; 97. DOI: 10.1103/PhysRevLett.97.053901

[7] Swartzlander GA Jr, Ford EL, Abdul-Malik RS, Close LM, Peters MA. Astronomical demonstration of an optical vortex coronagraph. Optical Express. 2008;**16**:10200–10207. DOI: 10.1364/OE.16.010200

[8] Gibson G, Courtial J, Padgett M, Vasnetsov M, Pasco V. Free-space information transfer using light beams carrying orbital angular momentum. Optical Express. 2004;**12**:5448. 10.1364/OPEX.12.005448

[9] Bouchal Z, Čelechovský R. Mixed vortex states of light as information carriers. New Journal of Physics. 2004;**6**:131. DOI:10.1088/1367-2630/6/1/131

[10] Spektor B, Normatov A, Shamir J. Singular beam microscopy. Applied Optics. 2008;**47**: A78–A87. DOI: 10.1364/AO.47.000A78

[11] Turitsyn SK, Mezentsev VK, Dubov M, Rubenchik AM, Fedoruk MP. Sub-critical regime of femtosecond inscription. Optical Express. 2007;**15**:14750–14764. DOI: 10.1364/OE.15.014750

[12] Gahagan KT, Swartzlander GA. Optical vortex trapping of particles. Optics Letter. 1996;**21**:827–829. DOI: 10.1364/OL.21.000827

[13] Garcés-Chávez V, Volke-Sepulveda K, Chávez-Cerda S, Sibbett W, Dholakia K. Transfer of orbital angular momentum to an optically trapped low-index particle. Physics Review A. 2002;**66**:063402. DOI: 10.1103/PhysRevA.66.063402

[14] Reicherter M, Haist T, Wagemann E, Tiziani H. Optical particle trapping with computer-generated holograms written on a liquid-crystal display. Optical Letters. 1999;**24**:608–610. DOI: 10.1364/OL.24.000608

[15] Ladavac K, Grier D. Micromechanical pump assembled and driven by holographic optical vortices. Optical Express. 2004;**12**:1144–1149. DOI: 10.1364/OPEX.12.001144

[16] Friese M, Rubinsztein-Dunlop H, Gold J, Hagberg P, Hanstorp D. Optically driven micromachine elements. Applied Physics Letters. 2001;**78**:547–549. DOI: 10.1063/1.1339995

[17] Masajada J, Dubik B. Optical vortex generation by three plane wave interference. Optics Communications. 2001;**198(1)**:21–27. DOI: 10.1016/S0030-4018(01)01499-7

[18] Yang Y, Wang W, Moitra P, Kravchenko I, Briggs D, Valentine J. Dielectric meta-reflectarray for broadband linear polarization conversion and optical vortex generation. Nano Letters. 2014;**14(3)**:1394–1399. DOI: 10.1021/nl4044482

[19] Brasselet E, Murazawa N, Misawa H, Juodkazis S. Optical vortices from liquid crystal droplets. Physical Review Letters. 2009;**103(10)**:103903. DOI: 10.1103/PhysRevLett.103.103903

[20] Curtis Jennifer E, Brian AK, Grier DG. Dynamic holographic optical tweezers. Optics Communications. 2002;**207(1)**:169–175.DOI: 10.1016/S0030-4018(02)01524-9

[21] Ojeda-Castaneda J, Gomez-Reino C. Selected Papers on Zone Plates (SPIE Optical Engineering Press, 1996), vol. MS128, 512–518. DOI:10.1109/TMTT.1961.1125382

[22] Macia E. The role of aperiodic order in science and technology. Reports on Progress in Physics. 2006;**69**:397–441. DOI: 10.1088/0034-4885/69/2/R03

[23] Saavedra G, Furlan WD, Monsoriu JA, Fractal zone plates. Optical Letters. 2003;**28**:971–973. DOI: 10.1080/09500340500356973

[24]] Furlan WD, Saavedra G, Monsoriu JA. White-light imaging with fractal zone plates. Optical Letters. 2007;**32**:2109–2111. DOI: 10.1364/OL.32.002109

[25] Monsoriu JA, Calatayud A, Remon L, Furlan WD, Saavedra G, Andrés P. Bifocal Fibonacci diffractive lenses. IEEE Photonics Journal. 2013;**5**:3400106. DOI: 10.1109/JPHOT.2013.2248707

[26] Ferrando V, Calatayud A, Andres P, Torroba R, Furlan WD, Monsoriu JA. Imaging properties of kinoform Fibonacci lenses. IEEE Photonics Journal. 2014;**6**: 6500106. DOI: 10.1109/JPHOT.2014.2304560

[27] Ferrando V, Giménez F, Furlan WD, Monsoriu JA. Bifractal focusing and imaging properties of Thue–Morse Zone Plates. Optics Express. 2015;**23(15)**:19846–19853. DOI: 10.1364/OE.23.019846

[28] Gimenez F, Furlan WD, Calatayud A, Monsoriu JA. Multifractal zone plates. Optical Society of America. 2010;**27**:1851–1855. DOI: 10.1364/JOSAA.27.001851

[29] Giménez F, Monsoriu JA, Furlan WD, Pons A. Fractal photon sieve. Optics Express. 2006;**14(25)**:11958–11963.DOI: 10.1364/OE.14.011958

[30] Giménez F, Furlan WD, Calatayud A, Monsoriu JA. Multifractal zone plates. OSA. 2010;**27(8)** :1851–1855. DOI: 10.1364/JOSAA.27.001851

[31] Tao SH, Yuan XC, Lin J, Burge R. Sequence of focused optical vortices generated by a spiral fractal zone plates. Applied Physics Letters. 2006;**89**: 031105. DOI: 10.1063/1.2226995

[32] Furlan WD, Giménez F, Calatayud A, Monsoriu JA. Devil's vortex-lenses. Optics Express. 2009;**17**:21891–21896. DOI: 10.1364/OE.17.021891

[33] Monsoriu JA, Furlan WD, Saavedra G, Giménez F. Devil's lenses. Optics Express. 2007;**15 (21)**:13858–13864. DOI: 10.1364/OE.15.013858

[34] Furlan WD, Giménez F, Calatayud A, Remon L, Monsoriu JA. Volumetric multiple optical traps produced by Devil's lenses. Journal of the European Optical Society-Rapid Publications. 2010;**5**:10037s. DOI: 10.2971/jeos.2010.10037s

[35] Calatayud A, Rodrigo JA, Remón L, Furlan WD, Cristóbal G, Monsoriu JA. Experimental generation and characterization of Devil's vortex-lenses. Applied Physics B. 2012;**106**:915–919. DOI: 10.1007/s00340-012-4913-0

[36] Vyas S, Senthikumaran P. Interferometric optical vortex array generator. Applied Optics. 2007;**46**:2893–2898. DOI: 10.1364/AO.46.002893

[37] García-Martínez P, Sánchez-López MM, Davis JA, Cottrell DM, Sand D, Moreno I. Generation of Bessel beam arrays through Dammann gratings. Applied Optics. 2012;**51**:1375–1381. DOI: 10.1364/AO.51.001375

[38] Calabuig A, Sánchez-Ruiz S, Martínez-León L, Tajahuerce E, Fernández-Alonso M, Furlan, WD, Pons-Martí, A. Generation of programmable 3D optical vortex structures through devil's vortex-lens arrays. Applied Optics. 2013;**52(23)**:5822–5829. DOI: 10.1364/AO.52.005822

[39] Maciá E. Exploiting aperiodic designs in nanophotonic devices. Reports in Progress on Physics. 2012;75:1–42. DOI: 10.1088/0034-4885/75/3/036502

[40] Sah Y, Ranganath G. Optical diffraction in some Fibonacci structures. Optics Communications. 1995;**114**:18–24. DOI: 10.1016/0030-4018(94)00600-Y

[41] Zhou Y, Panetta K, Agaian S, Chen CP. Image encryption using P-Fibonacci transform and decomposition. Optics Communications. 2012; **285(5)**:594-608. DOI: 10.1016/j.optcom.2011.11.044

[42] Lusk D, Abdulhalim I, Placido F. Omnidirectional reflection from Fibonacci quasi-periodic one-dimensional photonic crystal. Optics Communications. 2001;**198(4)**:273–279. DOI: 10.1016/S0030-4018(01)01531-0

[43] Basin SL. The Fibonacci sequence as it appears in nature. Fibonacci Quarterly. 1963;**1 (1)**:53–56.

[44] Garland TH. Fascinating Fibonaccis: Mystery and Magic in Numbers. Dale Seymour Publications. Palo Alto, CA 94303-0879, 1987. DOI: 10.1111/j.1949-8594.1990

[45] Monsoriu JA, Calatayud A, Remón L, Furlan WD, Saavedra G, Andrés P. Bifocal Fibonacci diffractive lenses. In IEEE Photonics Journal 2013. Vol. 5, No. 3, pp. 34001061–34001067. Institute of Electrical and Electronics Engineers (IEEE). DOI: 10.1109/JPHOT.2013.2248707

[46] Calatayud A, Ferrando V, Remón L, Furlan WD, Monsoriu JA. Twin axial vortices generated by Fibonacci lenses. Optics Express. 2013;**21(8)**:10234–10239. DOI: 10.1364/OE.21.010234

[47] Tsao CW, Cheng YH, Hsueh WJ. Localized modes in one-dimensional symmetric Thue-Morse quasicrystals. Optics Express. 2014;**22**:24378–24383. DOI: 10.1364/OE.22.024378

[48] Hsueh WJ, Chang CH, Lin CT. Exciton photoluminescence in resonant quasi-periodic Thue-Morse quantum wells. Optical Letters. 2014;**39**:489–492. DOI: 10.1364/OL.39.000489

[49] Monsoriu JA, Depine RA, Silvestre E. Non-Bragg band gaps in 1D metamaterial aperiodic multilayers Optics Express. 2007;**2**:07002. DOI: 10.1364/OE.14.012958

[50] Huang H, Liu D, Zhang H, Kong X. Electronic transport and shot noise in Thue-Morse sequence graphene superlattice. Applied Physics. 2013;**113**:043702 . DOI: 10.1142/S021798 4916501815

[51] Ferrando V, Giménez F, Furlan WD, Monsoriu JA. Bifractal focusing and imaging properties of Thue–Morse Zone Plates. Optics Express. 2015;**23**(15):19846–19853. DOI: 10.1364/OE.23.019846.

Chapter 11

Partially Coherent Vortex Beam: From Theory to Experiment

Xianlong Liu, Lin Liu, Yahong Chen and
Yangjian Cai

Additional information is available at the end of the chapter

http://dx.doi.org/10.5772/66323

Abstract

Partially coherent vortex beam exhibits some unique and interesting properties, for example, correlation singularities (i.e., ring dislocations) exist in its correlation function, and one can determine the magnitude of the topological charge of the vortex phase from the number of the ring dislocations. Modulating the coherence of a vortex beam provides a convenient way for shaping its focused beam spot, which is useful for material processing and optical trapping. Furthermore, a partially coherent vortex beam has an advantage over a partially coherent beam without vortex phase for reducing turbulence-induced scintillation, which will be useful in free-space optical communications. We introduce recent theoretical and experimental developments on partially coherent vortex beams.

Keywords: partially coherent vortex beam, generation, propagation, application

1. Introduction

Coherence is an important property of a light beam, which has been investigated widely in the past few decades [1]. Coherence can be regarded as a consequence of correlations between the components of the fluctuating electric field at two or more points. Light beam with low coherence is called partially coherent beam, and such beam has an advantage over a coherent beam in many applications, such as optical imaging [2–4], optical trapping [5, 6], free-space optical communications [7, 8], laser radar systems [9, 10] and remote sensing [11]. Before 2000, most literatures on partially coherent beam were focused on the conventional partially coherent beam named Gaussian Schell-model (GSM) beam [12–16], whose intensity and degree of coherence satisfy Gaussian distributions. Since 2000, partially coherent beams with prescribed phase, state of polarization and degree of coherence were investigated widely due to their extraordinary properties and potential applications [17–31].

Phase is another important property of a light beam, which is characterized by the wavefront on propagation. Conventional Gaussian beam carries customary quadratic phase with spherical wavefront. Vortex beam, such as Laguerre-Gaussian beam, carries a vortex phase with helical wavefront. The intensity in the vortex beam center is zero while the phase is undefined, and this point is called phase singularity. In 1992, Allen et al. found that the vortex beam carries an orbital angular momentum (OAM) of $l\hbar$ with l being the topological charge [32]. Since then, numerous efforts have been devoted to vortex beams, and such beams have been applied in many applications [33–42], such as atom and particle trapping, optical tweezer, quantum information process, and laser cooling.

Vortex beam with low coherence is called partially coherent vortex beam, which was first proposed by Gori et al. [43]. Later, various partially coherent vortex beams were introduced [44–54]. Partially coherent vortex beam differs in many aspects from a coherent vortex beam, and it exhibits some unique interesting properties, for example, correlation singularities (i.e., ring dislocations) exist in its correlation function (i.e., cross-spectral density or degree of coherence) [47, 55–59]. Here the correlation singularity is defined as the point where the value of the cross-spectral density or degree of coherence equal zero, while the corresponding phase is undefined. Recently, more and more attention is being paid to partially coherent vortex beams [60–73], more interesting and useful results are being revealed. In this chapter, we will introduce recent theoretical and experimental developments on partially coherent vortex beams.

2. Theoretical models for various partially coherent vortex beams

There are different types of partially coherent vortex beams, such as partially coherent beam with helicoidal modes [43], partially coherent vortex beam with a separable phase [44, 45], Gaussian Schell-model vortex (GSMV) beam [46], partially coherent LG_{0l} beam [47, 48], partially coherent LG_{pl} beam [49], partially coherent Bessel-Gaussian beam [50], special correlated partially coherent vortex beam [51, 52] and vector partially coherent vortex beam [53, 54]. Here we only introduce some models which can be realized in experiment easily.

A scalar partially coherent beam can be characterized by the cross-spectral density (CSD) in the space-frequency domain or mutual intensity in the space-time domain [1]. For a GSMV beam, its CSD in the source plane is expressed as follows [46]:

$$W(r_1, r_2, \varphi_1, \varphi_2) = \exp\left[-\frac{r_1^2 + r_2^2}{4\sigma_0^2} - \frac{r_1^2 + r_2^2 - 2r_1 r_2 \cos(\varphi_1 - \varphi_2)}{2\delta_0^2} + il(\varphi_1 - \varphi_2)\right], \quad (1)$$

where r and φ are the radial and azimuthal (angle) coordinates, σ_0 and δ_0 denotes the beam width and coherence width of a GSMV beam, respectively, and l denotes the topological charge. When $\delta_0 = \infty$, the GSMV beam reduces to a coherent Gaussian vortex beam.

http://dx.doi.org/10.5772/66323

The CSD of a partially coherent LG_{pl} beam in the source plane is expressed as follows [49]:

$$W(r_1, r_2, \varphi_1, \varphi_2) = \left(\frac{\sqrt{2}r_1}{\omega_0}\right)^l \left(\frac{\sqrt{2}r_2}{\omega_0}\right)^l L_p^l\left(\frac{2r_1^2}{\omega_0^2}\right) L_p^l\left(\frac{2r_2^2}{\omega_0^2}\right) \exp\left(-\frac{r_1^2+r_2^2}{\omega_0^2}\right) \exp\left(il\varphi_1 - il\varphi_2\right) \times \exp\left[-\frac{r_1^2+r_2^2-2r_1r_2\cos(\varphi_1-\varphi_2)}{2\delta_0^2}\right], \tag{2}$$

where L_p^l denotes the Laguerre polynomial with mode orders p and l. When $p = 0$, Eq. (2) reduces to the CSD of a partially coherent LG_{0l} beam [47, 48]. When $p = 0$ and $l = 0$, Eq. (2) reduces to the CSD of the well-known GSM beam [12–16].

As a typical kind of special correlated partially coherent vortex beam, the CSD of a Laguerre-Gaussian correlated Schell-model vortex (LGCSMV) beam in the source plane is expressed as [51]:

$$W(\mathbf{r}_1, \mathbf{r}_2) = \exp\left[-\frac{\mathbf{r}_1^2+\mathbf{r}_2^2}{4\sigma_0^2} - \frac{(\mathbf{r}_1-\mathbf{r}_2)^2}{2\delta_0^2}\right] L_n^0\left[\frac{(\mathbf{r}_1-\mathbf{r}_2)^2}{2\delta_0^2}\right] \exp\left[il(\varphi_1-\varphi_2)\right]. \tag{3}$$

When $n = 0$, Eq. (3) reduces to the CSD of a GSMV beam.

A vector partially coherent beam can be characterized by the CSD matrix in space-frequency domain or the beam coherence-polarization matrix in the space-time domain [17]. The elements of the CSD matrix of a vector partially coherent vortex beam with uniform state of polarization named electromagnetic Gaussian Schell-model vortex (EGSMV) beam in the source plane are given as [53]:

$$W_{\alpha\beta}(\mathbf{r}_1, \mathbf{r}_2) = A_\alpha A_\beta B_{\alpha\beta} \exp\left[-\frac{\mathbf{r}_1^2}{4\sigma_\alpha^2} - \frac{\mathbf{r}_2^2}{4\sigma_\beta^2} - \frac{(\mathbf{r}_1-\mathbf{r}_2)^2}{2\delta_{\alpha\beta}^2}\right] \exp\left[il(\varphi_1-\varphi_2)\right], \tag{4}$$

where A_x and A_y are the amplitudes of x and y components of the electric field, respectively. σ_i is the r.m.s width of the intensity distribution along the i direction, δ_{xx}, δ_{yy} and δ_{xy} are the r.m.s widths of autocorrelation function of the x components of the electric field, of the y components of the electric field and of the mutual correlation function of x and y components of the electric field, respectively. $B_{xx} = B_{yy} = 1$, $B_{xy} = |B_{xy}| \exp(i\phi_{xy})$ is the complex correlation coefficient between the x and y components of the electric field with ϕ_{xy} being the phase difference between the x and y components. The nine real parameters A_x, A_y, σ_x, σ_y, δ_{xx}, δ_{yy}, δ_{xy}, $|B_{xy}|$, ϕ_{xy} of an EGSMV beam are shown to satisfy several intrinsic constraints and obey some simplifying assumptions [17].

The elements of the CSD matrix of a vector partially coherent vortex beam with non-uniform state of polarization named radially polarized partially coherent vortex beam in the source plane are expressed as [54]:

$$W_{xx}(r_1,\varphi_1,r_2,\varphi_2) = \frac{r_1 r_2 \cos\varphi_1 \cos\varphi_2}{4\sigma_0^2} T(r_1,\varphi_1,r_2,\varphi_2), \tag{5}$$

$$W_{xy}(r_1,\varphi_1,r_2,\varphi_2) = \frac{r_1 r_2 \cos\varphi_1 \sin\varphi_2}{4\sigma_0^2} T(r_1,\varphi_1,r_2,\varphi_2), \tag{6}$$

$$W_{yx}(r_1,\varphi_1,r_2,\varphi_2) = W_{xy}^*(r_2,\varphi_2,r_1,\varphi_1), \tag{7}$$

$$W_{yy}(r_1,\varphi_1,r_2,\varphi_2) = \frac{r_1 r_2 \sin\varphi_1 \sin\varphi_2}{4\sigma_0^2} T(r_1,\varphi_1,r_2,\varphi_2), \tag{8}$$

with

$$T(r_1,\varphi_1,r_2,\varphi_2) = \exp\left[-\frac{r_1^2 + r_2^2 - 2r_1 r_2 \cos(\varphi_1 - \varphi_2)}{2\delta_0^2} - \frac{r_1^2 + r_2^2}{4\sigma_0^2} - il\varphi_1 + il\varphi_2\right]. \tag{9}$$

3. Propagation of partially coherent vortex beams

Propagation of a partially coherent vortex beam in free space can be studied with the help of the well-known Huygens-Fresnel integral, and propagation of a partially coherent vortex beam through a paraxial ABCD optical can be studied with the help of the following generalized Collins formula [74]:

$$\begin{aligned} W(\rho_1,\phi_1,\rho_2,\phi_2) = &\frac{1}{\lambda^2 BB^*} \exp\left(\frac{ikD^*\rho_2^2}{2B^*} - \frac{ikD\rho_1^2}{2B}\right) \int_0^{2\pi}\int_0^{2\pi}\int_0^{\infty}\int_0^{\infty} W(r_1,\varphi_1,r_2,\varphi_2) \\ &\times \exp\left\{ik\left[\frac{A^* r_2^2}{2B^*} - \frac{A r_1^2}{2B} + \frac{\rho_1 r_1}{B}\cos(\phi_1 - \varphi_1)\right]\right. \\ &\times \exp\left[-ik\frac{\rho_2 r_2}{B^*}\cos(\phi_2 - \varphi_2)\right] r_1 r_2 dr_1 dr_2 d\varphi_1 d\varphi_2 , \end{aligned} \tag{10}$$

where $W(\rho_1,\phi_1,\rho_2,\phi_2)$ is the cross-spectral density in the receiver plane, $k = 2\pi/\lambda$ is the wave number with λ being the optical wavelength, A, B, C, and D are the elements of the transfer matrix of the paraxial optical system, the asterisk denotes the complex conjugate.

Propagation of a partially coherent vortex beam through turbulent atmosphere can be studied with the help of the following generalized Huygens-Fresnel integral [75]

$$\begin{aligned} W(\boldsymbol{\rho}_1,\boldsymbol{\rho}_2) = &\int_{-\infty}^{\infty}\int_{-\infty}^{\infty}\int_{-\infty}^{\infty}\int_{-\infty}^{\infty} W(\mathbf{r}_1,\mathbf{r}_2)\exp\left[-\frac{ik(\boldsymbol{\rho}_1 - \mathbf{r}_1)^2}{2z} + \frac{ik(\boldsymbol{\rho}_2 - \mathbf{r}_2)^2}{2z}\right] \\ \exp&\left\{-\frac{\pi^2 k^2 z}{3}\left[(\boldsymbol{\rho}_1 - \boldsymbol{\rho}_2)^2 + (\boldsymbol{\rho}_1 - \boldsymbol{\rho}_2)\cdot(\mathbf{r}_1 - \mathbf{r}_2) + (\mathbf{r}_1 - \mathbf{r}_2)^2\right]\int_0^{\infty}\kappa^3 \Phi_n(\kappa)d\kappa\right\} d^2\mathbf{r}_1 d^2\mathbf{r}_2. \end{aligned} \tag{11}$$

Here $\Phi_n(\kappa)$ is the one-dimensional power spectrum of the refractive-index fluctuations of the atmospheric turbulence, $\boldsymbol{\kappa}$ is the spatial frequency.

http://dx.doi.org/10.5772/66323

The average intensity and the degree of coherence of a partially coherent vortex beam in the receiver plane are obtained as:

$$I(\rho) = W(\rho,\rho), \quad \mu(\rho_1,\rho_2) = \frac{W(\rho_1,\rho_2)}{\sqrt{I(\rho_1)I(\rho_2)}}. \tag{12}$$

Coherent vortex beam displays dark hollow beam profile in the source plane or on propagation in free space. For a partially coherent vortex beam, it also displays dark hollow beam profile in the source plane, while its beam profile varies on propagation due to the degradation caused by the source spatial coherence, and one can shape the beam spot of a partially coherent vortex beam in the focal plane through varying the initial coherence width, for example, the beam profile of the focused beam spot gradually transforms from a dark hollow beam profile to a flat-topped beam profile and finally to a Gaussian beam profile when the coherence width gradually decreases (see **Figure 1**). Furthermore, when the initial coherence width is fixed, one also can shape the beam spot of a partially coherent vortex beam through varying its initial topological charge because the topological charge plays a role of anti-degradation caused by the coherence [46].

For a vector partially coherent beam with nonuniform state of polarization (i.e., radially polarized partially coherent beam), it is known that such beam always displays radial polarization on propagation (see **Figure 2(a1)–(e1)**) although its degree of polarization varies. For a radially polarized partially coherent vortex beam, one finds from [54] that the vortex phase induces changes of not only the degree of polarization but also the state of polarization (see **Figure 2(a2)–(e2)** and **(a3)–(e3)**) besides rotation of the beam spot, that is, radial polarization disappears and elliptical polarization appears on propagation. The state of polarization displays left-handed elliptical polarization around the beam center and right-handed elliptical polarization outside of the beam center for $l > 0$, and the handedness of the polarization ellipse

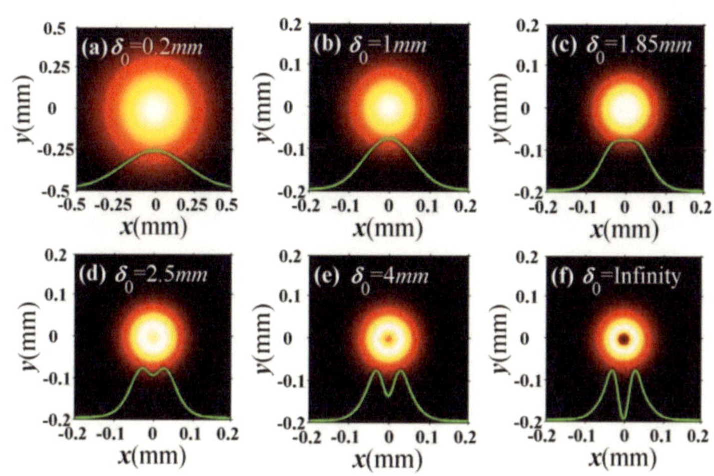

Figure 1. Average intensity of a partially coherent LG_{0l} beam with $l = 2$ in the focal plane for different values of the initial coherence width δ_0.

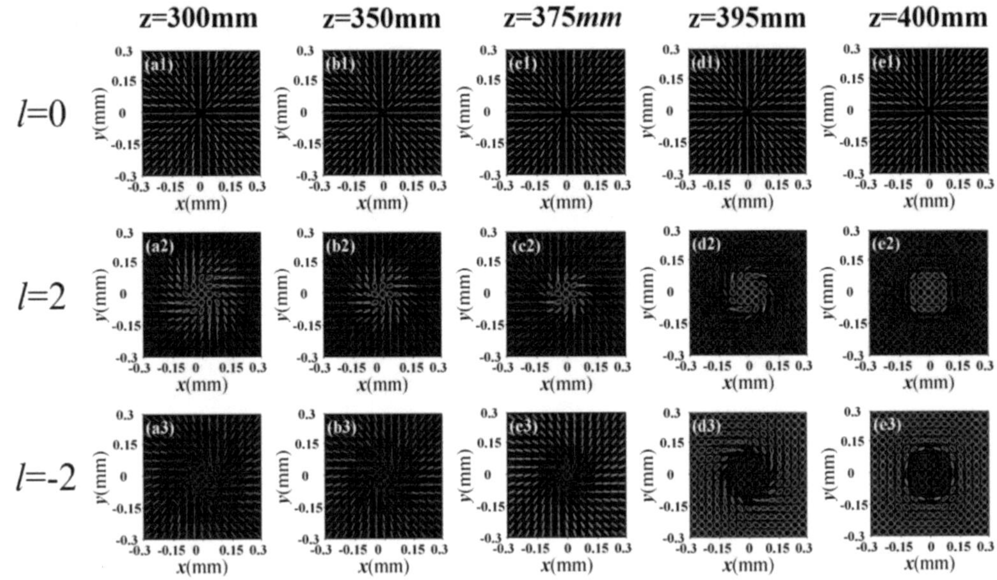

Figure 2. Changes of the state of polarization of a focused radially polarized partially coherent vortex beam on propagation for different values of the topological charge *l*.

is reversed for $l<0$. Furthermore, the polarization ellipse rotates clockwise for $l>0$ and anticlockwise for $l<0$ on propagation. Thus, modulating the magnitude and sign of the topological charge of the vortex phase provides a convenient way for modulating the polarization properties of a vector partially coherent vortex beam. What is more, the phenomenon of vortex phase-induced changes of the state of polarization of a radially polarized partially coherent beam may be used to detect a phase object.

Coherent vortex beam carries phase singularity in the source plane and on propagation. Phase singularity is defined as the point where the intensity is zero while the phase is undefined. When the spatial coherence of a vortex beam is reduced, the dark hollow beam profile disappears on propagation due to the degradation caused by the coherence (see **Figure 3(a)–(c)**). Thus, a partially coherent vortex beam does not carry phase singularity on propagation, while an interesting correlation singularity named ring dislocation appears (see **Figure 3(d)–(f)**). Here the correlation singularity is defined as the point where the amplitude of the cross-spectral density $W(\rho,-\rho)$ or degree of coherence $\mu(\rho,-\rho)$ is zero while the corresponding phase is undefined. It was demonstrated in [47] both theoretically and experimentally that correlation singularity (i.e., ring dislocation) exists in a partially coherent vortex beam on propagation. One finds from **Figure 3(d)–(f)** that the ring dislocation becomes more obvious with the decrease of the spatial coherence.

The study of optical beam propagation in turbulent atmosphere is a venerable subject. It is known that the turbulence induces scintillation (i.e., intensity fluctuations), beam wander and deformation of laser beam, which impedes the applications of free-space optical communications, optical imaging and remote sensing. Propagation properties of a partially coherent beam

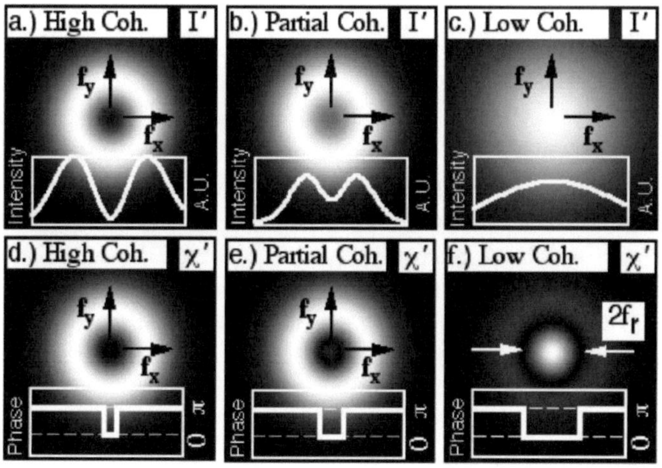

Figure 3. Far-field intensity distribution and corresponding amplitude and phase distribution of the correlation function of a partially coherent LG_{0l} beam.

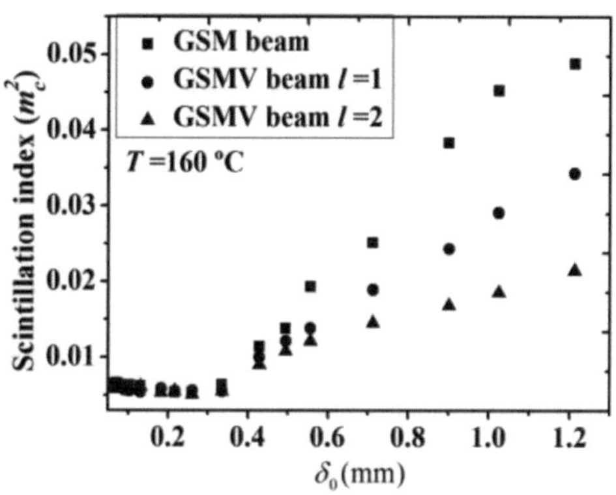

Figure 4. Experimental results of the scintillation index of a GSM or GSMV beam ($l = 1, 2$) at the centroid versus the initial coherence width after propagating through thermal turbulence.

in turbulent atmosphere have been investigated in detail in the past few decades, and it was found that a GSM beam has an advantage over a coherent Gaussian beam for reducing turbulence-induced scintillation and degradation [7, 8]. The scintillation index of a beam in turbulent atmosphere is defined as follows:

$$m_c^2 = \langle I^2(\rho_x, \rho_y)\rangle / \langle I(\rho_x, \rho_y)\rangle^2 - 1, \tag{13}$$

where $I(\rho_x, \rho_y)$,$\langle I(\rho_x, \rho_y)\rangle$ and $\langle I^2(\rho_x, \rho_y)\rangle$ represent the instantaneous intensity, the average intensity and the intensity correlation function of the beam in the receiver plane.

Is it possible to further reduce turbulence-induced scintillation compared to GSM beam? Recently, propagation properties of a partially coherent vortex beam in turbulent atmosphere have been investigated both theoretically and experimentally [60–62]. It was shown in [62] that a GSMV beam has an advantage over a GSM beam for further reducing turbulence-induced scintillation (see **Figure 4**). From **Figure 4**, one sees that the scintillation index of a GSMV beam or a GSM beam decreases with the decrease of initial coherence width δ_0, and the scintillation index of a GSMV beam is always smaller than that of a GSM beam unless the coherence width is very small ($\delta_0 < 0.35$mm). When the coherence width is very small, the scintillation index of a GSMV beam is almost the same with that of a GSM beam, and the phenomenon can be explained by the fact that the influence of the coherence on the scintillation index plays a dominant role, while the influence of the vortex phase is negligible when the coherence width is very small. Thus, the scintillation index of the GSMV beam with extremely low coherence is similar to that of a GSM beam.

4. Generation of partially coherent vortex beams

Up to now, many different methods have been developed to generate a coherent vortex beam, such as spiral phase plate [76], transverse mode selection [77], holographic grating [78], spatial light modulator [79], helical optical fiber [80] and uniaxial crystal [81], while only few papers were devoted to generation of partially coherent vortex beams [46–48, 51, 53, 54, 73].

One can generate a GSMV beam in experiment with the help of a rotating ground-glass disk, Gaussian amplitude filter and a spiral phase plate [46]. As shown in **Figure 5**, a focused laser beam generated by a He-Ne laser is reflected by a mirror and then illuminates a RGGD, producing a partially coherent beam with Gaussian statistics. The thin lens L_2 is used to collimate the transmitted light, and the GAF is used to transform the intensity of the transmitted light into a Gaussian profile. The transmitted light behind the GAF is a GSM beam. The coherence width of the GSM beam is determined by the focused beam spot size on the RGGD, which is controlled by the varying the distance between lens L_1 and RGGD. After passing

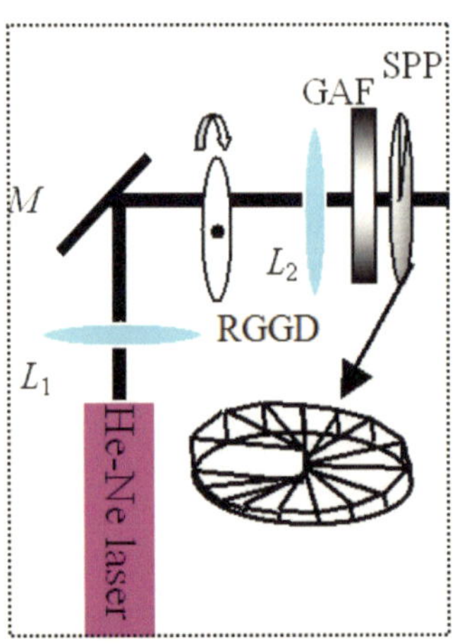

Figure 5. Experimental setup for generating a GSMV beam. L_1, L_2, thin lenses; M, mirror; RGGD, rotating ground-glass disk; GAF, Gaussian amplitude filter; SPP, spiral phase plate.

through a SPP located just behind the GAF, the GSM beam becomes a GSMV beam. It is true that the beam spot of the generated GSMV beam in the focal plane is shaped by varying its coherence width (see **Figure 6**).

One can generate a partially coherent LG_{pl} beam in experiment with the help of a rotating ground-glass disk, Gaussian amplitude filter and a SLM [48]. As shown in **Figure 7**, the RGGD and GAF are used to generate a GSM beam. The generated GSM beam goes toward a spatial light modulator (SLM), which acts as a grating with fork pattern designed by the method of computer-generated holograms. The first-order diffraction pattern of the beam reflected from the SLM is regarded as a partially coherent LG_{pl} beam and is selected out by a circular aperture.

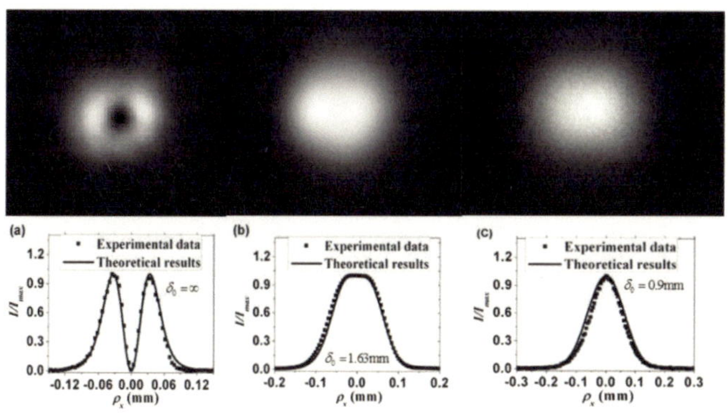

Figure 6. Experimental results of the focused intensity distribution and the corresponding cross line (dotted curve) of the generated GSMV beam for three different coherence widths. The solid curves are calculated by theoretical formulae.

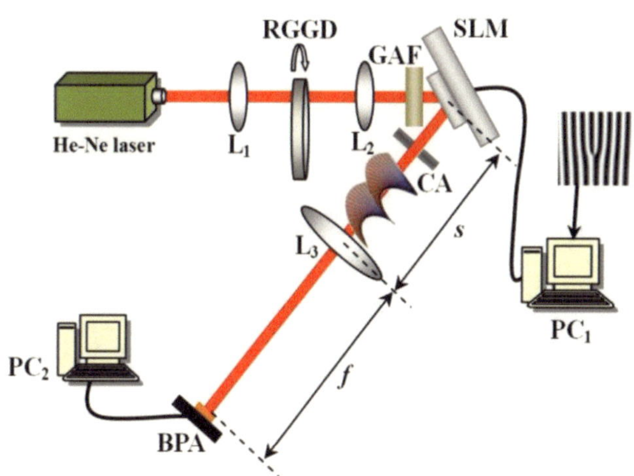

Figure 7. Experimental setup for generating a partially coherent LG_{pl} beam. L_1, L_2, L_3, thin lenses; RGGD, rotating ground-glass disk; GAF, Gaussian amplitude filter; SLM, spatial light modulator; CA, circular aperture; BPA, beam profile analyzer.

Figure 8 shows the experimental setup for generating a LGCSMV beam [25]. A beam emitted from the He-Ne laser passes through a beam expander, and then it goes toward a SLM. The first order of the beam from the SLM is a dark hollow beam and is selected out by a circular aperture. The generated dark hollow beam illuminates a RGGD, producing an incoherent

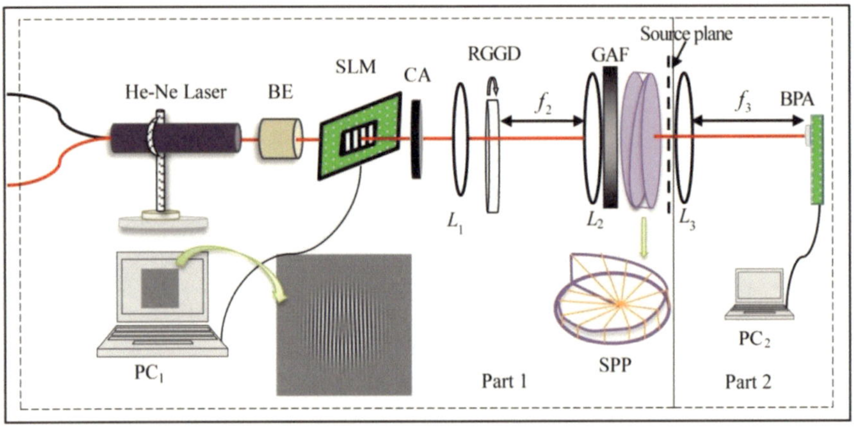

Figure 8. Experimental setup for generating a LGCSMV beam. BE, beam expander; SLM, spatial light modulator; CA, circular aperture; L_1, L_2, L_3, thin lenses; RGGD, rotating ground-glass disk; GAF, Gaussian amplitude filter; SPP, spiral phase plate; BPA, beam profile analyzer.

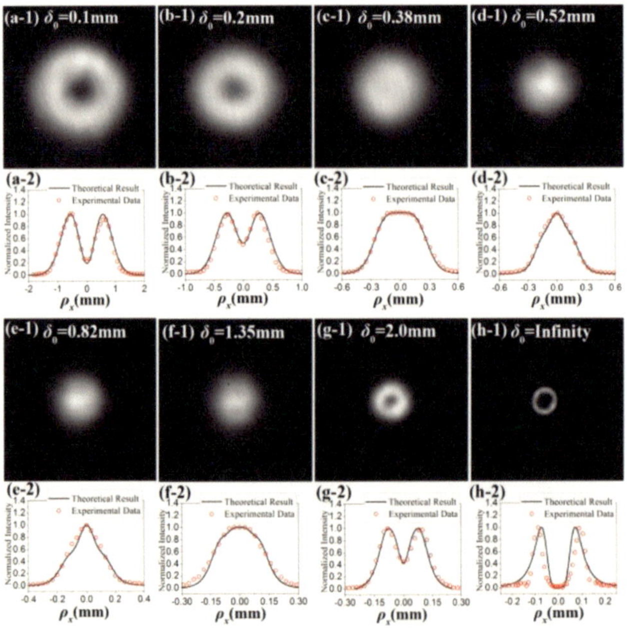

Figure 9. Experimental results of the intensity distribution and the corresponding cross line of the generated LGCSM vortex beam in the focal plane for different values of the coherence width. The solid curve denotes the theoretical results.

beam with dark hollow beam profile. After passing through free space with length f_2, the thin lens L_2, and the GAF, the generated incoherent dark hollow beam becomes a LGCSM beam. After passing through a SPP, the generated LGCSM beam becomes a LGCSMV beam. Due to the influence of the special correlation function, the LGCSMV beam exhibits interesting focusing properties, for example, the focused beam spot displays dark hollow beam profile when the coherence width is very large or very small, and displays flat-topped beam profile of Gaussian beam profile when the coherence width takes a middle value (see **Figure 9**), which are much different from the focusing properties of the conventional partially coherent vortex beam (see **Figures 1** and **6**).

In a similar way, experimental generation of an EGSMV beam and a radially polarized partially coherent vortex beam were reported in [53,54], respectively. It was shown that the vortex phase induces not only the rotation of the beam spot, but also the changes of the beam shape, the degree of polarization and the state of polarization. Furthermore, it was revealed that the vortex phase plays a role of resisting the coherence-induced degradation of the intensity distribution and the coherence-induced depolarization.

More recently, a new experimental technique is developed in [81] to generate partially coherent vortex beams with arbitrary azimuthal index using only a spatial light modulator (see **Figure 10**). This technique is based on digitally simulating the intrinsic randomness of broadband light passing through a spiral phase plate, and it provides control over the transverse coherence length, which will be useful for study of vector singularities in partially coherent fields or in the fields of optical communications and imaging systems where coherence plays a key role.

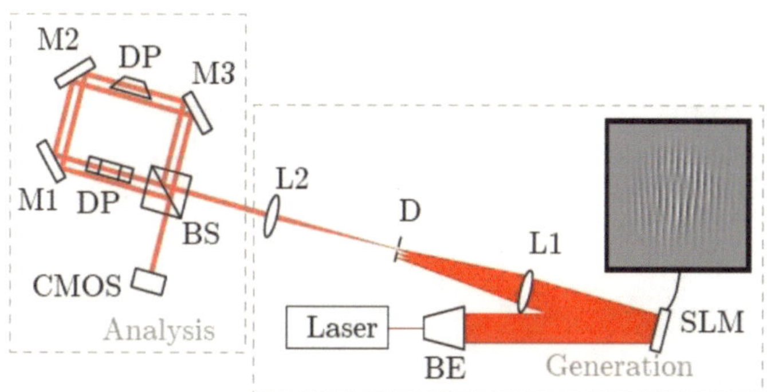

Figure 10. Experimental setup for digital generation of partially coherent vortex beam. HeNe Laser; BE: beam expander; L1–L2: lenses; D: iris diaphragm; SLM: spatial light modulator; BS: beam splitter; DP: Dove prism; M1–M3: mirrors; CMOS: camera.

5. Determination of the topological charge of a partially coherent vortex beam

It is known that a vortex beam carries an OAM of $l\hbar$ with l being the topological charge [32]. Different methods have been developed to determine or measure the topological charge (i.e.,

OAM) of a coherent vortex beam [82–88]. However, these methods are mainly based on measuring the intensity distribution of a vortex beam passing through different diffractive elements or intensity distribution produced by the interference of a vortex beam with a reference plane wave. For a partially coherent vortex beam, the conventional methods for measuring the topological charge will be invalid because the intensity distribution is seriously deformed, while fortunately, some methods have been proposed to measure or determine the topological charge of a partially coherent vortex beam based on the measurement of the correlation function recently [65–71].

For a coherent vortex beam, it is known from [88] that the number of dark rings in the Fourier transform of the intensity of a coherent vortex beam equals to the magnitude of the topological charge, thus one can determine the magnitude of the topological charge once we obtain the information of the intensity of a vortex beam. With the decrease of initial coherence width, the hollow profile of the intensity distribution of a vortex beam in the focal plane or in the far field disappears gradually and finally becomes a Gaussian beam profile (see **Figure 11(a-1)–(a-4)**), and the dark rings of the Fourier transform of the intensity distribution disappear (see **Figure 11(b-1)–(b-4)**). Then how to determine the topological charge of a partially coherent vortex beam? It is shown in [47] that the correlation function of a partially coherent vortex beam (i.e., partially coherent LG_{0l} beam) displays correlation singularity (i.e., ring dislocation) in the far field. Here the correlation function denotes the cross-spectral density $W(\rho,-\rho)$ or the degree of coherence $\mu(\rho,-\rho)$. Later in [65,66], it was revealed that for the number of ring dislocations of a partially coherent LG_{0l} beam in the focal plane or in the far field equals to the magnitude of the topological charge (see **Figure 12**). From **Figure 12**, one sees that the ring

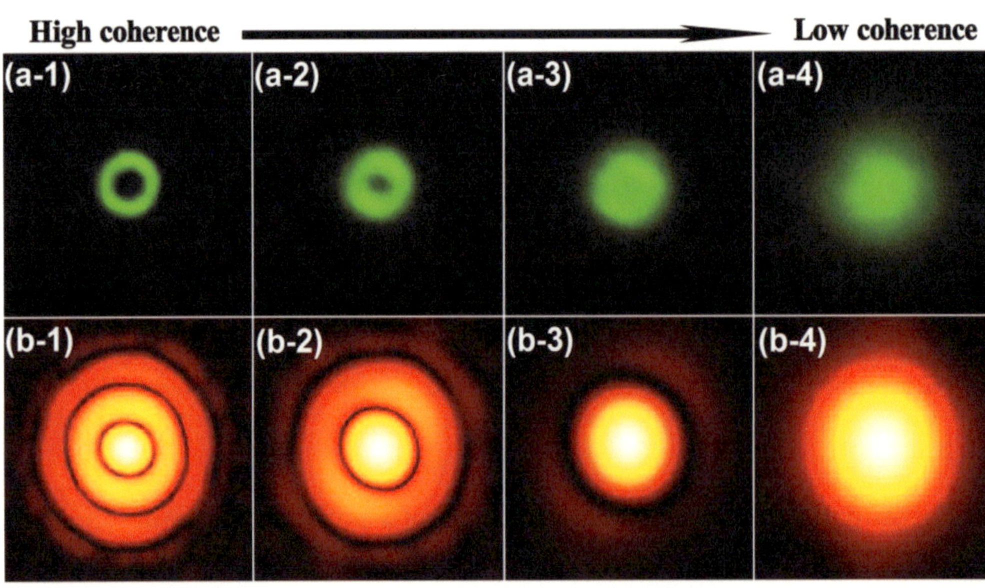

Figure 11. (a-1)–(a-4) Intensity distribution and (b-1)–(b-4) the corresponding Fourier transform of a partially coherent LG_{0l} beam with $l = 3$ in the focal plane for different state of coherence.

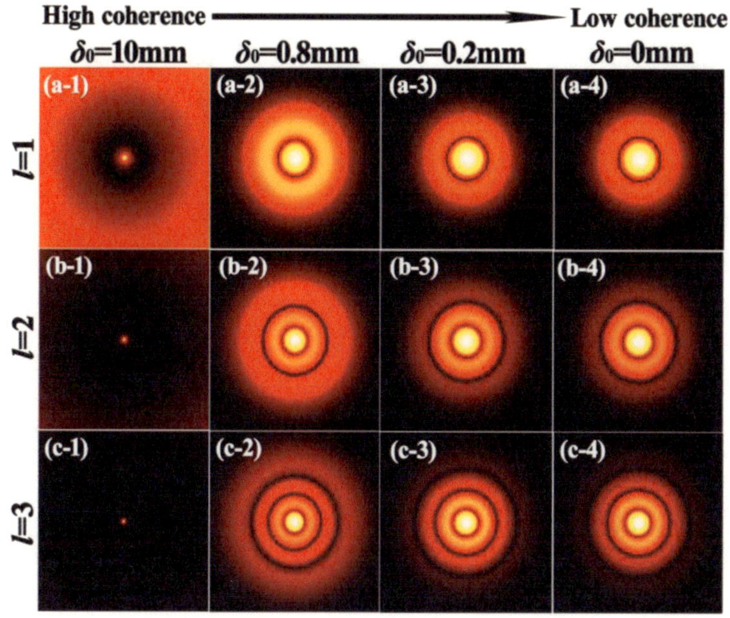

Figure 12. Distribution of the modulus of the degree of coherence of a partially coherent LG_{0l} beam with different values of the topological charge l in the focal plane for different state of coherence.

dislocations in the correlation function does not exist when initial coherence width is large, and the rings dislocations gradually appear with the decrease of the coherence, and such rings dislocations even exist in an incoherent LG_{0l} beam. Thus, one can determine the magnitude of the topological charge of a partially coherent or incoherent LG_{0l} beam through measuring its correlation function, and this phenomenon is confirmed experimentally in [67]. For a partially coherent LG_{pl} beam, it was demonstrated both theoretically [68] and experimentally [69] that the number of the ring dislocations equal to $2p+|l|$ (see **Figure 13**). Recently, it was predicted in [70] that one may determine p and l through measuring the double-correlation function, which is defined as $W(\rho, 2\rho)$ or $\mu(\rho, 2\rho)$.

Abovementioned literatures are confined to measure the magnitude of the topological charge of a partially coherent vortex beam. In fact, the sign of the topological charge of vortex phase also plays an important role in practical applications, for example, the sign of the vortex phase provides an additional degree of freedom for optical storage and communication [89, 90]. Recently, a simple method for simultaneous determination of the sign and the magnitude of the topological charge of a partially coherent LG_{0l} beam was proposed. This method is based on the measurement of the modulus of the degree of coherence of a partially coherent LG_{0l} beam after propagating through a couple of cylindrical lenses (see **Figure 14**). It was found that the distribution of the modulus of the degree of coherence becomes anisotropic, and it rotates anti-clockwise (or clockwise) during propagation when the sign of the topological charge is positive (or negative), furthermore, the modulus of the degree of coherence displays fringes distribution within certain propagation distances and the number of the bring fringes equals to

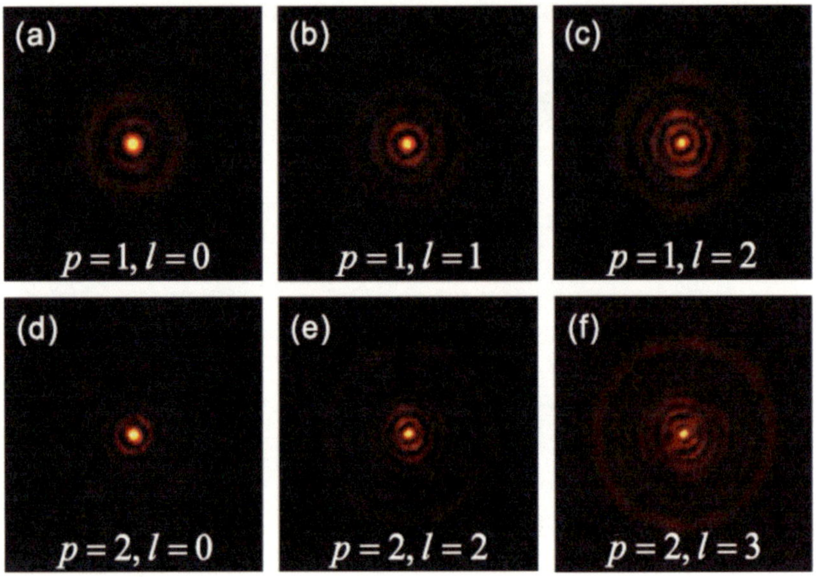

Figure 13. Experimental results of the distribution of the modulus of the degree of coherence of a partially coherent LG_{pl} beam in the focal plane for different values of p and l.

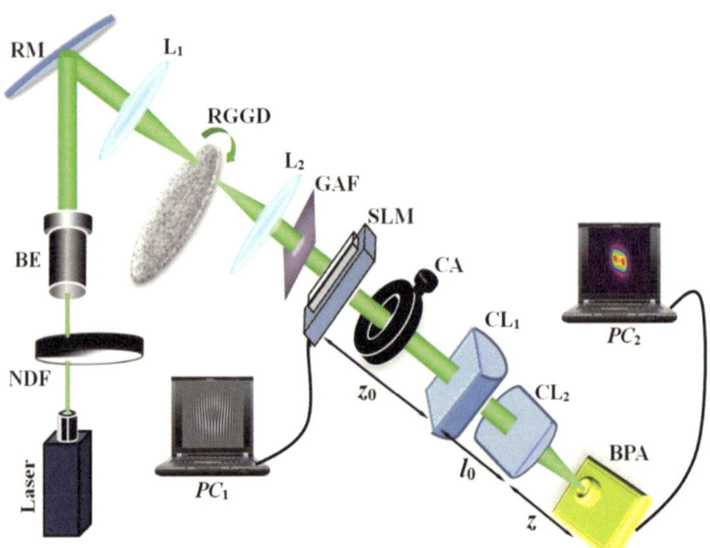

Figure 14. Experimental setup for determining the magnitude and the sign of the topological charge. NDF, neutral-density filter; BE, beam expander; RM, reflecting mirror; L_1, L_2, thin lenses; RGGD, rotating ground-glass disk; GAF, Gaussian amplitude filter; SLM, spatial light modulator; CA, circular aperture; CL_1, CL_2, cylindrical lenses; BPA, beam profile analyzer.

$2|l|+1$ (see **Figure 15**). One can extend this method to determine the sign and the magnitude of the topological charge of a partially coherent LG_{pl} beam based on the measurement of the double-correlation function.

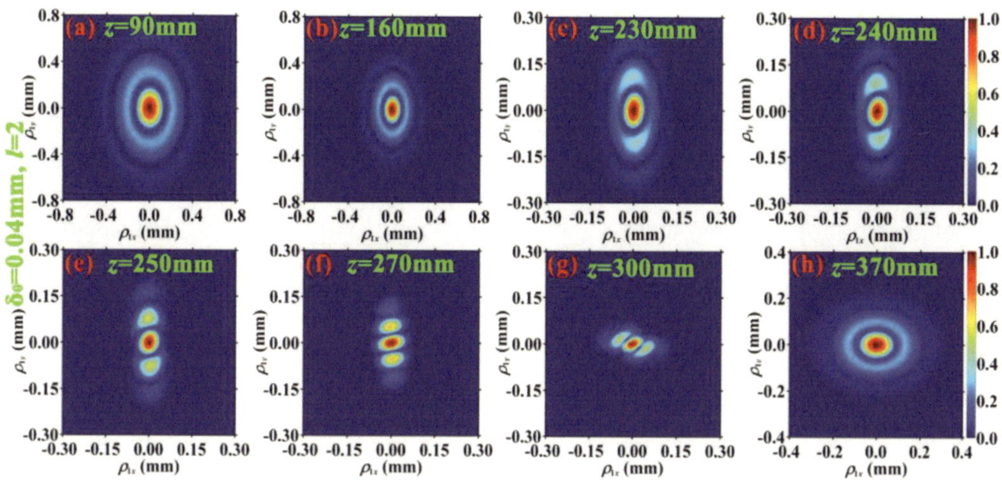

Figure 15. Distribution of the modulus of the degree of coherence of a partially coherent LG_{0l} beam after passing through a couple of cylindrical lenses at different propagation distances with l=2 and $\delta_0 = 0.04$mm.

6. Application of partially coherent vortex beams

Due to their extraordinary propagation properties, partially coherent vortex beams are useful in many applications, such as material processing, optical trapping, free-space optical communications and optical imaging.

It was shown in [46] that one can shape the beam profile of a focused partially coherent vortex beam through varying its initial spatial coherence width, and one can obtain flat-topped beam profile, dark hollow beam profile and Gaussian beam profile in the focal plane. The formed flat-topped beam profile is useful in material processing [91], and in trapping a Rayleigh particle whose refractive index is larger than that of the ambient [92], and the formed dark hollow beam profile is useful in trapping a Rayleigh particle whose refractive index is smaller than that of the ambient [72].

It is known that atmosphere turbulence induces scintillation of laser beam, which impedes the application of free-space optical communications. It was shown in [93] that partially coherent beam can be used to reduce turbulence-induced scintillation, and is useful in free-space optical communications [7, 8]. In [62], it was demonstrated experimentally a partially coherent vortex beam has an advantage over a partially coherent beam without vortex phase for reducing turbulence-induced scintillation, thus is expect to be useful in free-space optical communication.

Finally, we know that both partially coherent beam and vortex beam are useful in super-resolution imaging [94, 95], one may expect that partially coherent vortex beam has an advantage over partially coherent beam and vortex beam in super-resolution imaging. What is more, partially coherent vortex beam carries correlation singularities, and one may apply correlation singularities for information encoding, transfer and decoding.

7. Summary

We have presented on a review on recent theoretical and experimental developments on partially coherent vortex beam. The theoretical models, propagation properties, and generation methods for various partially coherent vortex beams have been illustrated in detail. Partially coherent vortex beams display many unique and interesting properties, and are useful in some applications, such as material processing, optical trapping, free-space optical communications and optical imaging. We believe this field will grow further and expand rapidly, and more and more interesting results and potential applications will be revealed.

Acknowledgements

This work is supported by the National Natural Science Fund for Distinguished Young Scholar under Grant no. 11525418, the National Natural Science Foundation of China under Grant no. 11404234 and 11274005, and the Project of the Priority Academic Program Development (PAPD) of Jiangsu Higher Education Institutions.

Author details

Xianlong Liu, Lin Liu, Yahong Chen and Yangjian Cai*

*Address all correspondence to: yangjiancai@suda.edu.cn

College of Physics, Optoelectronics and Energy & Collaborative Innovation Center of Suzhou Nano Science and Technology, Soochow University, Suzhou, China

References

[1] L. Mandel and E. Wolf, editors. *Optical Coherence and Quantum Optics*. Cambridge: Cambridge University Press; 1995.

[2] D. Kermisch. Partially coherent image processing by laser scanning. J. Opt. Soc. Am. 1975;**65**(8):887–891.

[3] Y. Cai and S. Zhu. Ghost imaging with incoherent and partially coherent light radiation. Phys. Rev. E. 2005;**71**(5):056607.

[4] T. E. Gureyev, D. M. Paganin, A. W. Stevenson, S. C. Mayo, and S. W. Wilkins. Generalized eikonal of partially coherent beams and its use in quantitative imaging. Phys. Rev. Lett. 2004;**93**(6):068103.

[5] C. Zhao, Y. Cai, and O. Korotkova. Radiation force of scalar and electromagnetic twisted Gaussian Schell-model beams. Opt. Express. 2009;**17**(24):21472–21487.

[6] Y. Dong, F. Wang, C. Zhao, and Y. Cai. Effect of spatial coherence on propagation, tight focusing and radiation forces of an azimuthally polarized beam. Phys. Rev. A. 2012;**86** (1):013840.

[7] J. C. Ricklin and F. M. Davidson. Atmospheric turbulence effects on a partially coherent Gaussian beam: implications for free-space laser communication. J. Opt. Soc. Am. A. 2002;**19**(9):1794–1802.

[8] F. Wang, X. Liu, and Y. Cai. Propagation of partially coherent beam in turbulent atmosphere: a review (invited review). Prog. Electromagn. Res. 2015;**150**:123–143.

[9] Y. Cai, O. Korotkova, H. T. Eyyuboğlu, and Y. Baykal. Active laser radar systems with stochastic electromagnetic beams in turbulent atmosphere. Opt. Express. 2008;**16** (20):15834–15846.

[10] O. Korotkova, Y. Cai, and E. Watson. Stochastic electromagnetic beams for LIDAR systems operating through turbulent atmosphere. Appl. Phys. B. 2009;**94**(4):681–690.

[11] G. Wu and Y. Cai. Detection of a semirough target in turbulent atmosphere by a partially coherent beam. Opt. Lett. 2011;**36**(10):1939–1941.

[12] E. Wolf and E. Collett. Partially coherent sources which produce the same far-field intensity distribution as a laser. Opt. Commun. 1978;**25**(3):293–296.

[13] F. Gori. Collet–Wolf sources and multimode lasers. Opt. Commun. 1980;**34**(3):301–305.

[14] P. de Santis, F. Gori, G. Guattari, and C. Palma. An example of a Collett–Wolf source. Opt. Commun. 1979;**29**(3):256–260.

[15] A. T. Friberg and R. J. Sudol. Propagation parameters of Gaussian Schell-model beams. Opt. Commun. 1982;**41**(6):383–387.

[16] E. Tervonen, A. T. Friberg, and J. Turunen. Gaussian Schell-model beams generated with synthetic acousto-optic holograms. J. Opt. Soc. Am. A. 1992;**9**(5):796–803.

[17] E. Wolf, editor. *Introduction to the Theory of Coherence and Polarization of Light*. Cambridge: Cambridge University; 2007.

[18] Y. Cai. Generation of various partially coherent beams and their propagation properties in turbulent atmosphere: a review. Proc. of SPIE. 2011;**7924**:792402.

[19] Y. Cai, Y. Chen, and F. Wang. Generation and propagation of partially coherent beams with nonconventional correlation functions: a review [invited]. J. Opt. Soc. Am. A. 2014;**31**(9):2083–2096.

[20] Y. Cai, F. Wang, C. Zhao, S. Zhu, G. Wu, and Y. Dong. Partially coherent vector beams: from theory to experiment. In: Q. Zhan, editor. *Vectorial Optical Fields: Fundamentals and Applications*. 1st ed. Singapore: World Scientific; 2013. pp. 221–273.

[21] F. Gori, M. Santarsiero, G. Piquero, R. Borghi, A. Mondello, and R. Simon. Partially polarized Gaussian schell model beams. J. Opt. A, Pure Appl. Opt. 2001;**3**(1):1–9.

[22] E. Wolf. Unified theory of coherence and polarization of random electromagnetic beams. Phys. Lett. A. 2003;**312**(5):263–267.

[23] O. Korotkova and E. Wolf. Changes in the state of polarization of a random electromagnetic beam on propagation. Opt. Commun. 2005;**246**(1):35–43.

[24] H. Roychowdhury, S. A. Ponomarenko, and E. Wolf. Change in the polarization of partially coherent electromagnetic beams propagating through the turbulent atmosphere. J. Mod. Opt. 2005;**52**(11):1611–1618.

[25] H. Lajunen and T. Saastamoinen. Non-uniformly correlated partially coherent pulses. Opt. Lett. 2013;**21**(1):190–195.

[26] S. Sahin and O. Korotkova. Light sources generating far fields with tunable flat profiles. Opt. Lett. 2012;**37**(4):2970–2972.

[27] F. Wang, X. Liu, Y. Yuan, and Y. Cai. Experimental generation of partially coherent beams with different complex degrees of coherence. Opt. Lett. 2013;**38**(11):1814–1816.

[28] C. Liang, F. Wang, X. Liu, Y. Cai, and O. Korotkova. Experimental generation of cosine Gaussian-correlated Schell-model beams with rectangular symmetry. Opt. Lett. 2014;**39** (4):769–772.

[29] O. Korotkova. Random sources for rectangularly-shaped far fields. Opt. Lett. 2014;**39** (1):64–67.

[30] Y. Chen, F. Wang, L. Liu, C. Zhao, Y. Cai, and O. Korotkova. Generation and propagation of a partially coherent vector beam with special correlation functions. Phys. Rev. A. 2014;**89**(1):013801.

[31] Y. Chen, J. Gu, F. Wang, and Y. Cai. Self-splitting properties of a Hermite-Gaussian correlated Schell-model beam. Phys. Rev. A. 2015;**91**(1):013823.

[32] L. Allen, M. W. Beijersbergen, R. Spreeuw, and J. P. Woerdman. Orbital angular momentum of light and the transformation of Laguerre-Gaussian laser modes. Phys. Rev. A. 1992;**4**(11):8185–8189.

[33] L. Allen, M. J. Padgett, and M. Babiker. The orbital angular momentum of light. Prog. Opt. 1999;39:291–372.

[34] H. He, M. E. J. Friese, N. R. Heckenberg, and H. Rubinsztein-Dunlop. Direct observation of transfer of angular momentum to absorptive particles from a laser beam with a phase singularity. Phys. Rev. Lett. 1995;**75**(5):826–829.

[35] T. Kuga, Y. Torii, N. Shiokawa, T. Hirano, Y. Shimizu, and H. Sasada. Novel optical trap of atoms with a doughnut beam. Phys. Rev. Lett. 1997;**78**(25):4713–4716.

[36] D. G. Grier. A revolution in optical manipulation. Nature. 2003;**424**(6950):810–816.

[37] G. Molina-Terriza, J. P. Torres, and L. Torner. Twisted photons. Nat. Phys. 2007;**3**(5):305–310.

[38] A. Mair, A. Vaziri, G. Weighs, and A. Zeilinger. Entanglement of the orbital angular momentum states of photons. Nature. 2001;**412**(6844):313–316.

[39] J. Ng, Z. Lin, and C. T. Chan. Theory of optical trapping by an optical vortex beam. Phys. Rev. Lett. 2010;**104**(10):103601.

[40] J. Wang, J. Yang, I. M. Fazal, N. Ahmed, Y. Yan, H. Huang, Y. Ren, Y. Yue, S. Dolinar, M. Tur, and A. E. Willner. Terabit free-space data transmission employing orbital angular momentum multiplexing. Nat. Photonics. 2012;**6**(7):488–496.

[41] E. Nagali, F. Sciarrino, F. De Martini, L. Marrucci, B. Piccirillo, E. Karimi, and E. Santamato. Quantum information transfer from spin to orbital angular momentum of photons. Phys. Rev. Lett. 2009;**103**(1):013601.

[42] M. P. J. Lavery, F. C. Speirits, S. M. Barnett, and M. J. Padgett. Detection of a spinning object using light's orbital angular momentum. Science. 2013;**341**(6145):537–540.

[43] F. Gori, M. Santarsiero, R. Borghi, and S. Vicalvi. Partially coherent sources with helicoidal modes. J. Mod. Opt. 1998;**45**(3):539–554.

[44] V. G. Boggatyryova, V. C. Felde, P. V. Polyanskii, S. A. Ponomarenko, M. S. Soskin, and E. Wolf. Partially coherent vortex beams with a separable phase. Opt. Lett. 2003;**28** (11):878–880.

[45] S. A. Ponomarenko. A class of partially coherent beams carrying optical vortices. J. Opt. Soc. Am. A. 2001;**18**(1):150–156.

[46] F. Wang, S. Zhu, and Y. Cai. Experimental study of the focusing properties of a Gaussian Schell-model vortex beam. Opt. Lett. 2011;**36**(16):3281–3283.

[47] D. Palacios, I. Maleev, A. Marathay, and G. A. Swartzlander. Spatial correlation singularity of a vortex field. Phys. Rev. Lett. 2004;**92**(14):143905.

[48] C. Zhao, Y. Dong, Y. Wang, F. Wang, Y. Zhang, Y. Cai. Experimental generation of a partially coherent Laguerre-Gaussian beam. Appl. Phys. B. 2012;**109**(2):345–349.

[49] F. Wang, Y. Cai, and O. Korotkova. Partially coherent standard and elegant Laguerre-Gaussian beams of all orders. Opt. Express. 2009;**17**(25):22366–22379.

[50] Z. Zhang, J. Pu, and X. Wang. Focusing of partially coherent Bessel-Gaussian beams through a high-numerical-aperture objective. Opt. Lett. 2008;**33**(1):49–51.

[51] Y. Chen, F. Wang, C. Zhao, and Y. Cai. Experimental demonstration of a Laguerre-Gaussian correlated Schell-model vortex beam. Opt. Express. 2014;22(5):5826–5838.

[52] Y. Zhang, L. Liu, C. Zhao, and Y. Cai. Multi-Gaussian Schell-model vortex beam. Phys. Lett. A. 2014;**378**(9):750–754.

[53] X. Liu, F. Wang, L. Liu, C. Zhao, and Y. Cai. Generation and propagation of an electromagnetic Gaussian Schell-model vortex beam. J. Opt. Soc. Am. A. 2015;**32**(11):2058–2065.

[54] L. Guo, Y. Chen, X. Liu, L. Liu, and Y. Cai. Vortex phase-induced changes of the statistical properties of a partially coherent radially polarized beam. Opt. Express. 2016;**24**(13):13714–13728.

[55] I. D. Maleev, D. M. Palacios, A. S. Marathay, and G. A. Swartzlander. Spatial correlation vortices in partially coherent light: theory. J. Opt. Soc. Am. B. 2004;**21**(11):1895–1900.

[56] T. Van Dijk and T. D. Visser. Evolution of singularities in a partially coherent vortex beam. J. Opt. Soc. Am. A. 2009;**26**(4):741–744.

[57] I. D. Maleev and G. A. Swartzlander. Propagation of spatial correlation vortices. J. Opt. Soc. Am. B. 2008;**25**(6):915–922.

[58] W. Wang, M. Takeda. Coherence current, coherence vortex, and the conservation law of coherence. Phys. Rev. Lett. 2006;**96**(22):223904.

[59] W. Wang, Z. Duan, S. G. Hanson, Y. Miyamoto, and M. Takeda. Experimental study of coherence vortices: Local properties of phase singularities in a spatial coherence function. Phys. Rev. Lett. 2006;**96**(7):073902.

[60] Y. Zhang, M. Tang, and C. Tao. Partially coherent vortex beams propagating in a turbulent atmosphere. Chin. Opt. Lett. 2005;**3**(10):559–561.

[61] F. Wang, Y. Cai, H. Eyyuboglu, and Y. Baykal. Average intensity and spreading of partially coherent standard and elegant Laguerre-Gaussian beams in turbulent atmosphere. Prog. Electromagn. Res. 2010;**103**:33–56.

[62] X. Liu, Y. Shen, L. Liu, F. Wang, and Y. Cai. Experimental demonstration of vortex phase-induced reduction in scintillation of a partially coherent beam. Opt. Lett. 2013;**38**(24):5323–5326.

[63] H. Pires, J. Woudenberg, and M. Van Exter. Measurements of spatial coherence of partially coherent light with and without orbital angular momentum. J. Opt. Soc. Am. A. 2010;**27**(12):2630–2637.

[64] H. Pires, J. Woudenberg, and M. Van Exter. Measurement of the orbital angular momentum spectrum of partially coherent beams. Opt. Lett. 2010;**35**(6):889–891.

[65] C. Zhao, F. Wang, Y. Dong, Y. Han and Y. Cai. Effect of spatial coherence on determining the topological charge of a vortex beam. Appl. Phys. Lett. 2012;**101**(26):261104.

[66] Y. Yang, M. Mazilu, and K. Dholakia. Measuring the orbital angular momentum of partially coherent optical vortices through singularities in their cross-spectral density functions. Opt. Lett. 2012;**37**(23):4949–4951.

[67] A. Y. Escalante, B. Perez-Garcia, R. I. Hernandez-Aranda, and G. A. Swartzlander. Determination of angular momentum content in partially coherent beams through cross correlation measurements. Proc. SPIE. 2013;**8843**:884302.

[68] Y. Yang, M. Chen, M. Mazilu, A. Mourka, Y. Liu and K. Dholakia. Effect of the radial and azimuthal mode indices of a partially coherent vortex field upon a spatial correlation singularity. New J. Phys. 2013;**15**(11):113053.

[69] R. Liu, F. Wang, D. Chen, Y. Wang, Y. Zhou, H. Gao, P. Zhang, and F. Li. Measuring mode indices of a partially coherent vortex beam with Hanbury Brown and Twiss type experiment. App. Phys. Lett. 2016;**108**(5):051107.

[70] Y. Yang and Y. Liu. Measuring azimuthal and radial mode indices of a partially coherent vortex field. J. Opt. 2016;**18**(1):015604.

[71] J. Chen, X. Liu, J. Yu, and Y. Cai. Simultaneous determination of the sign and the magnitude of the topological charge of a partially coherent vortex beam. App. Phys. B. 2016;**122**:201

[72] C. Zhao and Y. Cai. Trapping two types of particles using a focused partially coherent elegant Laguerre-Gaussian beam. Opt. Lett. 2011;**36**(12):2251–2253.

[73] B. Perez-Garcia, A. Yepiz, R. I. Hernandez-Aranda, A. Forbes, and G. A. Swartzlander. Digital generation of partially coherent vortex beams. Opt. Lett. 2016; **41**(15):3471–3474.

[74] Q. Lin and Y. Cai. Tensor ABCD law for partially coherent twisted anisotropic Gaussian-Schell model beams. Opt. Lett. 2002;**27**(4):216–218.

[75] S. C. H. Wang and M. A. Plonus. Optical beam propagation for a partially coherent source in the turbulent atmosphere. J. Opt. Soc. Am. 1979;**69**(9):1297–1304.

[76] K. Sueda, G. Miyaji, N. Miyanaga, M. Nakatsuka. Laguerre-Gaussian beam generated with a multilevel spiral phase plate for high intensity laser pulses. Opt. Express. 2004;**12** (15):3548–3553.

[77] M. Vaupel and C. O. Weiss. Circling optical vortices. Phys. Rev. A. 1995;**51**(5):4078–4085.

[78] N. R. Heckenberg, R. McDuff, C. P. Smith, and A. G. White. Generation of optical phase singularities by computer-generated holograms. Opt. Lett. 1992;**17**(3):221–223.

[79] J. B. Bentley, J. A. Davisa, M. A. Bandres, and J. C. Guti_errez-Vega. Generation of helical Ince Gaussian beams with a liquid-crystal display. Opt. Lett. 2006;**31**(5):649–651.

[80] C. N. Alexeyev and M. A. Yavorsky. Generation and conversion of optical vortices in long-period helical core optical fibers. Phys. Rev. A. 2008;**78**(4):043828.

[81] A. Volyar, V. Shvedov, T. Fadeyeva, A. S. Desyatnikov, D. N. Neshev, W. Krolikowski, and Yu. S. Kivshar. Generation of single-charge optical vortices with an uniaxial crystal. Opt. Express. 2006;**14**(9):3724–3729.

[82] V. Denisenko, V. Shvedov, A. S. Desyatnikov, D. N. Neshev, W. Krolikowski, A. Volyar, M. Soskin, and Y. S. Kivsar. Determination of topological charges of polychromatic optical vortices. Opt. Express. 2009;**17**(26):23374–23379.

[83] G. C. G. Berkhout and M. W. Beijersbergen. Method for probing the arbital angular momentum of optical vortices in electromagnetic waves from astronomical objects. Phys. Rev. Lett. 2008;**101**(10):100801.

[84] J. M. Hickmann, E. J. S. Fonseca, W. C. Soares, and S. Chávez-Cerda. Unveiling a truncated optical lattice associated with a triangular aperture using light's orbital angular momentum. Phys. Rev. Lett. 2010;**105**(5):053904.

[85] R. Liu, J. Long, F. Wang, Y. Wang, P. Zhang, H. Gao and F. Li. Characterizing the phase profile of a vortex beam with angular-double-slit interference. J. Opt. 2013;**15**(12):125712.

[86] D. Fu, D. Chen, R. Liu, Y. Wang, H. Gao, F. Li, and P. Zhang. Probing the topological charge of a vortex beam with dynamic angular double slits. Opt. Lett. 2015;**40**(5):788–791.

[87] G. C. G. Berkhout, M. P. J. Lavery, J. Courtial, M. W. Beijersbergen, and M. J. Padgett. Efficient sorting of orbital angular momentum states of light. Phys. Rev. Lett. 2010;**105**(15):153601.

[88] S. Prabhakar, A. Kumar, J. Banerji, and R. P. Singh. Revealing the order of a vortex through its intensity record. Opt. Lett. 2011;**36**(22):4398–4400.

[89] N. Bozinovic, Y. Yue, Y. X. Ren, M. Tur, P. Kristensen, H. Huang, A. E. Willner, and S. Ramachandran. Terabit-scale orbital angular momentum mode division multiplexing in fibers. Science. 2013;**340**(6140):1545–1548.

[90] Y. Ming, J. Tang, Z. Chen, F. Xu, L. Zhang, and Y. Lu. Generation of n00n state with orbital angular momentum in a twisted nonlinear photonic crystal. IEEE. J. Quantum Electron. 2015;**21**(3):6601206.

[91] D. W. Coutts. Double-pass copper vapor laser master-oscillator power-amplifier systems: Generation of flat-top focused beams for fiber coupling and percussion drilling. IEEE J. Quantum Electron. 2002;**38**(9):1217–1224.

[92] C. Zhao, Y. Cai, X. Lu, and H. T. Eyyuboğlu. Radiation force of coherent and partially coherent flat-topped beams on a Rayleigh particle. Opt. Express. 2009;**17**(3):1753–1765.

[93] Y. Baykal and M. A. Plonus. Intensity fluctuations due to a spatially partially coherent source in atmospheric turbulence as predicted by Rytov's method. J. Opt. Soc. Am. A. 1985;**2**(12):2124–2132.

[94] F. Tamburini, G. Anzolin, G. Umbriaco, A. Bianchini, and C. Barbieri. Overcoming the Rayleigh criterion limit with optical vortices. Phys. Rev. Lett. 2006;**97**(16):163903.

[95] J. E. Oh, Y. W. Cho, G. Scarcelli, and Y. H. Kim. Sub-Rayleigh imaging via speckle illumination. Opt. Lett. 2013;**38**(5):682–684.

Chapter 12

Vortices and Singularities in Electric Dipole Radiation near an Interface

Xin Li, Henk F. Arnoldus and Zhangjin Xu

Additional information is available at the end of the chapter

http://dx.doi.org/10.5772/66459

Abstract

An oscillating electric dipole in free space emits its energy along straight lines. We have considered the effect of a nearby interface with a material medium. Interference between the directly emitted radiation and the reflected radiation leads to intricate flow line patterns. When the interface is a plane mirror, numerous interference vortices appear, and when the distance between the dipole and the mirror is not too small, these vortices lie on four strings. At the center of each vortex is a singularity, and these singularities are due to the fact that the magnetic field vanishes at these locations. When the interface is a boundary between dielectric media, reflection leads again to interference. The pattern for the transmitted radiation depends on whether the medium is thicker or thinner than the material in which the dipole is embedded. For thicker dielectrics, the field lines bend toward the normal, reminiscent of, but not the same as, the behavior of optical rays. For thinner media, oscillation of energy across the interface appears, and above a crossing point, there is a tiny vortex. We have also considered the case of a dipole in between two parallel mirrors.

Keywords: vortex, singularity, Poynting vector, dipole radiation, interface, mirror

1. Introduction

The common conception about the propagation of light is that the radiation travels along straight lines. Such a picture certainly seems to hold for a laser beam and for sunlight. The success of ray diagrams for the construction of images by lenses and mirrors also supports this picture. Reflection by and transmission through an interface is another example of a process that can be described by a ray picture of light. However, light is electromagnetic radiation, and ultimately any optical phenomenon must be accounted for by a solution of Maxwell's equations for the electric and magnetic fields. In the geometrical optics limit of light propagation [1], spatial variations on the scale of a wavelength or less are neglected, and optical rays are defined as the orthogonal trajectories of the wave fronts of a propagating wave. From a different point of view, we can define the direction of light propagation as the direction of the energy flow in the radiation field. This direction is determined by the electromagnetic Poynting vector [2]. It can be shown that in the geometrical optics limit for propagation in vacuum the field lines of the Poynting vector are straight lines and are identical to the optical rays, defined with the help of the propagation of wave fronts.

The concept of optical rays breaks down when spatial variations on the scale of a wavelength or less are of concern, or when coherence in the radiation leads to macroscopic constructive and destructive interference (as for a diffraction grating). We shall consider a small particle, like an atom, molecule, or nano-particle, irradiated by a monochromatic laser beam, oscillating with angular frequency ω. We shall assume that the beam is linearly polarized. The electric field of the laser beam induces an electric dipole moment in the particle, and this dipole moment oscillates with angular frequency ω along the same direction as the electric field of the immersing beam. The oscillating dipole moment emits electromagnetic radiation (expressions for the electric and magnetic fields are given below), and the Poynting vector is easily calculated. It appears that the field lines of the Poynting vector are straight lines, coming out of the dipole and running radially outward to infinity. This is illustrated in **Figure 1**.

The energy flow lines for a free (linear) dipole are straight at all distances. Any deviation from this radially outgoing pattern is due to the environment of the particle. For instance, when the particle is embedded in an absorbing medium, the imaginary part of the permittivity is responsible for a bending of the field lines in the near field toward the dipole axis [3]. Here we shall consider the case where the particle is located near an interface. Some of the emitted radiation by the dipole will be incident on the interface, and here reflection and transmission takes place. The reflected light will interfere with the incident light, and in the far field this leads to maxima and minima in the radiated power per unit solid angle. The structure of the

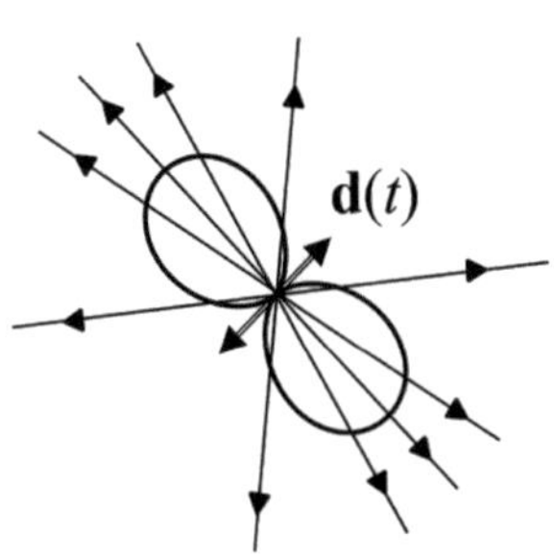

Figure 1. The electric dipole moment $\mathbf{d}(t)$ oscillates along the direction indicated by the double-headed arrow. The field lines of the Poynting vector come out of the dipole, and run radially outward. The solid curve is a polar diagram of the power per unit solid angle. No radiation is emitted along the dipole axis, and the maximum intensity is emitted perpendicular to the dipole axis.

http://dx.doi.org/10.5772/66459

angular power distribution of the transmitted light in the far field depends on the value of the critical angle and the distance between the particle and the surface. Usually, when a traveling plane wave is incident upon an interface, the transmitted wave is again traveling, and is bent toward the normal. When the angle of incidence approaches 90°, the transmitted wave is still traveling, and the angle of transmission is called the critical angle (for transmission). In the dipole spectrum, evanescent waves are present, and they are still transmitted as traveling waves, provided the wavelength is not too small. Therefore, above the critical transmission angle, all transmitted light comes from evanescent dipole waves, and this can lead to a large lobe in the power distribution above the critical angle [4]. Rather than considering the effects in the far field, we shall here present results for the power flow in the near field. We shall illustrate that interference gives rise to interesting flow patterns, including singularities, vortices, and strings of vortices. We also show that transmission in the near field exhibits interesting features.

2. Dipole radiation in free space

The oscillating dipole moment can be written as

$$\mathbf{d}(t) = d_o \hat{\mathbf{u}} \cos(\omega t), \tag{1}$$

where d_o is the amplitude of the oscillation, and $\hat{\mathbf{u}}$ is a unit vector representing the direction of oscillation. Let $\mathbf{r}$ be the position vector of a field point, with respect to the location of the dipole, and let r be the length of $\mathbf{r}$. Then, $\hat{\mathbf{r}} = \mathbf{r}/r$ is the unit vector into the direction of the field point. The dimensionless distance between the dipole and the field point is defined as $q = k_o r$, with $k_o = \omega/c$ the wave number of the light. On this scale, a distance of 2π corresponds to one wave length. The emitted electric field has the form

$$\mathbf{E}(\mathbf{r}, t) = \mathrm{Re}[\mathbf{E}(\mathbf{r})e^{-i\omega t}], \tag{2}$$

with $\mathbf{E}(\mathbf{r})$ the complex amplitude, and a similar expression holds for the magnetic field $\mathbf{B}(\mathbf{r}, t)$. We define the constant

$$\zeta = \frac{k_o^3 d_o}{4\pi\varepsilon_o}. \tag{3}$$

The dimensionless complex amplitudes $\mathbf{e}(\mathbf{r})$ and $\mathbf{b}(\mathbf{r})$ of the fields are introduced as

$$\mathbf{E}(\mathbf{r}) = \zeta\, \mathbf{e}(\mathbf{r}), \tag{4}$$

$$\mathbf{B}(\mathbf{r}) = \frac{\zeta}{c}\mathbf{b}(\mathbf{r}). \tag{5}$$

We then obtain for an electric dipole [5]

$$\mathbf{e}(\mathbf{r}) = \left\{ \hat{\mathbf{u}} - (\hat{\mathbf{r}} \cdot \hat{\mathbf{u}})\hat{\mathbf{r}} + [\hat{\mathbf{u}} - 3(\hat{\mathbf{r}} \cdot \hat{\mathbf{u}})\hat{\mathbf{r}}] \frac{i}{q} \left(1 + \frac{i}{q}\right) \right\} \frac{e^{iq}}{q}, \tag{6}$$

$$\mathbf{b}(\mathbf{r}) = (\hat{\mathbf{r}} \times \hat{\mathbf{u}}) \left(1 + \frac{i}{q}\right) \frac{e^{iq}}{q}. \tag{7}$$

The time-averaged Poynting vector for radiation in free space is given by

$$\mathbf{S}(\mathbf{r}) = \frac{1}{2\mu_o} \mathrm{Re}[\mathbf{E}(\mathbf{r}) \times \mathbf{B}(\mathbf{r})^*]. \tag{8}$$

For an electric dipole, we split off a factor:

$$\mathbf{S}(\mathbf{r}) = \frac{\zeta^2}{2\mu_o c} \boldsymbol{\sigma}(\mathbf{r}), \tag{9}$$

so that

$$\boldsymbol{\sigma}(\mathbf{r}) = \mathrm{Re}[\mathbf{e}(\mathbf{r}) \times \mathbf{b}(\mathbf{r})^*], \tag{10}$$

With the above expressions for $\mathbf{e}(\mathbf{r})$ and $\mathbf{b}(\mathbf{r})$, we find

$$\boldsymbol{\sigma}(\mathbf{r}) = \frac{1}{q^2} \hat{\mathbf{r}} \sin^2\alpha, \tag{11}$$

and here α is the angle between the dipole axis (e.g., the direction of vector $\hat{\mathbf{u}}$) and the observation direction $\hat{\mathbf{r}}$. Since the vector $\boldsymbol{\sigma}(\mathbf{r})$ is proportional to $\hat{\mathbf{r}}$, the field lines of the vector field are straight, and run radially outward from the site of the dipole. This is shown in **Figure 1**. No power is emitted along the dipole axis ($\alpha = 0$), and the power per unit solid angle is maximum in the direction perpendicular to the dipole axis ($\alpha = \pi/2$).

3. Dipole radiation near a mirror

The simplest interface is a flat, infinite, and perfect mirror. We take the surface of the mirror as the xy-plane, and the dipole is located on the positive z-axis, at a distance H above the mirror. The dipole direction vector $\hat{\mathbf{u}}$ makes an angle γ with the positive z-axis, and we take $\hat{\mathbf{u}}$ in the yz-plane. Therefore,

$$\hat{\mathbf{u}} = \mathbf{e}_y \sin\gamma + \mathbf{e}_z \cos\gamma. \tag{12}$$

The electric field above the mirror is equal to the field of the dipole plus the electric field of an image dipole located at a distance H below the mirror on the z-axis, as illustrated in **Figure 2**, and the same holds for the magnetic field. The dipole moment direction of the image dipole is

http://dx.doi.org/10.5772/66459

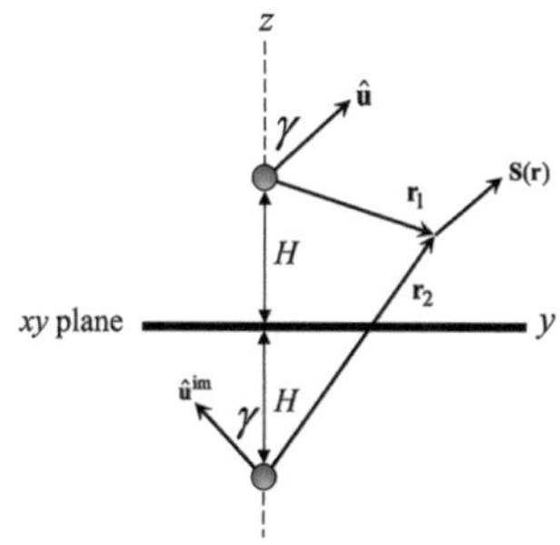

Figure 2. The electric dipole is located at a distance H above a flat mirror (xy-plane), and the image dipole is located at a distance H below the surface. Vectors $\mathbf{r}_1$ and $\mathbf{r}_2$ are the position vectors of the field point with respect to the location of the dipole and the image dipole, respectively. The field point is represented by the position vector $\mathbf{r}$ with respect to the origin of coordinates. This vector is not shown here.

$$\hat{\mathbf{u}}^{\mathrm{im}} = -\ \mathbf{e}_y \sin\gamma + \mathbf{e}_z \cos\gamma. \tag{13}$$

The complex amplitudes of the electric and magnetic fields of the dipole are given by Eqs. (6) and (7), but with $\mathbf{r}$ replaced by the position vector of the field point relative to the dipole location. We indicate this vector by $\mathbf{r}_1$. Let $\mathbf{r}$ be the position vector of the field point where we wish to evaluate the fields. We see from **Figure 2** that

$$\mathbf{r}_1 = \mathbf{r}–H\mathbf{e}_z. \tag{14}$$

The coordinates of the field point only come in through $\hat{\mathbf{r}}$ and q in Eqs. (6) and (7), so we replace $\hat{\mathbf{r}}$ by $\hat{\mathbf{r}}_1$, and q by $q_1 = k_o r_1$. Similarly, the fields of the image dipole are found by replacing $\mathbf{r}$ by

$$\mathbf{r}_2 = \mathbf{r} + H\mathbf{e}_z, \tag{15}$$

q by $q_2 = k_o r_2$, and $\hat{\mathbf{u}}$ by $\hat{\mathbf{u}}^{\mathrm{im}}$. With the fields $\mathbf{e}(\mathbf{r})$ and $\mathbf{b}(\mathbf{r})$ constructed, the Poynting vector follows from Eq. (10). It is easy to verify that for $z = 0$, e.g., just above the mirror surface, the Poynting vector is along the surface.

4. Computation of field lines

With the above method, the Poynting vector $\boldsymbol{\sigma}(\mathbf{r})$ can be computed for a given field point $\mathbf{r}$. We shall use dimensionless Cartesian coordinates $\overline{x} = k_o x$, $\overline{y} = k_o y$ and $\overline{z} = k_o z$, and the dimensionless distance $h = k_o H$. The Poynting vector is a function of $\overline{x}$, $\overline{y}$ and $\overline{z}$. Therefore, it is better to write $\boldsymbol{\sigma}(\mathbf{q})$ instead of $\boldsymbol{\sigma}(\mathbf{r})$, with $\mathbf{q} = k_o\mathbf{r}$. We shall do so from now on. The only free parameters are h and γ. A field line of the vector field $\boldsymbol{\sigma}(\mathbf{q})$ is a curve such that for each point $\mathbf{q}$ on the curve the vector $\boldsymbol{\sigma}(\mathbf{q})$ is on the tangent line. Let $\mathbf{q}(t)$ be a parameter representation of a field line, with t an arbitrary dummy variable. Then $\mathbf{q}(t)$ must be a solution of

$$\frac{\mathrm{d}}{\mathrm{d}t}\mathbf{q}(t) = \boldsymbol{\sigma}(\mathbf{q}(t)). \tag{16}$$

In Cartesian coordinates this becomes:

$$\frac{\mathrm{d}}{\mathrm{d}t}\overline{x}(t) = \boldsymbol{\sigma}_x(\overline{x}(t), \overline{y}(t), \overline{z}(t)), \tag{17}$$

and similarly for $\overline{y}(t)$ and $\overline{z}(t)$, so this is a set of three differential equations. The independent variable t does not appear explicitly, and such equations are called autonomous. The solution is determined by an initial point $(\overline{x}_o, \overline{y}_o, \overline{z}_o)$.. The field line through this point is found by solving Eq. (16). We set $t = 0$ at the initial point. The direction of the field line is the direction along the curve that follows from increasing t. So, the field line runs from the initial point into the direction that corresponds to the solution with $t > 0$. The solution with $t < 0$ is the part of the field line that runs toward the initial point.

Obviously, the differential Eq. (16) will in general need to be solved numerically. An interesting exception is the case for an arbitrary (elliptical) dipole in free space for which an analytical solution can be obtained, as reported in Ref. [6]. We use *Mathematica* to solve the set and produce the field line pictures. For two-dimensional problems, the routine StreamPlot only requires the expression for $\boldsymbol{\sigma}$ as a function of $\overline{x}$, $\overline{y}$ and $\overline{z}$. The initial points $(\overline{x}_o, \overline{y}_o, \overline{z}_o)$ are called StreamPoints and can be specified. Alternatively, one can let *Mathematica* select these initial points. This is much faster, and works well to get an initial picture. However, in this approach *Mathematica* cuts off field lines when they get too close together, and that does not necessarily look too good.

Finally, field lines are determined by the direction of vector $\boldsymbol{\sigma}$ at a point $\mathbf{q}$, and not by its direction. Therefore, the vector fields $\boldsymbol{\sigma}(\mathbf{q})$ and $f(\mathbf{q})\boldsymbol{\sigma}(\mathbf{q})$, with $f(\mathbf{q})$ an arbitrary positive function of $\mathbf{q}$, have the same field lines. This can also be seen by making the change of variables $t = f(\mathbf{q})t'$. We then get

$$\frac{\mathrm{d}}{\mathrm{d}t}\mathbf{q}(t) = f(\mathbf{q}(t))\sigma(\mathbf{q}(t)), \tag{18}$$

and this equation has the same solutions for the field lines as Eq. (16). It just gives a different parametrization of the curves. A popular choice is $f(\mathbf{q}) = 1/|\boldsymbol{\sigma}(\mathbf{q})|$, which makes the right-hand side of Eq. (18) a unit vector. For the mirror problem, the fields diverge in the neighborhood of the dipole, and a good choice seems to be $f(\mathbf{q}) = q_1^5$.

5. Field lines in the symmetry plane

The dipole direction vector $\hat{\mathbf{u}}$ and the direction of its image $\hat{\mathbf{u}}^{\mathrm{im}}$ are both in the yz-plane. If we take a field point in the yz-plane, then also $\hat{\mathbf{r}}$ is in the yz-plane, and therefore, $\mathbf{e}(\mathbf{r})$ from Eq. (6) is in the yz-plane, and the same holds for the complex amplitude of the image electric field. From

http://dx.doi.org/10.5772/66459

Eq. (7), we see that the complex amplitudes of the magnetic dipole field and the magnetic image field are along the x-axis if the field point is in the yz-plane. Therefore, the Poynting vector from Eq. (10) is in the yz-plane if the field point is in the yz-plane. Consequently, any field line through a point in the yz-plane stays in the yz-plane. This plane is the symmetry plane for the dipole near the mirror. For field lines off the symmetry plane the field lines will be 3D curves, and the flow pattern is reflection symmetric in the yz-plane. Field lines in 3D are difficult to visualize, so we shall only consider field lines in the symmetry plane.

Figure 3 shows the flow lines of energy for a dipole oscillating along the z-axis ($\gamma = 0$). The field line pattern is rotation symmetric around the z-axis. The field lines that run toward the surface smoothly bend away from the surface upon approach. At the origin of coordinates we have a singularity, since the field lines that come out of the dipole and run straight down split there. This point is marked with a white circle. **Figure 4** shows the flow pattern for a dipole oscillating parallel to the surface ($\gamma = \pi/2$). The pattern is reflection symmetric in the z-axis, so only the region $\overline{y} < 0$ is shown. Just left of the dipole are two singularities. It seems that a field

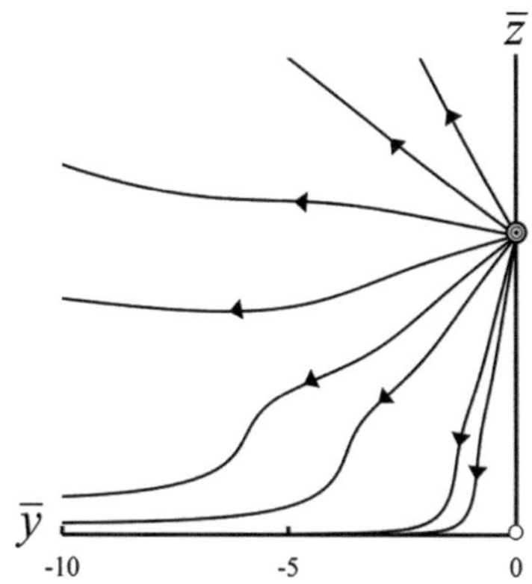

Figure 3. The dipole is located at a distance $h = 2\pi$ above the mirror, and the dipole oscillates along the z-axis.

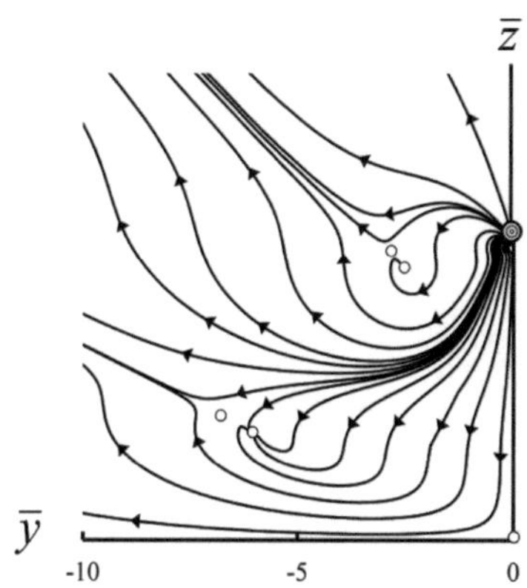

Figure 4. The dipole is located at a distance $h = 2\pi$ above the mirror, and the dipole oscillates parallel to the xy-plane.

line ends at the lower singularity, but this is essentially a minuscule vortex. The higher of the two is a singularity where field lines "collide." There is again a singularity right below the dipole on the mirror surface, and two other singularities appear at greater distances.

Figure 5 shows the field lines for a dipole oscillating under 45° with the z-axis ($\gamma = \pi/4$). Numerous singularities and vortices appear for this case. Vortex c is close to the dipole, and is similar to the small vortex in **Figure 4**. Vortices a and b have a greater extent, although they are still of subwavelength dimension. Some field lines come out of the dipole and swirl around vortex a, and some pass by vortex b. Interestingly, there are field lines that emanate from vortex a and end up at the center of vortex b. This seems to represent energy flowing from vortex a to vortex b, but that is not the case. Only the dipole is a source for the vector field of the Poynting vector. The singularity at the point labeled S seems like a bump in the road for the field lines that pass nearby. It can be shown analytically [7] that this singularity is a point on a singular circle in the plane of the mirror. The circle goes through the origin of coordinates and singularity S, and singularity S is located at $\overline{y} = -h \tan \gamma$. For $\gamma = \pi/4$ this is at $\overline{y} = -h$, and for the case of **Figure 5** this is at $\overline{y} = -h = -2\pi$. To the right of the z-axis, not shown in **Figure 5**, there seems to be no interesting structure.

When the dimensionless distance h between the dipole and the surface increases, so does the number of vortices. In **Figures 3–5**, this distance was taken as $h = 2\pi$, corresponding to one wavelength. In **Figure 6**, this distance is $h = 8\pi$, and angle γ is the same as in **Figure 5**. We notice

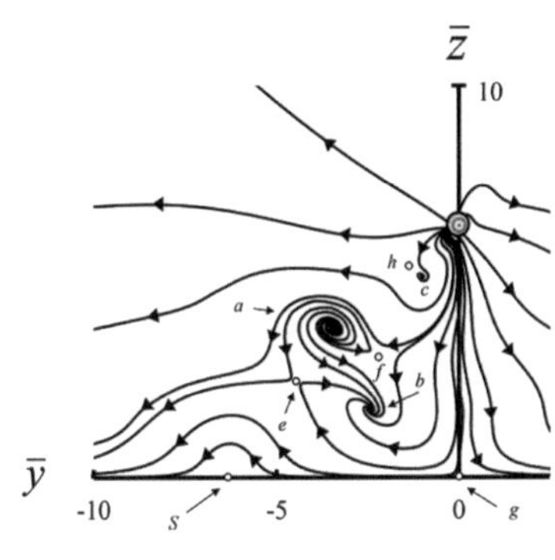

Figure 5. The dipole is located at a distance $h = 2\pi$ above the mirror, and the dipole oscillates under 45° with the z-axis.

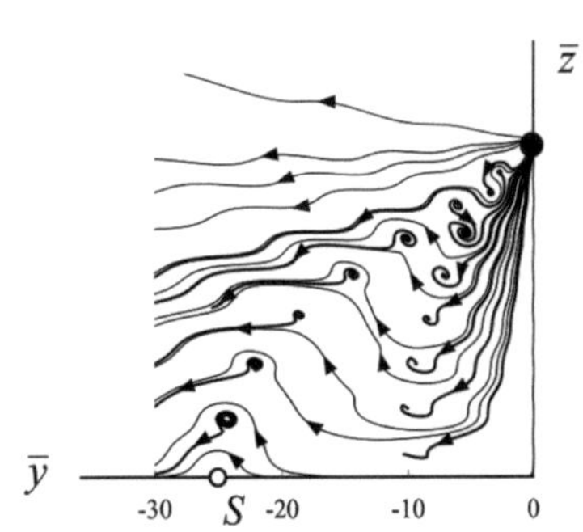

Figure 6. The dipole is located at a distance $h = 8\pi$ above the mirror, and the dipole oscillates under 45° with the z-axis.

http://dx.doi.org/10.5772/66459

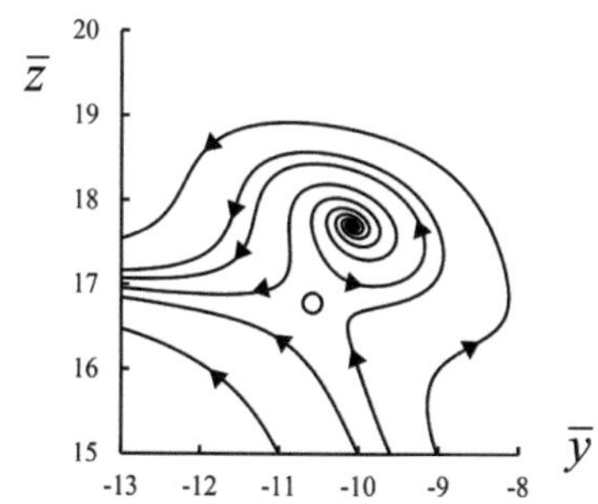

Figure 7. Enlargement of a vortex of **Figure 6**.

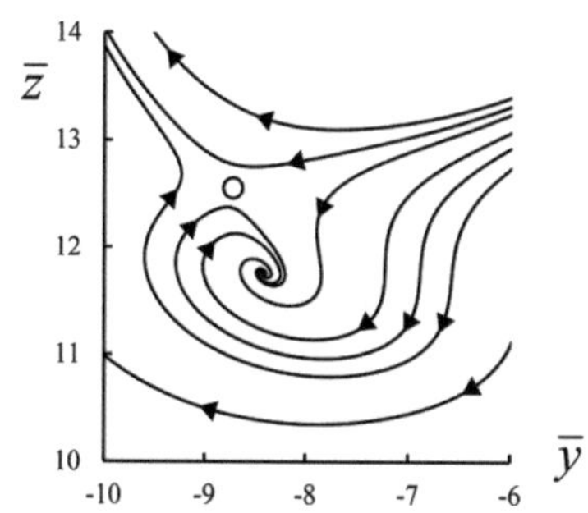

Figure 8. Enlargement of a vortex of **Figure 6**.

numerous vortices in the flow line pattern. The field lines rotate counterclockwise around the singularities that are close to the z-axis. **Figure 7** shows an enlargement of one of these vortices, and it follows from the pattern that there must be another singularity nearby. The vortices on the left have a clockwise rotation and **Figure 8** shows an enlargement. For the vortices on the left the field lines come out of the center of the vortex, and for the vortices on the right the field lines end at the center of the vortex, as in **Figures 7** and **8**, respectively. In **Figure 6**, the field lines that start or end at a vortex are drawn in bold. Many other field lines are present. They run from the dipole to infinity, either without coming in the neighborhood of the vortices (the four field lines on top of the picture) or they cross the array of vortices once or twice.

6. Location of the vortices

At a singularity, the Poynting vector vanishes. This can be due to $\mathbf{E}(\mathbf{r}) = 0$ or $\mathbf{B}(\mathbf{r}) = 0$ or $\mathbf{E}(\mathbf{r}) \times \mathbf{B}(\mathbf{r})^*$ is imaginary. We have found that for singularities that appear when field lines split or collide, $\mathbf{E}(\mathbf{r}) \times \mathbf{B}(\mathbf{r})^*$ is imaginary. Singularities at the center of a vortex are due to the vanishing of the magnetic field. For field lines in the xy-plane, the complex amplitude of the magnetic field is along the x-axis. For $\mathbf{B}(\mathbf{r})_x$ to be zero, both the real and imaginary part have to be zero at the same point. With the expressions for the magnetic field of source and image, we find that $\text{Re}[\mathbf{B}(\mathbf{r})_x] = 0$ leads to

$$\frac{q_1 \cos q_1 - \sin q_1}{q_1^3}[\overline{y}\cos\gamma + (h - \overline{z})\sin\gamma] + \frac{q_2\cos q_2 - \sin q_2}{q_2^3}[\overline{y}\cos\gamma + (h + \overline{z})\sin\gamma] = 0, \tag{19}$$

and similarly $\mathrm{Im}[\mathbf{B}(\mathbf{r})_x] = 0$ gives

$$\frac{q_1 \sin q_1 + \cos q_1}{q_1^3}[\overline{y}\cos\gamma + (h - \overline{z})\sin\gamma] + \frac{q_2\sin q_2 + \cos q_2}{q_2^3}[\overline{y}\cos\gamma + (h + \overline{z})\sin\gamma] = 0. \tag{20}$$

The solutions of Eq. (19) are curves in the yz-plane, and Eq. (20) also represents a set of curves in the yz-plane. At intersections between these sets of curves the magnetic field is zero, and this corresponds to the center of a vortex. **Figure 9** shows the curves for the same parameters as

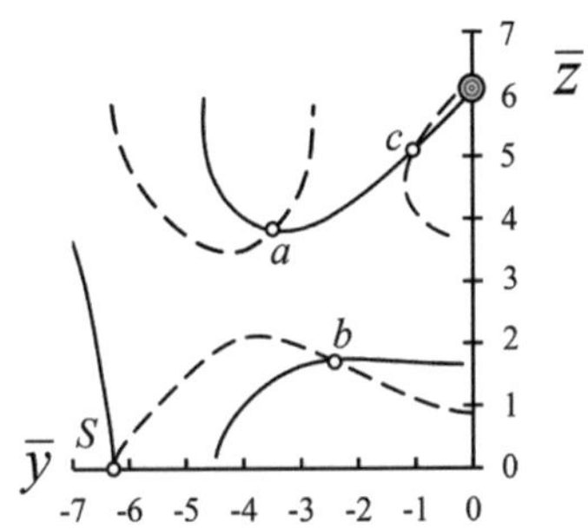

Figure 9. The vortices of **Figure 5** appear at intersections between the solid and dashed curves.

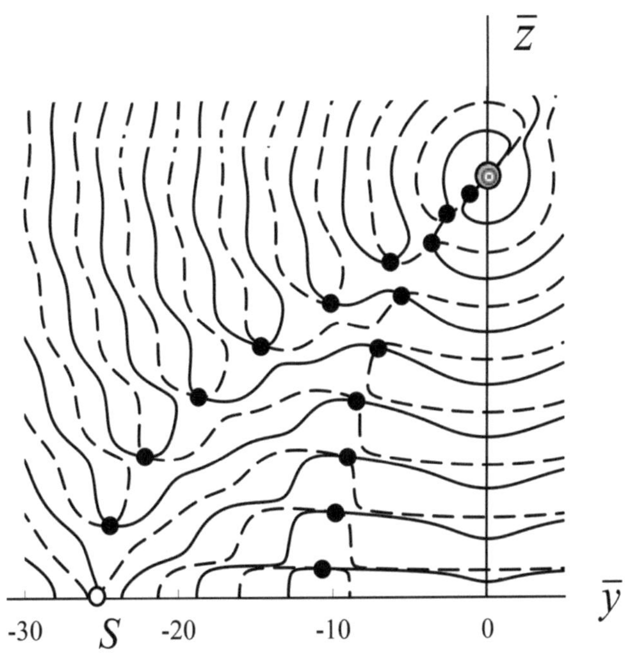

Figure 10. The vortices of **Figure 6** appear at intersections between the solid and dashed curves, and these intersections are indicated by black dots.

http://dx.doi.org/10.5772/66459

shown in **Figure 5**. The solid lines are the solutions of Eq. (19) and the dashed lines are the solutions of Eq. (20). The three intersections a, b, and c are the centers of the three vortices in **Figure 5**. Interestingly, at the center of the bump on the mirror, labeled S, the magnetic field also vanishes. This point is located at

$$\overline{y} = -h \tan \gamma, \quad \overline{z} = 0, \tag{21}$$

and it is easy to check that this is indeed a solution of both Eqs. (19) and (20). Here, the magnetic field is zero, but there is no vortex. For the case of the parameters for **Figure 6**, the solutions of Eqs. (19) and (20) are shown in Figure 10.

7. Vortex strings

In **Figure 10**, the intersections between the solid and dashed curves are not always precisely to determine and this gets worse with increasing h. By manipulating Eqs. (19) and (20), a different set of equations can be obtained. We find [8]

$$\begin{aligned}&[(q_1 q_2 + 1) \cos (q_2 - q_1) + (q_2 - q_1) \sin (q_2 - q_1)][\overline{y} \cos \gamma + (h + \overline{z}) \sin \gamma] \\ &\quad + \left(\frac{q_2}{q_1}\right)^3 (q_1^2 + 1)[\overline{y} \cos \gamma + (h - \overline{z}) \sin \gamma] = 0,\end{aligned} \tag{22}$$

$$(q_1 q_2 + 1) \sin (q_2 - q_1) = (q_2 - q_1) \cos (q_2 - q_1). \tag{23}$$

The solid curve in **Figure 11** is the solution of Eq. (22) and the dashed curves are the solutions of Eq. (23). Vortices appear at the intersections. The parameters are the same as for **Figure 10**. Interestingly, Eq. (23) is independent of the orientation angle γ of the dipole. When γ varies, the solid curve rotates with it, but the dashed curves stay the same. The solid curve starts at the location of the dipole and runs to point S on the mirror, as can easily be checked from Eq. (21).

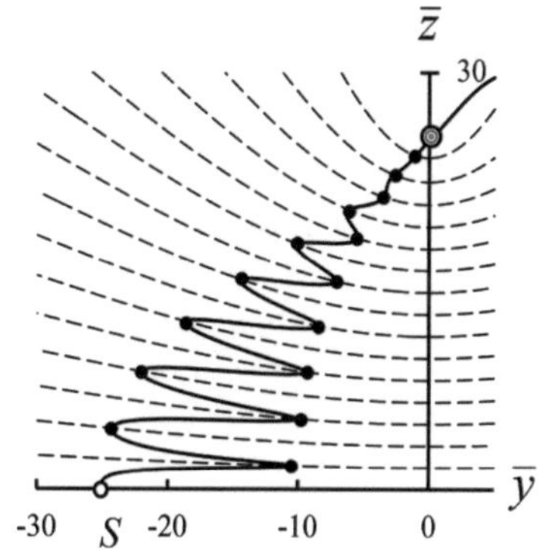

Figure 11. The black dots are the intersections between the solid curve and the dashed curves. These points correspond to the location of vortices, and here we have used the same parameters γ and h as for **Figure 10**.

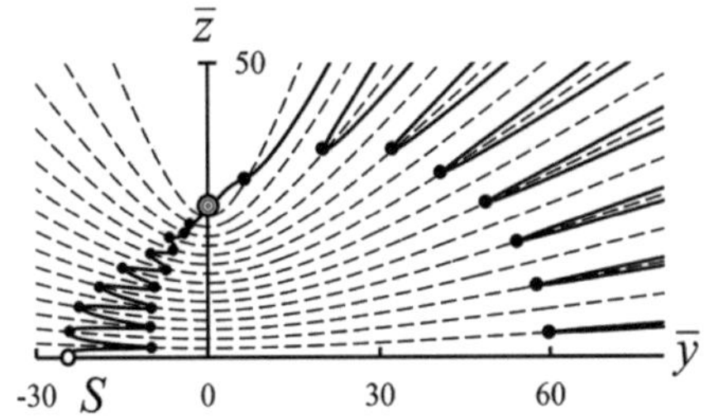

Figure 12. The figure shows a larger view of the picture in **Figure 11**.

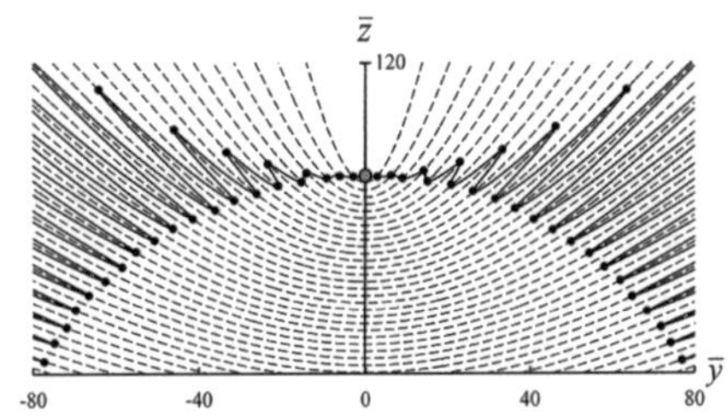

Figure 13. The figure shows vortex strings for $\gamma = \pi/2$ and $h = 30\pi$.

The vortices in **Figure 11** appear to lie on two 'strings'. The left string starts at the dipole and runs to point *S* on the mirror. The second string is in between the left string and the *z*-axis. It can be seen from **Figure 6** that all the vortices on the left string have a counterclockwise rotation, as in **Figure 7**. On the right string the vortices have a clockwise rotation, as in **Figure 8**. **Figure 12** shows a larger view of the same graph as in **Figure 11**. The solid line passes the dipole and continues in the upper right part of the graph. It appears that far away there are also intersections between the solid curve and the dashed curves, and these correspond also to the location of vortices. The vortices appear to lie on a third string. It can be shown that these vortices have a counterclockwise rotation.

The left string ends at point S on the mirror. The location of this point depends on the dipole angle γ, according to Eq. (21). When γ increases, the point moves to the left, and for $\gamma \to \pi/2$ the point moves to infinity. **Figure 13** shows the string pattern for $\gamma = \pi/2$ and $h = 30\pi$. The field line pattern must be reflection symmetric in the *z*-axis, and so is the string pattern. We see that a fourth string of vortices appears, and from symmetry it follows that these have a clockwise rotation. When we look again at **Figure 12**, the solid curve must intersect the dashed curves outside the picture in the upper right corner. Therefore, there is a fourth string very far away, and outside the picture. Apparently, there are always four vortex strings in electric dipole radiation near a mirror.

8. Dielectric interface

An interesting generalization of the free dipole near the mirror is the case of a dipole embedded in a dielectric medium, and near an interface with another dielectric material. For this

http://dx.doi.org/10.5772/66459

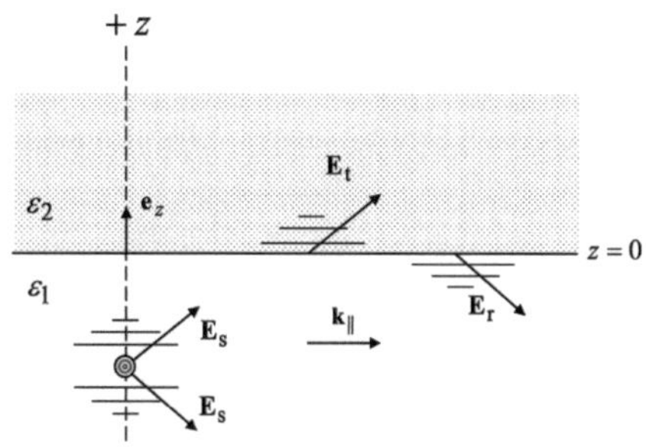

Figure 14. The figure shows the setup for a dipole embedded in a dielectric material and near an interface with another dielectric material.

problem, we reverse the z-axis, as compared to **Figure 2**, and we place the dipole on the negative z-axis, at a distance H below the interface. This is illustrated in **Figure 14**. The dielectric constant of the embedding medium is ε_1 and the substrate has dielectric constant ε_2. The corresponding indices of refraction are $n_1 = \sqrt{\varepsilon_1}$ and $n_2 = \sqrt{\varepsilon_2}$, respectively. The analysis of this system is considerably more complicated than the mirror problem. For simplicity of notation, we shall assume that the dipole oscillates along the z-axis, e.g., $\hat{\mathbf{u}} = \mathbf{e}_z$. The approach here is to represent the source fields (6) and (7) by an angular spectrum of plane waves. The reflection and transmission of each partial wave is accounted for by the appropriate Fresnel coefficients. A partial wave can be traveling or evanescent, and in **Figure 14** this is schematically indicated by vectors and dashed parallel lines, respectively. In each triad of partial source, reflected and transmitted waves the parallel component of each of the three wave vectors is the same. Upon reflection and transmission only the z-component of a wave vector can change. Set $\alpha = k_{||}/k_o$, which is the dimensionless magnitude of the parallel component of each wave vector. The dimensionless z-components of the wave vectors can be expressed in terms of the functions

$$v_i = \sqrt{n_i^2 - \alpha^2}\,, \quad i = 1, 2. \tag{24}$$

The wave vector of an incident partial wave of the source field has a z-component of $k_o v_1$. Similarly, the reflected and transmitted waves have wave vectors with z-components $-k_o v_1$ and $k_o v_2$, respectively. For $\alpha < n_i$, v_i is real, and the corresponding wave is traveling. For $\alpha > n_i$, v_i is positive imaginary, and the corresponding wave is evanescent. The Fresnel reflection and transmission coefficients for an incident p-polarized plane wave with parameter α are

$$R_p(\alpha) = \frac{\varepsilon_2 v_1 - \varepsilon_1 v_2}{\varepsilon_2 v_1 + \varepsilon_1 v_2}, \tag{25}$$

$$T_p(\alpha) = \frac{n_2}{n_1} \frac{2\varepsilon_1 v_1}{\varepsilon_2 v_1 + \varepsilon_1 v_2}. \tag{26}$$

For the dipole oscillating along the z-axis, all partial waves are p polarized.

The setup is rotation symmetric around the z-axis, so we only need to consider the solution in the yz-plane, with $y > 0$. Then the electric field is in the yz-plane, the magnetic field is along the

x-axis, and the Poynting vector is in the yz-plane. We obtain for the complex amplitudes of the dimensionless electric and magnetic reflected fields [9]

$$\mathbf{e}_r(\mathbf{r}) = \frac{1}{\varepsilon_1}\int_0^\infty d\alpha\, \alpha^2 R_p(\alpha)\left[\mathbf{e}_z \frac{i\alpha}{v_1} J_0(\alpha\overline{y}) - \mathbf{e}_y J_1(\alpha\overline{y})\right] e^{iv_1(h-\overline{z})}, \tag{27}$$

$$\mathbf{b}_r(\mathbf{r}) = -\mathbf{e}_x \int_0^\infty d\alpha \frac{\alpha^2}{v_1} R_p(\alpha) J_1(\alpha\overline{y}) e^{iv_1(h-\overline{z})}, \tag{28}$$

and for the transmitted fields we find

$$\mathbf{e}_t(\mathbf{r}) = \frac{1}{n_1 n_2}\int_0^\infty d\alpha \frac{\alpha^2}{v_1} T_p(\alpha)[\mathbf{e}_z i\alpha J_0(\alpha\overline{y}) + \mathbf{e}_y v_2 J_1(\alpha\overline{y})] e^{i(v_1 h + v_2\overline{z})}, \tag{29}$$

$$\mathbf{b}_t(\mathbf{r}) = -\mathbf{e}_x \frac{n_2}{n_1}\int_0^\infty d\alpha \frac{\alpha^2}{v_1} T_p(\alpha) J_1(\alpha\overline{y}) e^{i(v_1 h + v_2\overline{z})}. \tag{30}$$

Here, J_0 and J_1 are Bessel functions. These four fields are integral representations of the solutions, and, obviously, these integrals need to be computed numerically. On the z-axis we have $\overline{y} = 0$, and since $J_1(0) = 0$ we see that the magnetic fields vanish on the z-axis. The magnetic source field, Eq. (7), also vanishes on the z-axis. Therefore, the Poynting vector is zero on the z-axis, and so the z-axis is a singular line of the flow lines pattern.

For the case of **Figure 15**, we have $n_1 = 1$ and $n_2 = 2$. The energy flows from a thinner to a thicker medium. Upon transmission, the field lines of the Poynting vector bend toward the normal, just like optical rays would do for this case. However, the angle of incidence and the angle of transmission for the Poynting vector are not related by Snell's law, as the

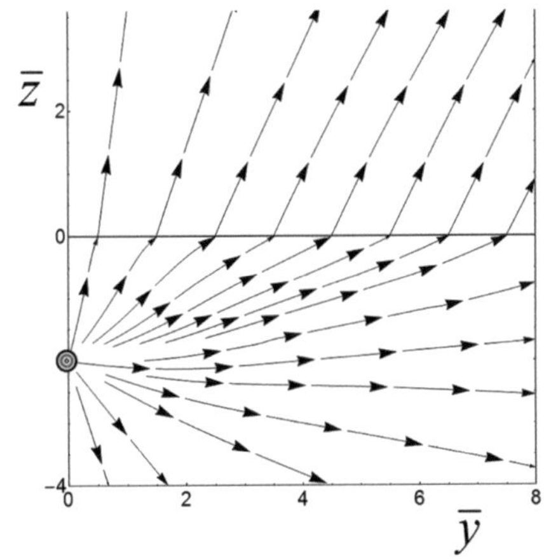

Figure 15. The figure shows the transmission through an interface into a thicker medium.

http://dx.doi.org/10.5772/66459

corresponding angles for optical rays are. The critical angle for this interface is 30°, and we see that away from the z-axis the field lines leave the interface under approximately 30°. This critical angle corresponds to an angle of incidence of 90°, and we see from the figure that the field lines do not approach the interface under 90°. The reason that the field lines leave the interface under approximately the critical angle is that the Fresnel transmission coefficient has a sharp maximum at this angle. The angular spectrum is a superposition of waves with all angles of incidence, and extends into the evanescent region as well. The transmission coefficient favors partial waves that approach the interface under about 90°. These are the partial waves that are on the borderline of the traveling and the evanescent regions of the angular spectrum.

More interesting is the case for transmission into a thinner medium, as illustrated in **Figure 16**. The indices of refraction are $n_1 = 2$ and $n_2 = 1$. On crossing the interface, the field lines bend away from the normal, just like optical rays. We see here that the field lines are more curved than in **Figure 15**. Some of the field lines that entered the n_2 medium bend down, and then cross the interface again. The field lines run through the n_1 material for a short distance and then bend up and cross the interface again into the n_2 region. This oscillation of energy back and forth through the interface persists over long distances. **Figure 17** shows an enlargement of the first dip of the field lines below the interface. We see that a vortex appears in medium n_2,

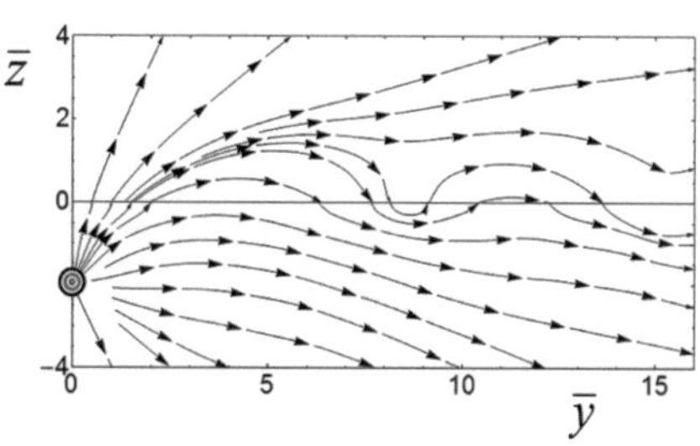

Figure 16. The figure shows the transmission through an interface to a thinner medium.

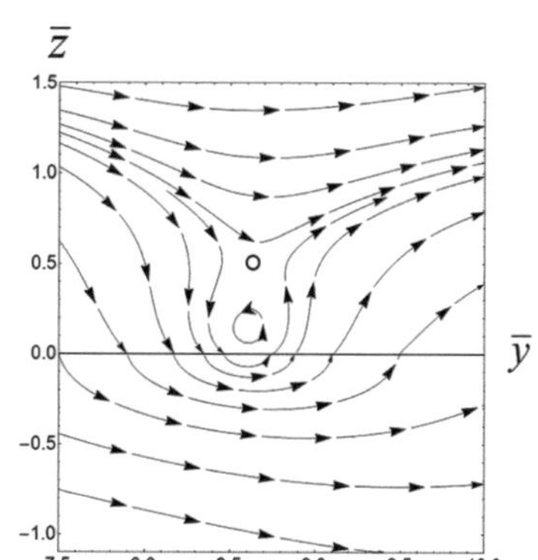

Figure 17. Enlargement of a part of **Figure 16.** Just above the first dip of the field lines under the interface a vortex appears. Above the vortex is a singularity, indicated by a white circle.

just above the dip of the field lines under the interface. And just above the vortex, we necessarily have a singularity because near this point the field lines run in opposite directions.

9. Dipole in between mirrors

An interesting variation of the mirror problem is the case for a dipole in between parallel mirrors, as depicted in **Figure 18**. A second mirror is placed at a distance D above the first mirror. Now, the dipole has a mirror image in both mirrors. In order to satisfy the boundary conditions at both mirrors, a mirror image must have again an image in the other mirror, and so on. This leads to an infinite sequence of images. Let us label the images with m. We then find that the images are located at

$$z_m = (m + \frac{1}{2})D + (-1)^m (H - \frac{1}{2}D), \tag{31}$$

with m integer. For $m = 0$, this is the actual dipole in **Figure 18**. The image with $m = -1$ is the image dipole from **Figure 2**. Images with m even have a dipole orientation vector $\hat{\mathbf{u}}$ and images with m odd have $\hat{\mathbf{u}}^{\mathrm{im}}$ as orientation vector.

This can be combined as

$$\hat{\mathbf{u}}_m = (-1)^m \mathbf{e}_y \sin\gamma + \mathbf{e}_z \cos\gamma, \tag{32}$$

for the orientation of image m. The dimensionless distance between the mirrors is denoted by $\delta = k_o D$.

Figure 19 shows the energy flow pattern for a horizontal dipole midway between the mirrors. The distance between the mirrors is $\delta = 4\pi$. Most of the radiation is emitted in the vertical direction. It travels to the mirrors, and there the field lines bend and continue horizontally. No radiation is emitted along the dipole axis, and we see that the line $\bar{z} = 2\pi$ is a singular line. Field lines approach this line from above and below, and then stop at this line. Consequently, the Poynting vector has to be zero on this line. For **Figure 20**, we have $\gamma = \pi/4$ and $\delta = \pi$, and the dipole is midway between the mirrors. We notice the appearance of two vortices very close to the dipole (subwavelength distance), and the rotation is counterclockwise for both. Some of

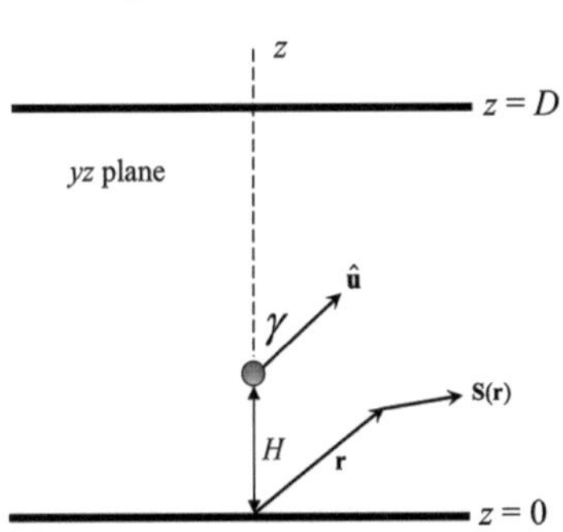

Figure 18. The figure shows the setup for the dipole in between mirrors.

the field lines run downward to the right swing up turn around (outside the figure) and end up running to the left. Similarly, some field lines start toward top-left, and then turn around, swing by the dipole, and then run off to the right. In **Figure 21**, we have again a horizontal

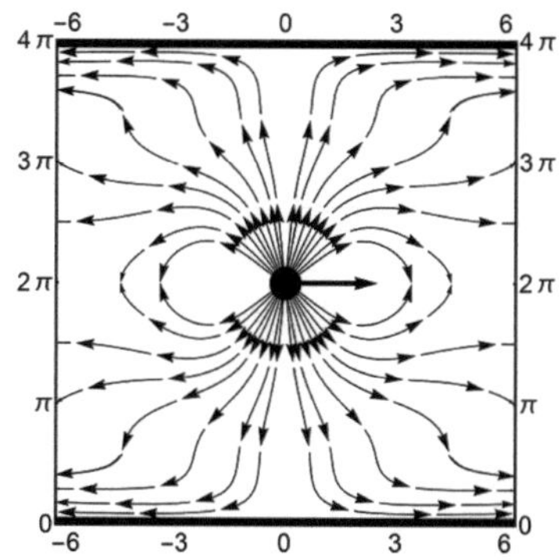

Figure 19. The figure shows the flow lines of energy for a horizontal dipole midway between the mirrors.

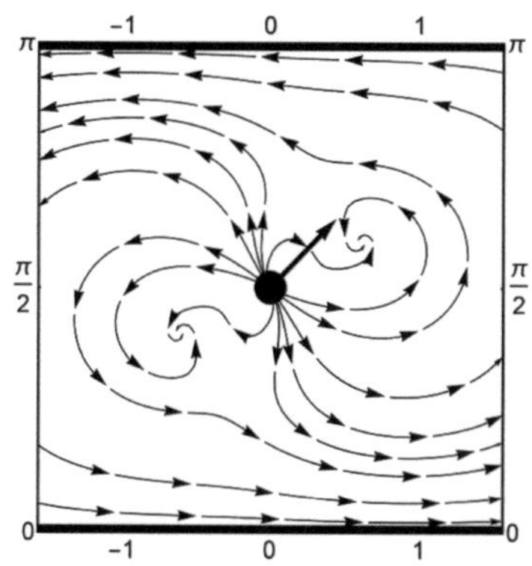

Figure 20. The figure shows the flow lines of energy for a dipole midway between the mirrors, and oscillating under 45° with the z-axis.

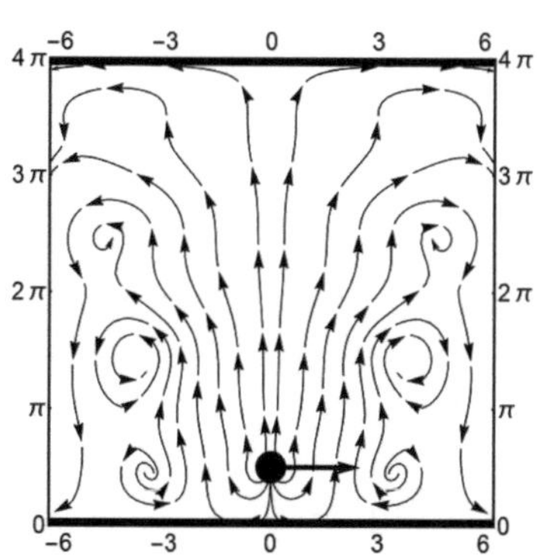

Figure 21. The figure shows the flow lines of energy for a dipole close to the lower mirror and oscillating horizontally.

dipole, but it is now closer to the mirror on the bottom. Numerous vortices appear in the flow pattern.

10. Conclusions

An oscillating electric dipole in free space emits its energy along straight lines. Most radiation is emitted perpendicular to the dipole axis, and none comes out along the dipole axis. We have studied the effect of a nearby interface on this flow pattern. Reflection of radiation at the interface leads to interference between the directly emitted radiation and the reflected radiation. A mirror is impenetrable for radiation, and so all radiation bounces back at the interface. This also implies that the field lines of energy flow must be parallel to the mirror at the mirror surface. This effect is shown in **Figure 3** for a dipole oscillating perpendicular to the surface, and one wavelength away from the surface. The radiation comes out of the dipole, more or less as for emission in free space, but at the mirror surface the field lines bend, and the energy flows away along the mirror surface. For a dipole oscillating parallel to the surface, a typical flow pattern is shown in **Figure 4**. Again, at the mirror surface the field lines run away parallel to the surface, but in between the surface and the dipole several singularities appear, and there is also a vortex very close to the dipole. For the case shown in **Figure 5**, the dipole oscillates under 45° with the normal to the surface, and we see that two large vortices appear and one very small one. The rotation direction of the energy flow in the two large vortices is in opposite directions, and some energy flows from one vortex to the other. When the distance between the dipole and the surface is much larger than a wavelength, numerous tiny (subwavelength) vortices appear, and we found that the vortices are located on a set of four strings. This is shown most clearly in **Figure 13**.

When the surface is an interface between two dielectrics, we also need to consider the radiation transmitted into the substrate. Here, we only consider the simplest case of a dipole oscillating perpendicular to the interface. This can be generalized to arbitrary oscillation directions, and also to the case where the surface is an interface with a layer of material, and this layer is located on a substrate of yet another kind of material [10]. **Figure 15** illustrates a typical case of transmission into a thicker medium. The field lines bend toward the normal, just like optical rays would. However, the refraction angle for the flow lines does not follow Snell's law for optical rays. **Figure 16** shows field lines for transmission into a thinner medium. Now the field lines bend away from the normal, but some field lines bend so much that they return to the other side of the interface. There is oscillation of energy back and forth through the interface. An enlargement is shown in **Figure 17**, and we observe that a vortex appears just above the location where the energy goes back and forth through the interface.

We have also considered the case where the oscillating dipole is located in between two mirrors. For a horizontal dipole, the emitted radiation bends near the surfaces of the two mirrors, and then flows away horizontally, as shown in **Figure 19**. For the case in **Figure 20**, the dipole oscillates under 45° with the normal. Two vortices appear. Some of the radiation that is emitted to the bottom-right of the picture originally flows to the right along the surface of the lower mirror, but then turns around, swings by the dipole, and then continues to the left, along the surface of the top mirror. When the dipole is not located midway between the mirrors, as in **Figure 21**, numerous vortices appear, and the pattern repeats indefinitely to the left and right in the range outside the picture.

Author details

Xin Li[1], Henk F. Arnoldus[2*] and Zhangjin Xu[2]

*Address all correspondence to: hfa1@msstate.edu

1 Department of Physics, Millersville University, Millersville, PA, USA

2 Department of Physics and Astronomy, Mississippi State University, Mississippi State, MS, USA

References

[1] Born M, Wolf E. Principles of Optics. 6th ed. Pergamon: Oxford; 1980. Ch. 3.

[2] Griffiths D J. Introduction to Electrodynamics. 2nd ed. Upper Saddle River: Prentice Hall; 1989. p. 323.

[3] Li X, Pierce D M, Arnoldus H F. Redistribution of energy flow in a material due to damping. Opt. Lett. 2011; **36**: 349–351.

[4] Arnoldus H F, Foley J T. Transmission of dipole radiation through interfaces and the phenomenon of anti-critical angles. J. Opt. Soc. A. 2004; **21**: 1109–1117.

[5] Jackson J D. Classical Electrodynamics. 3rd ed. New York: Wiley; 1999. p. 411.

[6] Shu J, Li X, Arnoldus H F. Energy flow lines for the radiation emitted by a dipole. J. Mod. Opt. 2008; **55**: 2457–2471.

[7] Li X, Arnoldus H F. Electric dipole radiation near a mirror. Phys. Rev. A. 2010; **81**(053844): 1–10.

[8] Li X, Arnoldus H F. Vortex string in electric dipole radiation near a mirror. Opt. Commun. 2013; **305**: 76–81.

[9] Arnoldus H F, Berg M J, Li X. Transmission of electric dipole radiation through an interface. Phys. Lett. A. 2014; **378**: 755–759.

[10] Arnoldus H F, Berg M J. Energy transport in the near field of an electric dipole near a layer of material. J. Mod. Opt. 2015; **62**: 244–254.

Chapter 13

Spin-Wave Dynamics in the Presence of Magnetic Vortices

Sławomir Mamica

Additional information is available at the end of the chapter

http://dx.doi.org/10.5772/66099

Abstract

This chapter describes spin-wave excitations in nanosized dots and rings in the presence of the vortex state. The special attention is paid to the manifestation of the competition between exchange and dipolar interactions in the spin-wave spectrum as well as the correlation between the spectrum and the stability of the vortex. The calculation method uses the dynamic matrix for an all-discrete system, the numerical diagonalization of which yields the spectrum of frequencies and spin-wave profiles of normal modes of the dot. We study in-plane vortices of two types: a circular magnetization in circular dots and rings and the Landau state in square rings. We examine the influence of the dipolar-exchange competition and the geometry of the dot on the stability of the vortex and on the spectrum of spin waves. We show that the lowest-frequency mode profile proves to be indicative of the dipolar-to-exchange interaction ratio and the vortex stability is closely related to the spin-wave profile of the soft mode. The negative dispersion relation is also shown. Our results obtained for in-plane vortices are in qualitative agreement with results for core-vortices obtained from experiments, micromagnetic simulations, and analytical calculations.

Keywords: magnetic dot, in-plane vortex, spin waves, stability, dipolar-exchange competition

1. Introduction

One of the hottest topics nowadays are small magnetic dots and rings with a thickness in a range of few tens of nanometers and the diameter ranging from one hundred nanometers to a few micrometers. A strong interest in such systems originates from their potential applicability as well as rich physics [1]. The physical properties of magnetic nanodots are related mostly to the concurrence of two types of magnetic interactions, namely exchange and dipolar ones. Usually, the coexistence of long- and short-distance interactions leads to new phenomena, such as surface and subsurface localization of the spin waves in layered magnetic systems [2, 3], opening of the band gaps in magnonic crystals [4, 5], or splitting the spin-wave spectrum into subbands in patterned multilayers [6, 7]. In the case of exchange and dipolar interactions, the situation is even more interesting due to competitive effects of these two schematically shown in **Figure 1**.

The favorable alignment of two magnetic moments (also called them spins) coupled via exchange interaction depends on the sign of the so-called exchange integral, J, regardless of their mutual position. If $J > 0$ the spins are parallel (ferromagnetic, FM, coupling) while for $J < 0$ the spins are antiparallel (antiferromagnetic, AFM, coupling). Dipolar coupling, on the other hand, depends on the mutual positions of spins being FM if the spins are aligned one after another and AFM for spins alongside one another (see **Figure 1**). As a result, the ferromagnetic exchange interaction forces parallel configuration of spins leading to the magnetic monodomain whereas pure dipolar interaction leads to the in-plane alignment of spins and so-called labyrinth magnetic structures [8]. Additionally, the dipolar interaction is a long range one and consequently very sensitive for size and shape of the sample while the exchange interaction is local. Thus, the competition between these two also depends on the size and shape of the system.

The concurrence of these to competitive interactions is the origin of the variety of possible magnetic configurations and leads to the occurrence of magnetic vortices in nanosized dots and rings [9–12]. In the vortex configuration, a magnetization component lying in the plane of the dot forms a closure state. Depending on the shape of the system, this in-plane magnetization can be realized as a circular magnetization in circular dots and rings or as a Landau state (closure domain configuration) in square rings, as shown in **Figure 2a**. In square dots,

Figure 1. Exchange vs. dipolar interactions. Preferred configuration of magnetic moments depends on the sign of the exchange integral J for the exchange interaction while on the alignment of magnetic moments for the dipolar interaction.

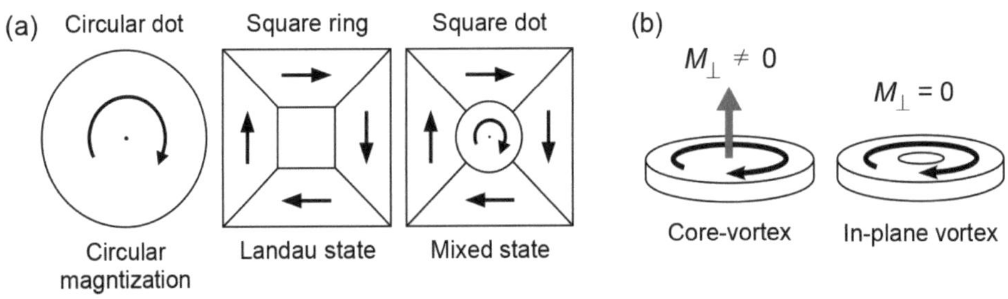

Figure 2. (a) Different preferred configurations of the in-plane magnetization component in dots of different shape. (b) Core-vortex vs. in-plane vortex.

according to the simulations [13], the magnetic configuration is a mixture of these two states: along borders Landau state appears, which is the effect of the minimization of the surface magnetic charges, while in the central part of the dot, the magnetization is circular as a result of the tendency to decrease the (local) exchange energy. The area of circular magnetization is relatively small; therefore, in large-square dots, the Landau state prevails in the major part of the dot. However, in small dots, the circular in-plane magnetization fails to fit the geometry of the system only in minor corner regions.

For strong exchange interaction, the circular in-plane configuration is not enough to minimize the exchange energy at the vortex center (which is not necessary the dot center, however, for the stable vortex its center is in close vicinity to the center of the dot). As a consequence, spins at the center are rotated from their in-plane alignment (forced by dipolar interactions) forming so-called vortex core, a tiny region with a nonzero out-of-plane component of magnetization (**Figure 2b**). In typical ferromagnets, such as cobalt or permalloy, the exchange interaction is strong thus in experiments the vortex core is observed [14–16]. In rings, the center of the vortex is removed from the sample, thus the magnetization lies in the plane of the dot throughout its volume [17] except rings with extremely small internal radius [18]. The potential applications of the magnetic vortex itself increase from the possibility of the switching of core polarity (up or down) and chirality (the direction of the in-plane magnetization: clockwise, CW, or counterclockwise, CCW), and these two can be switched independently [19, 20].

In square dots, beside the vortex core, domain walls appear as well at the borders between domains. Roughly speaking, there are two types of domain walls: with and without nonzero out-of-plane magnetization (Bloch and Néel type, respectively) [21]. Thus, in the first case, the total out-of-plane magnetization is not zero even without the vortex core. Consequently, the out-of-plane magnetization can differ from zero in square rings in which the core does not appear. As we will show later, the preferred type of domain walls depends on the competition between exchange and dipolar interactions.

There are two types of magnetic excitations in magnetic dots in the vortex state. First one is a gyrotropic mode, i.e., the precession of the vortex core around the dot center. This is a low-frequency excitation with the frequency usually in the range of hundreds of MHz, and it can be utilized to microwave generation [22, 23]. The second type are spin waves; high-frequency excitations with the frequency of several GHz [24]. The spin-wave excitations are normal modes of the confined magnetic system similar to the vibration of the membrane. They prove to be of a key importance for the vortex switching [25], can be used to generate the higher harmonics of the microwave radiation [26], and have a significant influence on the vortex stability [27, 28].

In this chapter, we study the stability of the magnetic vortex state and the spin-wave excitations spectrum in two-dimensional (2D) nanosized dots and rings in their dependence on the competition between dipolar and exchange interactions. We use a very efficient method based on the discrete version of the Landau-Lifshitz equation. Our theoretical approach is described in Section 2. In next sections, we present our results starting with the circular dot in which the in-plane circular vortex is assumed as a magnetic state. In Section 3, we analyze an exemplar spin-wave spectrum of the dot showing typical effects such as the negative dispersion relation and the influence of the lattice symmetry on the spin-wave spectrum. In Section 4, we examine the stability of the in-plane vortex vs. the dipolar-to-exchange interaction ration (d) and the size of the dot. The influence of the competition between dipolar end exchange interactions on the spin-wave spectrum of a dot is studied in Section 5. In next two sections, we consider the influence of the spin-wave profile of the soft mode on the vortex stability in circular (Section 6) and square rings (Section 7). Finally, we provide some concluding remarks in Section 8.

2. The model

The object of our study is a dot (ring) cut out of a 2D lattice of elementary magnetic moments (**Figure 3**). For circular dots, the external size L is defined as the number of lattice constants in the diameter of the circle used for cutting out the dot. The internal size of the ring, $L^{'}$, is the radius of the inner circle (in units of the lattice constant). For square rings, L means the number of lattice sites along the side of the square. Similarly, $L^{'}$ means the side of the removed square. In linear approximation used in this work, the magnetic moment $\boldsymbol{M_R}$, where $\boldsymbol{R}$ is the position vector, can be expressed as a sum of two components: static, $\boldsymbol{M}_{0,\boldsymbol{R}}$, and dynamic, $\boldsymbol{m_R}$, with the assumption that $|\boldsymbol{m_R}| \ll |\boldsymbol{M_R}|$, $|\boldsymbol{M}_{0,\boldsymbol{R}}| \simeq |\boldsymbol{M_R}|$, and $\boldsymbol{m_R} \perp \boldsymbol{M}_{0,\boldsymbol{R}}$. For any magnetic moment within the dot, we can define a local Cartesian coordinate system as follows: unit vector $\boldsymbol{i_R}$ is parallel to the static component $\boldsymbol{M}_{0,\boldsymbol{R}}$, unit vector $\boldsymbol{j_R}$ is oriented toward the vortex center lying in the plane of the dot, and unit vector $\boldsymbol{k_R}$ is the third Cartesian unit vector being perpendicular to the other two. In this coordinate system, a dynamic component of the magnetic moment is $\boldsymbol{m_R} = m_{j,\boldsymbol{R}}\boldsymbol{j_R} + m_{k,\boldsymbol{R}}\boldsymbol{k_R}$, where $m_{j,\boldsymbol{R}}$ and $m_{k,\boldsymbol{R}}$ we will refer to as in-plane and perpendicular coordinates of the magnetic moment, respectively. For in-plane vortices, the last component is always perpendicular to the plane of a dot.

The time evolution of any magnetic moment $\boldsymbol{M_R}$ is described by the damping-free Landau-Lifshitz (LL) equation, which in the linear approximation reads:

$$i\frac{\omega}{\gamma\mu_0}\boldsymbol{m_R} = \boldsymbol{M}_{0,\boldsymbol{R}} \times \boldsymbol{h_R} + \boldsymbol{m_R} \times \boldsymbol{H_R}, \tag{1}$$

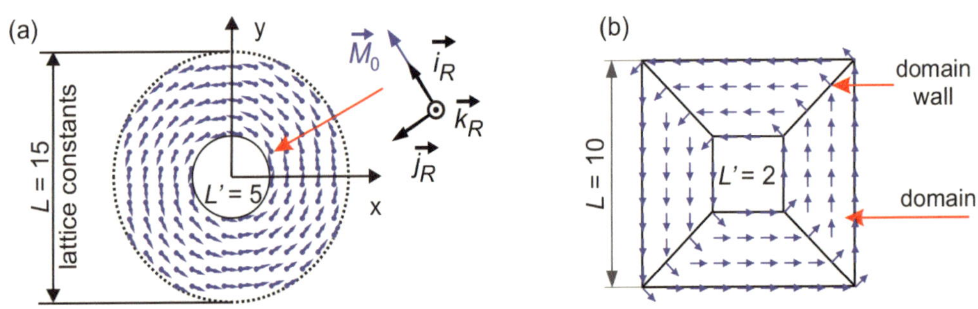

Figure 3. Schematic plots of two in-plane vortices typical for two types of rings: (a) a circular magnetization in a circular ring and (b) closure domains (Landau state) in a square ring. Both rings are based on a 2D square lattice with magnetic moments (represented by the arrows) arranged in the lattice sites. To the right in figure (a), the local coordinate system associated with the magnetic moment indicated by the arrow.

http://dx.doi.org/10.5772/66099

where i is the imaginary unit, γ is the gyromagnetic ratio, μ_0 is the vacuum permeability, and ω is the frequency of harmonic oscillations of $\boldsymbol{m}_R$. $\boldsymbol{H}_R$ and $\boldsymbol{h}_R$ are static and dynamic components of the effective field $\boldsymbol{H}_R^{\text{eff}} = \boldsymbol{H}_R + \boldsymbol{h}_R$ acting on the magnetic moment $\boldsymbol{M}_R$.

In this work, we consider exchange-dipolar systems only thus the effective field consists of two components:

$$\boldsymbol{H}_R^{\text{eff}} = \frac{2J}{\mu_0 (g\mu_B)^2} \sum_{\boldsymbol{R}' \in \text{NN}} \boldsymbol{M}_{R'} + \frac{1}{4\pi a^3} \sum_{\boldsymbol{R}' \neq \boldsymbol{R}} \left(\frac{3(\boldsymbol{R}'-\boldsymbol{R})\left(\boldsymbol{M}_{R'} \cdot (\boldsymbol{R}'-\boldsymbol{R})\right)}{|\boldsymbol{R}'-\boldsymbol{R}|^5} - \frac{\boldsymbol{M}_{R'}}{|\boldsymbol{R}'-\boldsymbol{R}|^3} \right).$$

The first term comes from the exchange interaction and can be derived from the Heisenberg Hamiltonian under the condition of uniform interactions. Since we restrict ourselves to nearest neighbor (NN) interactions the summation runs over NNs of the magnetic moment $\boldsymbol{M_R}$. Here J is the NN exchange integral, μ_B is the Bohr magneton, and g is the g-factor. The second term is a typical dipolar sum over all magnetic moments within the sample except $\boldsymbol{M_R}$. The position vectors $\boldsymbol{R}$ are expressed in the units of the lattice constant a.

From Eq. (1) one can derive the system of equations of motion for dynamic components of all magnetic moments as follows:

$$i\Omega m_{r,R} = - \sum_{\boldsymbol{R}' \in \text{NN}(\boldsymbol{R})} \boldsymbol{k}_R \cdot \boldsymbol{m}_{\boldsymbol{R}'} + m_{k,R} \sum_{\boldsymbol{R}' \in \text{NN}(\boldsymbol{R})} \boldsymbol{i}_R \cdot \boldsymbol{i}_{\boldsymbol{R}'}$$

$$-d\left(\sum_{\boldsymbol{R}' \neq \boldsymbol{R}} \left(\frac{3[(\boldsymbol{R}'-\boldsymbol{R}) \cdot \boldsymbol{k}_R][(\boldsymbol{R}'-\boldsymbol{R}) \cdot \boldsymbol{m}_R]}{|\boldsymbol{R}'-\boldsymbol{R}|^5} - \frac{\boldsymbol{k}_{\boldsymbol{R}} \cdot \boldsymbol{m}_{\boldsymbol{R}'}}{|\boldsymbol{R}'-\boldsymbol{R}|^3} \right) + m_{k,\boldsymbol{R}} \sum_{\boldsymbol{R}' \neq \boldsymbol{R}} \left(\frac{3[(\boldsymbol{R}'-\boldsymbol{R}) \cdot \boldsymbol{i}_R][(\boldsymbol{R}'-\boldsymbol{R}) \cdot \boldsymbol{i}_{\boldsymbol{R}'}]}{|\boldsymbol{R}'-\boldsymbol{R}|^5} - \frac{\boldsymbol{i}_{\boldsymbol{R}} \cdot \boldsymbol{i}_{\boldsymbol{R}'}}{|\boldsymbol{R}'-\boldsymbol{R}|^3} \right) \right)$$

$$i\Omega m_{k,\boldsymbol{R}} = \sum_{\boldsymbol{R}' \in \text{NN}(\boldsymbol{R})} \boldsymbol{j}_{\boldsymbol{R}} \cdot \boldsymbol{m}_{\boldsymbol{R}'} - m_{r,\boldsymbol{R}} \sum_{\text{R} \in \text{NN}(\boldsymbol{R})} \boldsymbol{i}_{\boldsymbol{R}} \cdot \boldsymbol{i}_{\boldsymbol{R}'}$$

$$+d\left(\sum_{\boldsymbol{R}' \neq \boldsymbol{R}} \left(\frac{3[(\boldsymbol{R}'-\boldsymbol{R}) \cdot \boldsymbol{j}_R][(\boldsymbol{R}'-\boldsymbol{R}) \cdot \boldsymbol{m}_{\boldsymbol{R}'}]}{|\boldsymbol{R}'-\boldsymbol{R}|^5} - \frac{\boldsymbol{j}_{\boldsymbol{R}} \cdot \boldsymbol{m}_{\boldsymbol{R}'}}{|\boldsymbol{R}'-\boldsymbol{R}|^3} \right) - m_{r,\boldsymbol{R}} \sum_{\boldsymbol{R}' \neq \boldsymbol{R}} \left(\frac{3[(\boldsymbol{R}'-\boldsymbol{R}) \cdot \boldsymbol{i}_R][(\boldsymbol{R}'-\boldsymbol{R}) \cdot \boldsymbol{i}_{\boldsymbol{R}'}]}{|\boldsymbol{R}'-\boldsymbol{R}|^5} - \frac{\boldsymbol{i}_{\boldsymbol{R}} \cdot \boldsymbol{i}_{\boldsymbol{R}'}}{|\boldsymbol{R}'-\boldsymbol{R}|^3} \right) \right), \quad (2)$$

where $\Omega = (g\mu_B\omega)/(2\gamma SJ)$ is the reduced frequency of a spin-wave excitation, S is the spin (we assume that all spins within the dot are the same thus any magnetic moment is equal $M_R = g\mu_B S$), and d is the only material parameter of the model referred to as a dipolar-to-exchange interaction ratio given by:

$$d = \frac{(g\mu_B)^2 \mu_0}{8\pi a^3 J}. \quad (3)$$

The above system of equations can be represented as an eigenvalue problem the matrix of which is called a *dynamic matrix*. The diagonalization of the dynamic matrix leads to the spectrum of frequencies and profiles of normal excitations of the dot. The spin-wave profile is a spatial distribution of the dynamic components of magnetic moments, i.e., the distribution of the amplitude of the magnetic moment precession. Dynamic components obtained from diagonalization are complex numbers with a phase shift $\pi/2$ between the real and imaginary part, which gives $T/4$ shift in time, where $T = 2\pi/\omega$ is a period of oscillations for a given mode.

Usually, the distribution of these components obtained for the same mode is similar and differ in the intensity only. Therefore, if the situation is clear, it is sufficient to provide one part (Re or Im) of the one component (in-plane or out-of-plane) to explain the character of the mode. The spin-wave profiles in circular dots are marked as (n, m) similarly to the vibrations of a membrane, i.e., accordingly to the number of nodal lines in the radial (n) and azimuthal (m) direction. The azimuthal modes occur in pairs: (n, −m) and (n, +m) with both modes of the same character; thus, in this work we denote them just (n, m), with m denoting |m|.

3. Spin-wave spectrum of a circular dot

In **Figure 4a** shows an example of the spin-wave spectrum obtained for a circular dot of the diameter $L = 101$. The dot is cut out from the square lattice and contains 8000 spins. A magnetic configuration is assumed to form an in-plane vortex. The spectrum is calculated for the dipolar-to-exchange interaction ratio $d = 0.42$. The shape of the spectrum is typical for exchange-dipolar systems: for low-frequency modes, the shape of the spectrum is determined by the dipolar interaction, while for the high frequencies by the exchange one. Of course, the spectrum is discrete, which is clearly seen in the inset where frequencies of 14 lowest modes

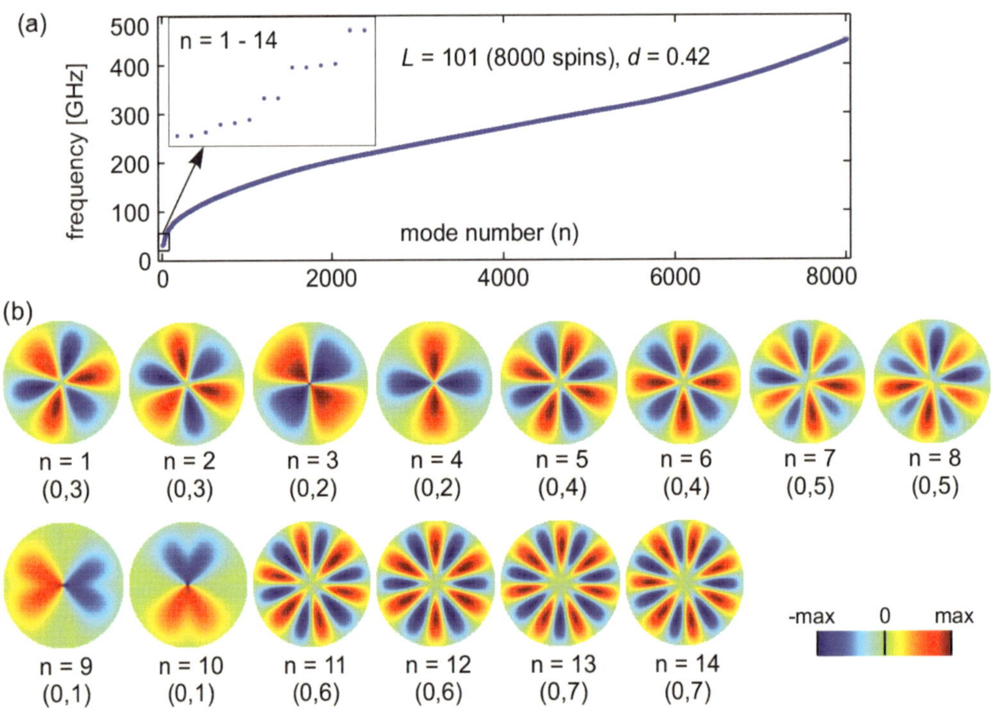

Figure 4. (a) An exemplar spin-wave spectrum calculated for the circular dot of the diameter $L = 101$ lattice constants consists of 8000 spins. The in-plane vortex configuration is assumed and the dipolar-to-exchange interaction ratio d is set to 0.42. The inset shows 14 lowest modes of the spectrum. (b) Spin-wave profiles of 14 lowest modes of the spectrum shown in (a).

are presented. Among these modes, one can distinguish a pairs of modes of the same frequencies. For example, modes 1 and 2, 7 and 8, 9 and 10, or 13 and 14 are degenerate in pairs. On the other hand, modes 3–6, 11 and 12 have unique frequencies.

To investigate this feature, we provide spin-wave profiles of the lowest eight modes in **Figure 4b**. As we can see that these 14 modes include seven pairs of modes with the same absolute value of the azimuthal number. Degenerate modes are of the odd azimuthal number: (0,3) modes 1 and 2, (0,5) modes 7 and 8, (0,1) modes 9 and 10, and (0,7) modes 13 and 14. In contrast, for even azimuthal numbers, degeneration is lifted. This originates from the discreteness of the lattice the dot is cut out from. If the symmetry of the profile matches the symmetry of the lattice, the degeneration is removed. For example, mode 3 has two nodal lines coincide with high-spin density lines (along the x and y axes in **Figure 3a**). Its counterpart, i.e., mode 4 is rotated by $\pi/4$ having antinodal lines along the high-spin density lines. This situation is analogous to the boundary of the Brillouin zone in the periodic system where the energy gap appears between two excitations: one having nodes in the potential wells and the other one having antinodes. Indeed, if the dot is based on the square lattice it can be considered as a system periodic in the azimuthal direction. A unit cell in this case corresponds to a quarter of the dot and is delimited by high-spin density lines. In such picture, one-half of the wavelength of modes (0,2) fits the unit cell with nodes or antinodes at the unit cell boundary. The same rule holds for hexagonal lattice where the degeneration is lifted if the azimuthal number is divisible by 3 [29]. It is worth to noting that there is also another type of degeneracy lifting caused by the coupling of the azimuthal modes with the gyrotropic mode [30, 31] which is not related to the discreteness of the dot and appears even for first-order azimuthal modes. In our work, this is not the case since we assume coreless vortex as a magnetic configuration.

For the dot under consideration, radial and azimuthal numbers are related to the wave vector in the corresponding direction. Thus, the spectrum shown in **Figure 4** exhibits negative dispersion relation for modes (0,1), (0,2), and (0,3), i.e., for this modes, the frequency decreases with an increase of azimuthal number. Such negative dispersion was also observed for core-vortices in circular dots experimentally [32, 33] and by means of analytical calculations [34, 35]. It was found that in a dot of a fixed thickness the increase in the diameter will cause the mode order to change, namely it will cause the negative dispersion to be stronger (the modes with higher azimuthal numbers will descend the spectrum). We show that this effect originates in the influence of the dipolar interaction regardless it is enhanced by the size of the system or by change of the dipolar-to-exchange interaction ratio.

4. Stability of the in-plane vortex

The dependence of the spin-wave spectrum on d is shown in **Figure 5** for the dot under consideration ($L = 101$, 8000 spins). For intermediate values of the dipolar-to-exchange interaction ratio, there are no zero-frequency modes in the spectrum which means that the assumed in-plane vortex is a (meta)stable magnetic configuration (see, e.g., our discussion in reference [36]). Going toward smaller values of d the exchange interaction gains the importance until

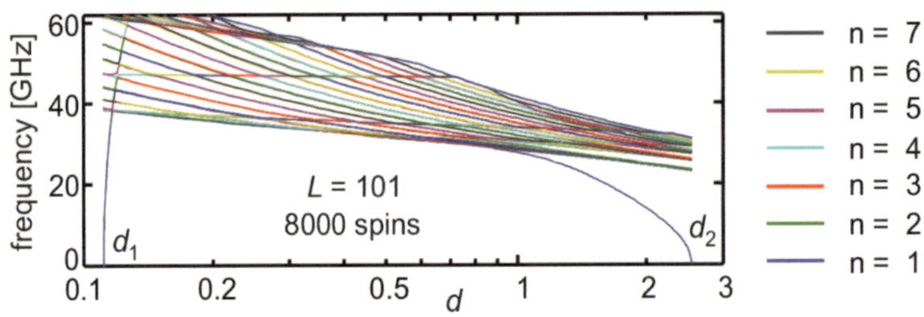

Figure 5. The frequency dependence on the dipolar-to-exchange interaction ratio d (in logarithmic scale) for 36 lowest modes in the spin-wave spectrum of the circular dot of the diameter $L = 101$ in the in-plane vortex state. The color assignment is indicated at the right; the colors repeat cyclically for successive modes. There are no zero-frequency modes between two critical values d_1 and d_2 which is indicative for the stability of the assumed magnetic configuration.

$d = d_1$. From this point, the frequency of the lowest mode is zero and the in-plane vortex is no more stable (or even metastable); the lowest mode becomes the nucleation mode responsible for the reorientation of the magnetic configuration. The profile of this mode reflects the tendency of the system to find a new stable state. Since this transition is forced by the exchange interaction we will call it the exchange-driven reorientation (transition). While d increases, which means the dipolar interaction gains the importance, another transition appears for $d = d_2$. In this case, the reorientation is caused by prevailing dipolar interaction; thus, it is referred to as dipolar-driven reorientation (transition). This behavior reflects the origin of the vortex state: the competition between dipolar and exchange interaction.

The importance of the dipolar interaction depends, besides its dependence on d, also on the size of the system. Therefore, the critical values of d should change with the dot size. **Figure 6a** shows critical values d_1 and d_2 vs. the number of spins, N, in which the dot consists of (which is equivalent to change of the dot diameter since the system is 2D). The critical value d_2 (for the dipolar driven reorientation) clearly depends on N, especially for small dots. Surprisingly, for the exchange-driven reorientation the critical value $d_1 = 0.1115$ and is constant in the whole range of the dot size, i.e., from 60 to 8000 spins ($L = 9$–101). (The same value and behavior of d_1 is reported in reference [37] where circular dots are studied by means of Monte Carlo simulations.)

To address this behavior of critical values in **Figure 6b**, we provide profiles of the lowest mode for two values of d, for $d \approx d_1$ (left profile) and for $d \approx d_2$ (right profile), for $L = 23$ (408 spins). Both profiles are localized at the vortex center but the localization near d_1 is much stronger than for d_2. The reorientation at d_1 is forced by the exchange interaction which is local and thus the dynamic interaction (between dynamic components of magnetic moments), confined to the very center of the dot, is not sensitive to the size of the dot. (It is not sensitive to the shape of the dot as well [28]). The second transition (d_2) is forced by the long-range dipolar interaction and the dynamic interaction of this type, although localized near the center, "feels" the dot size even for rather big dots. However, due to the localization this effect fades for larger dots, which is reflected in the d_2 curve in **Figure 6a**.

For typical ferromagnetic materials, the dipolar-to-exchange interaction ratio has very small value due to strong exchange. For example, using experimental data for ultrathin cobalt film

http://dx.doi.org/10.5772/66099

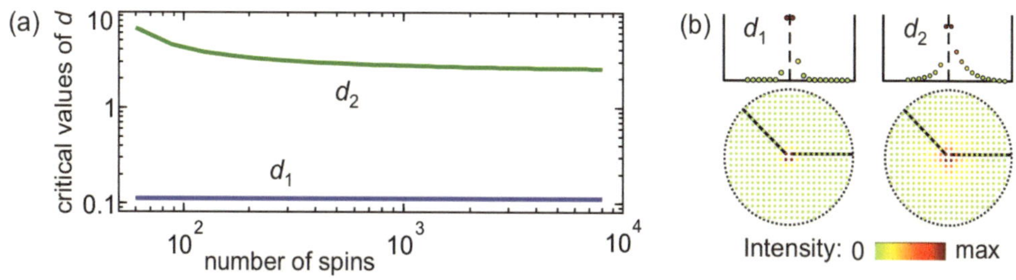

Figure 6. (a) Critical values d_1 and d_2 vs. the dot size (the number of magnetic moments within the dot, in logarithmic scale) for circular dot in the in-plane vortex state. (b) Exemplar profiles of the lowest mode in circular dots for $d \approx d_1$ (left profile) and for $d \approx d_2$ (right profile). Above each profile, its section along the indicated lines.

[38] from the relationship (3) we obtain $d_{Co} = 0.00043$ which is far below d_1. Consequently, in such materials, the in-plane vortex is unstable regardless the size of the dot (since d_1 is size independent).

5. Competition between interactions

As seen from **Figure 5**, for the majority of modes the frequency decreases with increasing *d* but with different rate. As a result, the order of modes in the spectrum changes with *d*; this effect is particularly intensive at the bottom of the spectrum. In particular, the mode of the lowest frequency has different symmetry of its profile in different ranges of *d* (compare **Figure 7**). The modes with the decreasing frequency can be divided into two groups: first one contains purely azimuthal modes (radial number equal zero). Within this group, the rate of the decreasing frequency grows with the increasing azimuthal number. However, above c.a. 55 GHz, this rate is visibly lower for another group of modes with the radial number 1. Within this second group, the situation repeats: for the mode (1,m) the frequency decrease rate is almost the same as for the mode (0,m) and it grows with increasing m. It shows that the impact of the dipolar-to-exchange interaction ratio on the mode frequency is determined mainly by its azimuthal number, the radial number being of little influence.

Besides the localized mode, there is one more mode in **Figure 5** the frequency of which acts in different way than the majority; in the broad range of *d* its frequency is almost constant. This mode, called fundamental mode, is an analogue of the uniform excitation [35]. Its profile is almost uniform within the dots without any nodal lines in azimuthal nor a radial direction thus it is labeled as (0,0). Highly uniform profile is the origin of the independence of the frequency of the fundamental mode on *d*.

As we already noticed that the mode order in the spin-wave spectrum is influenced by the dipolar-to-exchange interaction ratio and by the size of the dot, thus influences the character of the lowest mode. **Figure 7a** shows the dependence of the lowest mode frequency on *d* for different size of the dot. **Figure 7b** and c provides mode profiles for some values of *d* for two dot diameters: 51 and 101, respectively. Close to the critical value d_1 the profile is strongly localized at the vortex center regardless the dot size. This strong localization together with the

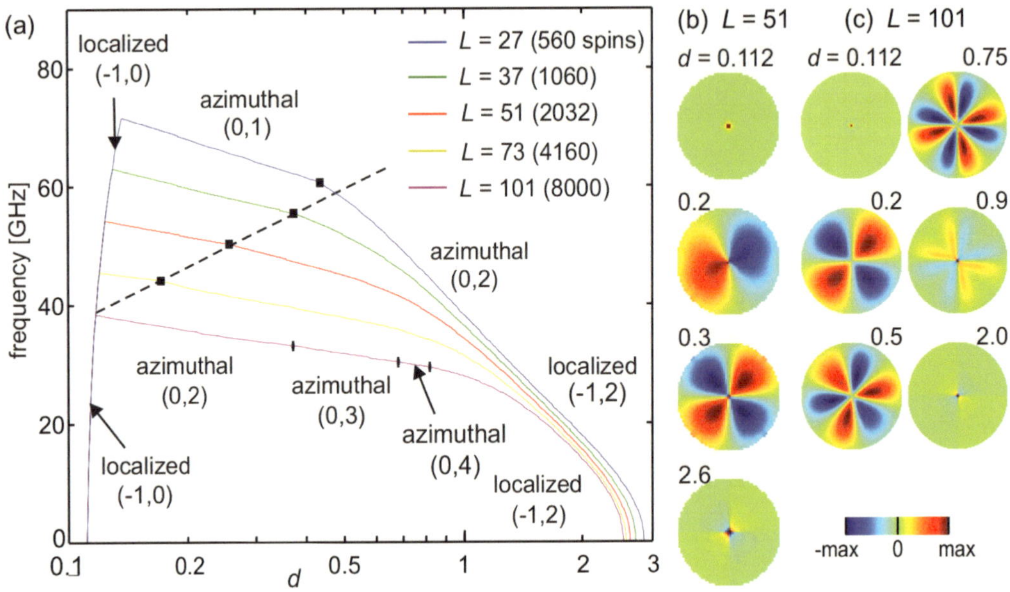

Figure 7. (a) The dependence of the lowest mode frequency vs. dipolar-to-exchange interaction ratio *d* (in logarithmic scale) in circular dots of different diameter *L* with the in-plane vortex as a magnetic configuration. On every curve, the crossing between first- and second-order azimuthal modes is marked with a black square (if exists). (b, c) Evolution of the lowest mode profile with *d* in dots of diameters 51 and 101, respectively.

short range of the exchange interaction (responsible for the magnetic reorientation below d_1) results in not only the independence of d_1 on the size and shape of the dot but also the frequency vs. *d* dependence is the same for the dot of any size. In this range of the dipolar-to-exchange interaction ratio, the lowest mode is a soft mode but with growing *d* its frequency increases rapidly and the mode ascend the spectrum very fast causing crossings with azimuthal modes of decreasing frequencies. After the first crossing, the azimuthal mode becomes the lowest in the spectrum.

For small dots ($L < 100$), the mode (0,1) is the lowest one after crossing with the localized mode. While *d* continues to increase till next crossing appears and (0,2) mode becomes the lowest. The point of crossing of these modes shifts to the smaller *d* with increasing size of the dot (see **Figure 7a**). Finally, for $L = 101$, the crossing between modes (0,1) and (0,2) takes place for lower *d* than the crossing with the localized mode. In a consequence, the first-order azimuthal mode is not the lowest one for any *d*. On the other hand, higher order modes may have the lowest frequency while *d* is growing (compare **Figure 7b** and **c**).

Here, we observe a general tendency of two interactions in question. The dipolar interaction favors higher order azimuthal modes. Thus, modes with the increasing azimuthal number m fall successively to the bottom of the spectrum as this type of interaction gains in importance regardless of whether their strengthen is due to the size (*L*) or material (*d*) of the dot. The exchange interaction in contrast favors modes with m = 1. Thus, the competition between the exchange and dipolar interaction manifests itself not only in the preferred magnetic configuration but also in the profile of the lowest frequency modes.

http://dx.doi.org/10.5772/66099

This rule changes if the vortex is close to unstable, i.e., close to the critical value of d. In this case, the soft mode is strongly localized at the vortex center. But even here this localized mode has nodal lines in the azimuthal direction for strong dipolar interaction and uniform for strong exchange interaction.

6. Circular rings

In circular rings, the central part of the dot is removed along with the vortex center. This causes significant reduction in the influence of the exchange interaction and consequently should result in the stabilization of the in-plane vortex for lower values of the dipolar-to-exchange interaction ratio. **Figure 8a** shows the typical dependence of the spin-wave spectrum in circular rings on d. The exemplar ring has external diameter $L = 25$ and internal one $L^{'} = 2$ which means that only four central magnetic moments are removed from the dot. The overall character of the picture is very similar to that for the dot shown in **Figure 5** with two exceptions: the range of the in-plane vortex stability and the behavior of the soft mode above d_1. (The decreasing of the frequency with growing d is much faster mostly due to the smaller external diameter.)

Just above d_1 the frequency of the soft mode increases steeply, as a consequence of increasing stability of the system, but before first crossing with the azimuthal mode the frequency slows down and finally becomes almost independent on d. The profile of this mode is shown in **Figure 8b** for $d = 0.01$; it is no more localized. Instead of this, the mode is a fundamental mode (0,0) being almost uniform within the ring. Due to the lack of the topological defect, there is no reason for the localization.

Other profiles provided in **Figure 8b** illustrate the change of the character of the lowest mode. Even if the external diameter of the ring is rather small, higher order azimuthal modes are the lowest for large enough d: (0,3) for $d = 1.3$ and (0,4) for $d = 2.0$. In full dots these modes could

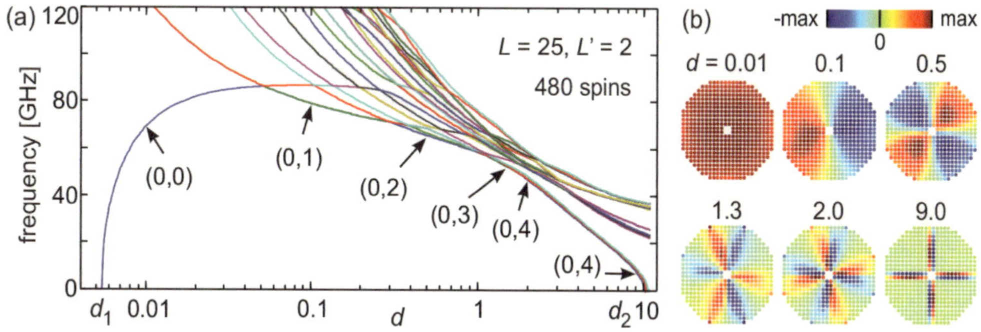

Figure 8. (a) The frequency dependence on the dipolar-to-exchange interaction ratio d (in logarithmic scale) for 25 lowest modes in the spin-wave spectrum of the circular ring of the external diameter $L = 25$ and the internal one $L^{'} = 2$ in the in-plane vortex state. (b) The evolution of the lowest mode profile. Profiles are calculated for six values of d marked with arrows in (a).

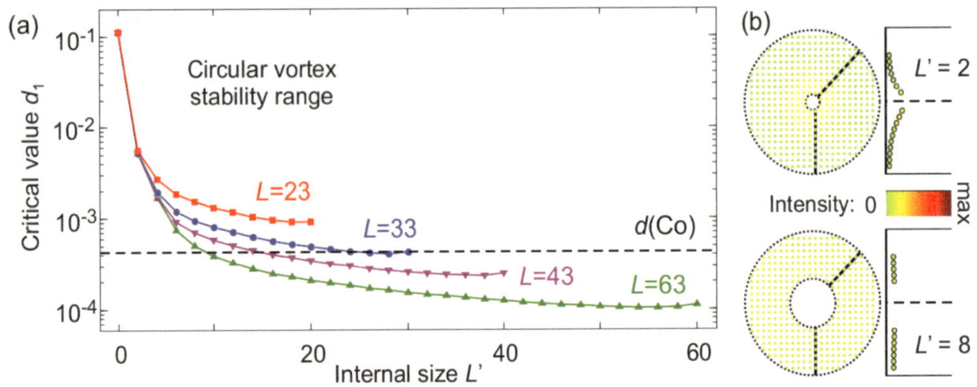

Figure 9. (a) Critical value d_1 vs. the internal diameter $L^{'}$ of the circular ring for four external diameters L. The dashed line in (a) indicates the value of d for Co/Cu(001) calculated from experimental results reported in reference [38]. (b) Spin-wave profiles of the soft mode in circular rings calculated for $d \approx d_1$ for two internal diameters $L^{'}$. The external diameter of rings is fixed at $L = 23$. To the right of each profile, its section along the indicated lines.

be the lowest for c.a. 4 times larger diameter which reflects the change of the balance between exchange and dipolar interaction after removing only few magnetic moments from the center of the dot.

The removing of these four central magnetic moments has also great impact on the stability of the in-plane vortex, as it should be expected. The critical value d_1 decreases from 0.115 for the full dot down to 0.0052 for the ring under consideration. However, this new critical value is still much larger than the value of d in common ferromagnetic materials. **Figure 9a** shows the change of the critical value d_1 with increasing the internal diameter of the ring for four external diameters: 23, 33, 43, and 63. In contrast to full dots in the rings d_1 visibly depends on both internal and external diameters (though for very small internal diameter the influence of the external size is weak). The increase in any diameter of the ring enhances the stability of the in-plane vortex. As a result, this magnetic configuration is stable even for such a material as cobalt if the ring is large enough ($d_1 < d_{Co}$).

The enhancement of the in-plane vortex stability due to the increasing of its internal diameter is rather obvious if we notice that the local exchange interaction between neighboring magnetic moments increases with decreasing distance from the vortex center (due to the change in the angle between them). In this context, the removal of the bigger circle from the center of the dot means the decreasing of the exchange interaction at the internal edge of the ring. Of course, this change in the exchange energy at the border should be visible in spin-wave profiles. To illustrate this effect, we calculate the profiles of the lowest mode for $d \approx d_1$ for the ring of the external diameter $L = 23$ and two different internal diameters, $L^{'} = 2$ and $L^{'} = 8$, shown in **Figure 9b**. Successive removing of the central part of the dot results in decreasing of the amplitude of the magnetic moments precession (smaller intensity of the profile) at the internal edge of the ring. On the other hand, the amplitude is slightly increased in the rest of the ring, especially at the outer edge. For larger hole in the ring, the profile is almost uniform in radial direction and d_1 is very little dependent on $L^{'}$. This nonzero intensity of the spin-wave profile

http://dx.doi.org/10.5772/66099

reaching the external edge of the ring explains the influence of the external diameter on the critical value d_1.

7. Square rings

In square rings, the in-plane vortex takes the form of the Landau state (closure domain configuration, see **Figure 2**). Unlike circular rings, here the magnetization along internal and external edges has the same conditions (no curvature). Another difference is the existence of domains walls. To see how these dissimilarities influence the in-plane vortex stability in **Figure 10a** we show the critical value d_1 vs. the internal size $L^{'}$ for square rings of different external size L. Similarly to the circular rings, the removal of the central part of the dot results in the drop of d_1, i.e., the in-plane vortex becomes stable for stronger exchange interaction. The critical value changes from 0.115 to 0.049. This time, in contrast to the previous case, this value is constant for broad range of the internal size of the ring. Additionally, d_1 does not depend on the external size of the ring as well. Therefore, the in-plane vortex (with the domain walls of Néel type) is not stable in square rings made from typical ferromagnetic materials.

To explain this behavior, **Figure 10b** shows spin-wave profiles of the lowest mode for $d \approx d_1$ for square rings of the external size $L = 22$ and three different internal sizes: $L^{'} = 0$ (full dot), $L^{'} = 2$, and $L^{'} = 16$. Removing of the central part of a dot, even just few magnetic moments, destroys the central localization of the lowest mode as it was in the case of circular dots, but now the localization is shifted to the corners of the resultant ring. Such corner-localized profile is not affected by the change of the size of the ring in large range of both, internal and external size. Again, the strongly localized spin-wave profile together with the local character of the exchange interaction causes the critical value of d for the exchange-driven reorientation to be independent on the size of the system. The high amplitude of the spin wave at the corners suggests also the increasing of the out-of-plane component of magnetization which means the formation of the Bloch-type domain walls.

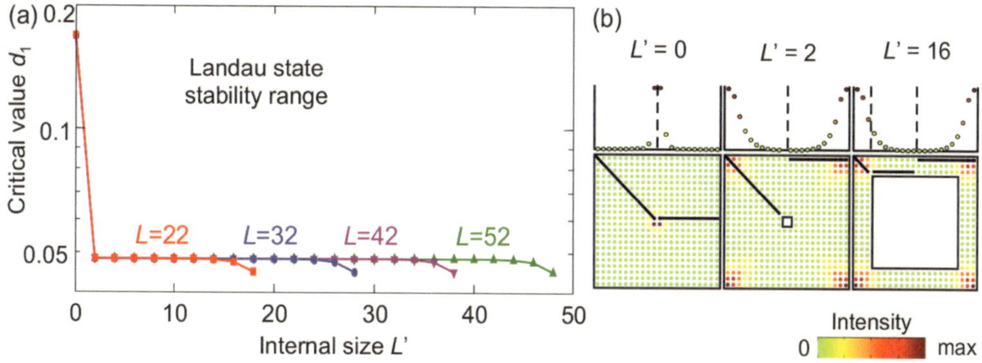

Figure 10. (a) Critical value d_1 vs. the internal size $L^{'}$ of the square ring for four external sizes L. (b) Spin-wave profiles of the soft mode in square rings calculated for $d \approx d_1$ for three internal sizes $L^{'}$. The external size of rings is fixed at $L = 22$. Above each profile, its section along the indicated lines.

8. Concluding remarks

In this chapter, we have shown our results concerning spin wave normal modes in nanosized dots and rings in the presence of the in-plane magnetic vortex. In experiments, in-plane vortices are observed in rings while in full dots made from typical ferromagnetic materials (e.g., cobalt or permalloy) the vortex core is formed at the vortex center [30]. Our results obtained for circular dots are consistent with this observation: in-plane vortex is stable in such a system for very weak exchange interaction, much weaker than in usual ferromagnets. We obtain the critical dipolar-to-exchange interaction ratio $d_1 = 0.1115$ (which corresponds to the exchange integral $J = 0.058$ eV) and this value is the same as received from Monte Carlo simulations in reference [37]. This critical value does not depend on the size of a dot which is also in agreement with simulations [37].

An interesting finding is the stability of the in-plane vortex in rings. In circular rings, the removal of a central part of a dot brings the dependence of d_1 on both diameters of the ring (external and internal) and, consequently, the in-plane vortex becomes stable even for strong exchange if the ring is large enough. In square rings, the situation is completely different: d_1 does not depend on any size of the ring (except extremely narrow rings). The critical value d_1 is reduced in comparison with full dots though not enough to stabilize the in-plane vortex. Therefore, in square rings made from usual ferromagnetic materials, the in-plane vortex is not stable (due to the preferred type of domain walls).

For the in-plane vortex configuration in full dots we found similar phenomena as reported from experiments, micromagnetic simulations, and analytical calculations, except those which arise from the existence of the gyrotropic motion of the vortex core, e.g., the splitting of the spin-wave frequency due to the coupling to the gyrotropic mode [31]. The qualitative agreement between results for in-plane and core vortices is an effect of the existence of the vortex center. Even without the out-of-plane component of the magnetization, the center of the vortex plays the role of the topological defect in the same manner as the vortex core. This defect acts as a nucleation center if d reaches its critical value and cause the localization of the soft mode. On the other hand, the properties of the spin waves in the presence of the vortex originate from the competition between exchange and dipolar interaction; thus the effects such as negative dispersion relation or diversity of the lowest mode profiles are similar for both types of vortices: with and without the core.

In our model, the dot is cut out from a discrete lattice which obviously has a consequence in the results. If the symmetry of azimuthal modes matches the symmetry of the lattice, the frequency of modes with opposite azimuthal numbers splits. Also the fundamental mode, an analogue of the uniform excitation, has nonuniform spin-wave profile whose symmetry reflects the symmetry of the lattice. (A similar effect was observed in micromagnetic simulations due to the artificial discretization of a sample [39–41].) In the case of circular dots and rings based on the discrete lattice, the edges are not smooth circles and cannot be smoothed as it is in continuous systems with artificial discretization, e.g., in micromagnetic simulations [42]. With the size of the ring, the edge smoothness increases but even for rather small dots (a dozen of lattice constants in the diameter) we obtain self-consistent results.

In this work, the method described in Section 2 is used for 2D dots and rings but its applicability is far beyond these simple systems. It can be used for 2D or 3D systems of an arbitrary shape, size, lattice, or magnetic configuration. Moreover, if the exchange interaction is neglected the method can be applied for nonperiodic systems too. Also interactions taken into account are not limited to dipolar and exchange only (the model with the anisotropy and the external field taken into account is derived in reference [43]). The main disadvantage of our approach is the lack of simulations; the assumption, instead of the simulation, of the magnetic configuration is useful for very simple magnetic configurations only. On the other hand, in comparison with time-domain simulations, the time of calculations is very short, and the spin-wave spectrum is obtained directly from diagonalization of the dynamic matrix (without the usage of the Fourier transformation). For simple magnetic configurations, our results are in perfect agreement with simulations [37, 13, 44]. In the case of more complicated systems, the simulations should be used for finding the stable magnetic configuration and for the simulated configuration the dynamical matrix method can be used to obtain the spin-wave spectrum.

Acknowledgements

The author thanks Jean-Claude Serge Lévy and Maciej Krawczyk for valuable discussions. The author received fund from Polish National Science Centre project DEC-2-12/07/E/ST3/00538 and from the EUs Horizon2020 research and innovation program under the Marie Sklodowska-Curie GA No644348.

Author details

Sławomir Mamica

Address all correspondence to: mamica@amu.edu.pl

Faculty of Physics, Adam Mickiewicz University in Poznan, Poznan, Poland

References

[1] Lau JW, Shaw JM. Magnetic nanostructures for advanced technologies: fabrication, metrology and challenges. J. Phys. D. 2011;**44**:303001. DOI: 10.1088/0022-3727/44/30/303001

[2] Puszkarski H, Lévy J-CS, Mamica S. Does the generation of surface spin-waves hinge critically on the range of neighbour interaction? Phys. Lett. A. 1998;**246**:347–352. DOI: 10.1016/S0375-9601(98)00518-0

[3] Mamica S, Puszkarski H, Lévy J-CS. The role of next-nearest neighbours for the existence conditions of subsurface spin waves in magnetic films. Phys. Status Solidi B. 2000;**218**: 561–569. DOI: 10.1002/1521-3951(200004)218:2<561::AID-PSSB561>3.0.CO;2-Q

[4] Mamica S, Krawczyk M, Kłos JW. Spin-wave band structure in 2D magnonic crystals with elliptically shaped scattering centres. Adv. Cond. Mat. Phys. 2012;**2012**:161387. DOI: 10.1155/2012/161387

[5] Romero Vivas J, Mamica S, Krawczyk M, Kruglyak VV. Investigation of spin wave damping in three-dimensional magnonic crystals using the plane wave method. Phys. Rev. B. 2012;**86**:144417. DOI: 10.1103/PhysRevB.86.144417

[6] Krawczyk M, Mamica S, Kłos JW, Romero Vivas J, Mruczkiewicz M, Barman A. Calculation of spin wave spectra in magnetic nanograins and patterned multilayers with perpendicular anisotropy. J. Appl. Phys. 2011;**109**:113903. DOI: 10.1063/1.3586249

[7] Pal S, Rana B, Saha S, Mandal R, Hellwig O, Romero Vivas J, Mamica S, Kłos JW, Mruczkiewicz M, Sokolovskyy ML, Krawczyk M, Barman A. Time-resolved measurement of spin-wave spectra in CoO capped [Co(t)/Pt(7 angstrom)](n-1) Co(t) multilayer systems. J. Appl. Phys. 2012;**111**:07C507. DOI: 10.1063/1.3672857

[8] Vedmedenko EY, Oepen HP, Ghazali A, Lévy J-CS, Kirschner J. Magnetic microstructure of the spin reorientation transition: a computer experiment. Phys. Rev. Lett. 2000;**84**:5884–5887. DOI: 10.1103/PhysRevLett.84.5884

[9] Vaz CAF, Kläui M, Bland JAC, Heyderman LJ, David C, Nolting F. Fundamental magnetic states of disk and ring elements. Nucl. Instrum. and Meth. B. 2006;**246**:13–19. DOI: 10.1016/j.nimb.2005.12.006

[10] Metlov KL, Lee YP. Map of metastable states for thin circular magnetic nanocylinders. Appl. Phys. Lett. 2008;**92**:112506. DOI: 10.1063/1.2898888

[11] Zhang W, Singh R, Bray-Ali N, Haas S. Scaling analysis and application: phase diagram of magnetic nanorings and elliptical nanoparticles. Phys. Rev. B. 2008;**77**:144428. DOI: 10.1103/PhysRevB.77.144428

[12] Chung S-H, McMichael RD, Pierce DT, Unguris J. Phase diagram of magnetic nanodisks measured by scanning electron microscopy with polarization analysis. Phys. Rev. B. 2010;**81**:024410. DOI: 10.1103/PhysRevB.81.024410

[13] Depondt P, Lévy J-CS, Mamica S. Vortex polarization dynamics in a square magnetic nanodot. J. Phys. Condens. Matter. 2013;**25**:466001. DOI: 10.1088/0953-8984/25/46/466001

[14] Shinjo T, Okuno T, Hassdorf R, Shigeto K, Ono T. Magnetic vortex core observation in circular dots of permalloy. Science. 2000;**289**:930–932. DOI: 10.1126/science.289.5481.930

[15] Wachowiak A, Wiebe J, Bode M, Pietzsch O, Morgenstern M, Wiesendanger R. Direct observation of internal spin structure of magnetic vortex cores. Science. 2002;**298**:577–580. DOI: 10.1126/science.1075302

[16] Miltat J, Thiaville A. Vortex cores – smaller than small. Science. 2002;**298**:555. DOI: 10.1126/science.1077704

[17] Li SP, Peyrade D, Natali M, Lebib A, Chen Y, Ebels U, Buda LD, Ounadjela K. Flux closure structures in cobalt rings. Phys. Rev. Lett. 2001;**86**:1102–1105. DOI: 10.1103/PhysRevLett.86.1102

[18] Mamica S. Stabilization of the in-plane vortex state in two-dimensional circular nanorings. J. Appl. Phys. 2013;**113**:093901. DOI: 10.1063/1.4794004

[19] Vavassori P, Metlushko V, Ilic B. Domain wall displacement by current pulses injection in submicrometer permalloy square ring structures. Appl. Phys. Lett. 2007;**91**:093114. DOI: 10.1063/1.2777156

[20] Jain S, Adeyeye AO. Probing the magnetic states in mesoscopic rings by synchronous transport measurements in ring-wire hybrid configuration. Appl. Phys. Lett. 2008;**92**: 202506. DOI: 10.1063/1.2936089

[21] Morrish AH. The Physical Principles of Magnetism. Wiley-IEEE Press. New York. 2001. ISBN: 978-0-7803-6029-7

[22] Pribiag VS, Krivorotov IN, Fuchs GD, Braganca PM, Ozatay O, Sankey JC, Ralph DC, Buhrman RA. Magnetic vortex oscillator driven by d.c. spin-polarized current. Nat. Phys. 2007;**3**:498–503. DOI: 10.1038/nphys619

[23] Guslienko KY. Spin torque induced magnetic vortex dynamics in layered nanopillars. J. Spintron. Magn. Nanomater. 2012;**1**:70–74. DOI: 10.1166/jsm.2012.1007

[24] Guslienko KY, Scholz W, Chantrell RW, Novosad V. Vortex-state oscillations in soft magnetic cylindrical dots. Phys. Rev. B. 2005;**71**:144407. DOI: 10.1103/PhysRevB.71.144407

[25] Bauer HG, Sproll M, Back CH, Woltersdorf G. Vortex core reversal due to spin wave interference. Phys. Rev. Lett. 2014;**112**:077201. DOI: 10.1103/PhysRevLett.112.077201

[26] Demidov VE, Ulrichs H, Urazhdin S, Demokritov SO, Bessonov V, Gieniusz R, Maziewski A. Resonant frequency multiplication in microscopic magnetic dots. Appl. Phys. Lett. 2011;**99**:012505. DOI: 10.1063/1.3609011

[27] Mozaffari MR, Esfarjani K. Spin dynamics characterization in magnetic dots. Phys. B. 2007;**399**:81–93. DOI: 10.1016/j.physb.2007.05.023

[28] Mamica S, Lévy J-CS, Depondt P, Krawczyk M. The effect of the single-spin defect on the stability of the in-plane vortex state in 2D magnetic nanodots. J. Nanopart. Res. 2011;**13**:6075–6083. DOI: 10.1007/s11051-011-0308-0

[29] Mamica S. Spin-wave spectra and stability of the in-plane vortex state in two-dimensional magnetic nanorings. J. Appl. Phys. 2013;**114**:233906. DOI: 10.1063/1.4851695

[30] Hoffmann F, Woltersdorf G, Perzlmaier K, Slavin AN, Tiberkevich VS, Bischof A, Weiss D, Back CH. Mode degeneracy due to vortex core removal in magnetic disks. Phys. Rev. B. 2007;**76**:014416. DOI: 10.1103/PhysRevB.76.014416

[31] Guslienko KY, Slavin AN, Tiberkevich V, Kim S-K. Dynamic origin of azimuthal modes splitting in vortex-state magnetic dots. Phys. Rev. Lett. 2008;**101**:247203. DOI: 10.1103/PhysRevLett.101.247203

[32] Buess M, Haug T, Scheinfein MR, Back CH. Micromagnetic dissipation, dispersion, and mode conversion in thin permalloy platelets. Phys. Rev. Lett. 2005;**94**:127205. DOI: 10.1103/PhysRevLett.94.127205

[33] Buess M, Knowles TPJ, Hollinger R, Haug T, Krey U, Weiss D, Pescia D, Scheinfein MR, Back CH. Excitations with negative dispersion in a spin vortex. Phys. Rev. B. 2005;**71**: 104415. DOI: 10.1103/PhysRevB.71.104415

[34] Ivanov BA, Zaspel CE. High frequency modes in vortex-state nanomagnets. Phys. Rev. Lett. 2005;**94**:027205. DOI: 10.1103/PhysRevLett.94.027205

[35] Zivieri R, Nizzoli F. Theory of spin modes in vortex-state ferromagnetic cylindrical dots. Phys. Rev. B. 2005;**71**:014411. DOI: 10.1103/PhysRevB.71.014411

[36] Mamica S, Lévy J-CS, Krawczyk M, Depondt P. Stability of the Landau state in square two-dimensional magnetic nanorings. J. Appl. Phys. 2012;**112**:043901. DOI: 10.1063/ 1.4745875

[37] Rocha JCS, Coura PZ, Leonel SA, Dias RA, Costa BV. Diagram for vortex formation in quasi-two-dimensional magnetic dots. J. Appl. Phys. 2010;**107**:053903. DOI: 10.1063/ 1.3318605

[38] Vollmer R, Etzkorn M, Kumar PSA, Ibach H, Kirschner J. Spin-wave excitation in ultrathin Co and Fe films on Cu(001) by spin-polarized electron energy loss spectroscopy (invited). J. Appl. Phys. 2004;**95**:7435. DOI: 10.1063/1.1689774

[39] Giovannini L, Montoncello F, Zivieri R, Nizzoli F. Spin excitations in nanometric magnetic dots: calculations and comparison with light scattering measurements. J. Phys. Condens. Matter. 2007;**19**:225008. DOI: 10.1088/0953-8984/19/22/225008

[40] Montoncello F, Giovannini L, Nizzoli F, Zivieri R, Consolo G, Gubbiotti G. Spin-wave activation by spin-polarized current pulse in magnetic nanopillars. J. Magn. Magn. Mater. 2010;**322**:2330–2334. DOI: 10.1016/j.jmmm.2010.02.033

[41] Wang R, Dong X. Sub-nanosecond switching of vortex cores using a resonant perpendicular magnetic field. Appl. Phys. Lett. 2012;**100**:082402. DOI: 10.1063/1.3687909

[42] Usov NA, Peschany SE. Magnetization curling in a fine cylindrical particle. J. Magn. Magn. Mater. 1993;**118**:L290. DOI: 10.1016/0304-8853(93)90428-5

[43] Mamica S. Vortices in two-dimensional nanorings studied by means of the dynamical matrix method. Low Temp. Phys. 2015;**41**:806–816. DOI: 10.1063/1.4932355

[44] Mamica S, Lévy J-CS, Krawczyk M. Effects of the competition between the exchange and dipolar interactions in the spin-wave spectrum of two-dimensional circularly magnetized nanodots. J. Phys. D. 2014;**47**:015003. DOI: 10.1088/0022-3727/47/1/015003